Moments of Valuation

Moments of Valuation

Exploring Sites of Dissonance

Edited by

Ariane Berthoin Antal, Michael Hutter,
and David Stark

OXFORD
UNIVERSITY PRESS

OXFORD
UNIVERSITY PRESS

Great Clarendon Street, Oxford, OX2 6DP,
United Kingdom

Oxford University Press is a department of the University of Oxford. It furthers the University's objective of excellence in research, scholarship, and education by publishing worldwide. Oxford is a registered trade mark of Oxford University Press in the UK and in certain other countries

First published 2015

Published in the United States of America by Oxford University Press
198 Madison Avenue, New York, NY 10016, United States of America

British Library Cataloguing in Publication Data
Data available

ISBN 978–0–19–870250–4

Acknowledgments

Many conversations and professional connections, often warmed by friendships, have led to this book. The institutional framework enabling the emergence of our ideas has been the research unit "Cultural Sources of Newness" at the WZB Berlin Social Science Center. During the lifespan of the unit's research program 2008–2014, several contributors came as visiting fellows, and the dialogue continued. David Stark visited in 2010 and eventually joined as an editor for this volume. In June 2013 we hosted a working conference at the WZB to enable the authors to share and develop their draft chapters. Nine months later, the volume is ready for printing, thanks to the careful attention paid to every line and each image by Julian Jürgenmeyer, Claudia Nentwich, Chris Reinelt, and Barbara Schlüter at the WZB. We are also grateful to David Musson for having believed in the book idea from the outset and to his staff for their support throughout the preparation process.

Ariane Berthoin Antal
Michael Hutter
David Stark

Contents

List of Figures

List of Tables

Notes on Contributors

Ariane Berthoin Antal is senior fellow at the WZB Berlin Social Science Center, where she currently leads the research stream on "Artistic Interventions in Organizations" in the research unit "Cultural Sources of Newness." She is Distinguished Research Professor at Audencia Nantes School of Management in France and Honorary Professor at the Technical University Berlin. Her books include the *Oxford Handbook of Organizational Learning and Knowledge* (co-edited with M. Dierkes, J. Child, and I. Nonaka, OUP, 2001) and *Learning Organizations. Extending the Field* (with P. Meusburger and L. Suarsana, Springer, 2014).

John Brewer is Eli and Eyde Broad Professor of Humanities and Social Sciences at the California Institute of Technology. He is the author or editor of more than a dozen books, including *The Sinews of Power: War, Money and the English State, 1688–1783* (Knopf and Unwin Hyman, New York and London, 1989); *The Pleasures of the Imagination: English Culture in the Eighteenth Century* (Harper Collins, London; Farrar, Straus and Giroux, New York, 1997; University of Chicago Press, 2000; new edition, Routledge, 2013); *A Sentimental Murder. Love and Madness in the Eighteenth Century* (Harper Collins, London; Farrar, Straus and Giroux, New York, 2004), and *The American Leonardo: A 20th Century Tale of Obsession, Art and Money* (Constable Robinson, London and Oxford University Press, New York, 2009).

Kimberly Chong is a senior teaching fellow in the Department of Anthropology and Sociology at the School of Oriental and African Studies, University of London. Her research interests include the anthropology of markets, restructuring and organizations, financial psychology, corporate ethics and science studies. Forthcoming publications include an article in *Socio-Economic Review.* She received a research grant from the Institute for New Economic Thinking to write a monograph, provisionally titled *Leveraging Value: An Ethnography of a Global Market Consultancy in China*.

Phillipa K. Chong is a visiting fellow in Sociology at Harvard University. Her research interests are in the areas of cultural sociology, sociology of knowledge, evaluation, and creativity. A current project compares conceptions of creativity in art and technology. She recently published about judgment in art in *Social Studies of Science.*

Liliana Doganova is a researcher at the Centre de Sociologie de l'Innovation, MINES-ParisTech, France. Her work explores entrepreneurial processes and valuation devices, such as business models and calculative techniques, with a particular focus on the biotech and cleantech sectors. She recently published *Valoriser la science. Les partenariats des start-up technologiques* and has contributed to journals such as *Research Policy* and *Public Policy*. Her current work focuses on the development of discounted cash flow analysis.

Ignacio Farías is a research fellow at the WZB Berlin Social Science Center. His main research areas are cultural sociology, urban studies and STS (Science and Technology Studies). His work has focused on mobilities and tourist practices, creative industries and studio processes, urban disasters and reconstruction planning, as well as current urban theory discussions on assemblage thinking. His work has appeared in *Sociological Review, European Journal of Social Theory, CITY, Space and Culture*, and *Mobilities*. He co-edited (with T. Bender) *Urban Assemblages: How ANT Changes Urban Studies.*

Antoine Hennion is professor at MINES-ParisTech and former Director of the Centre de Sociologie de l'Innovation (CSI) in Paris. At the CSI, he developed a problematization of mediation crossing cultural sociology and STS (Science and Technology Studies). He is now developing a pragmatist approach to diverse forms of attachment, from taste to issues about care, aging and disability. He has written extensively in the sociology of music, media and cultural industries, and on amateurs. Recent books are *La Passion musicale* (2007 [1993], English translation to be published in 2014), *La Grandeur de Bach* (with J.-M. Fauquet, 2000), *Figures de l'amateur* (with S. Maisonneuve, É. Gomart, 2000), and *Le Vin et l'environnement* (with G. Teil et al., 2011).

Michael Hutter is Director of the research unit "Cultural Sources of Newness" at the WZB Berlin Social Science Center. From 1987–2008, he held the chair for economic theory at Witten/Herdecke University, and from 2008–2013 he was research professor for knowledge and innovation at the Institute of Sociology, Technical University Berlin. Recent publications include *Beyond Price. Value in Economics, Culture and the Arts* (edited with D. Throsby, 2008) and *The Rise of the Joyful Economy* (Routledge, forthcoming).

Arlena Jung is a research fellow at the WZB Berlin Social Science Center. She has published articles on the medialization of science, boundary shifts in the science–policy–politics interface and the implications of different constructivist perspectives for observing transformation. She is the author of *Identität und Differenz, Sinnprobleme der differenzlogischen Systemtheorie* and is currently co-editing the special issue *Autonomie revisited—Beiträge zu einem umstrittenen Grundbegriff für die Beobachtung von Wissenschaft, Kunst und Politik in Zeitschrift für theoretische Soziologie* (with M. Franzen and D. Kaldeway, forthcoming).

Peter Karnøe is professor at the Center for Design, Innovation, and Sustainable Transition of Aalborg University Copenhagen, Denmark. His research has focused on technological entrepreneurship as distributed socio-technical practices and on emergent agency in innovation. He currently studies valuation work in cleantech innovation and in disputes about re-setting electricity markets with wind power. He has published in various journals including *Research Policy, Journal of Management Studies*, and *Industrial Marketing Management*.

Svetlana Kharchenkova is a PhD candidate at the Amsterdam Institute for Social Science Research (AISSR), University of Amsterdam, with a study devoted to the workings and development of the Chinese contemporary art market. Her research is grounded in cultural and economic sociology, and is part of the large-scale project

"The Globalization of High Culture: How Markets for Contemporary Art Develop in Brazil, Russia, India and China (BRIC)," funded by the Netherlands Organization for Scientific Research (NWO).

Rebecca-Lea Korinek is a research fellow in the project "Studying the Changing Orders of Political Expertise" at the WZB Berlin Social Science Center. Her research interests focus on comparative public administration, regulative policy and expertise, organization studies and science studies. Her dissertation compares changes in the practices used by national food safety agencies to construct their politico-epistemic authority. She has recently published in *der moderne staat (dms)—Zeitschrift für Public Policy, Recht und Management* and *Innovation: The European Journal of Social Sciences.*

Andrea Mennicken is Associate Professor in accounting at the London School of Economics and Political Science (LSE) and research associate at the Centre for Analysis of Risk and Regulation (CARR). She has published in *Accounting, Organizations and Society; Financial Accountability and Management; Foucault Studies,* among others. She co-edited with H. Vollmer *Zahlenwerk: Kalkulation, Organisation und Gesellschaft* (VS Verlag, 2007). Her research interests include social studies of valuation and accounting, auditing, transnational governance, processes of marketization, standardization, and public sector reforms.

Sophie Mützel is Professor of Sociology at the University of Lucerne, Switzerland. From 2009–2013 she was a member of the research unit "Cultural Sources of Newness" at the WZB Berlin Social Science Center. Her research areas include economic sociology, cultural sociology, and sociological theory. She uses texts as data to model changes in contextual meaning over time in different empirical fields. She has published book chapters on economic sociology and network research and articles in various journals, including *Current Sociology.* She recently co-edited (with J. Fuhse) a volume on relational sociology, *Relationale Soziologie. Zur kulturellen Wende der Netzwerkforschung* (VS Verlag, 2010).

Trevor Pinch is Goldwyn Smith Professor of Science and Technology Studies and Professor of Sociology at Cornell University. He holds degrees in physics and sociology. He has authored many books and numerous articles on aspects of the sociology of science, the sociology of technology, the sociology of economics, and sound studies. His books include *Analog Days: The Invention and Impact of the Moog Synthesizer.* He is co-editor of the *Oxford Handbook of Sound Studies.* He is also a performing musician with the Electric Golem.

Michael Power is Professor of Accounting at the London School of Economics and Political Science and former Director of the Centre for Analysis of Risk and Regulation (CARR). His research interests focus on auditing, accounting, risk management, regulation and organization theory. His major publications include *The Audit Society* (Oxford University Press, 1997) and *Organized Uncertainty: Designing a World of Risk Management* (Oxford University Press, 2007).

Claude Rosental is research professor of sociology at Centre National de la Recherche Scientifique (CNRS) and a member of the Institut Marcel Mauss and the Centre d'Etude des Mouvements Sociaux in Paris, France. His publications include works on the sociology of logic and public demonstrations and on sociological theory and methods. He is the author of *Weaving Self-Evidence: A Sociology of Logic* (Princeton University Press, 2008); *Les Capitalistes de la Science* (CNRS Editions, 2007), and co-editor of *La Cognition au Prisme des Sciences Sociales*, with B. Lahire (Editions des Archives Contemporaines, 2008).

David Stark is Arthur Lehman Professor of Sociology and International Affairs at Columbia University where he directs the Center on Organizational Innovation. He is the author of numerous books, including *The Sense of Dissonance: Accounts of Worth in Economic Life* (Princeton University Press, 2009) and *This Place, These People: Life and Shadow on the Great Plains* (with photographer Nancy Warner; Columbia University Press, 2013). His recent essays and articles have been published in *Sociologica* 2/2013, the *American Journal of Sociology* (2006 and 2010) and the *American Sociological Review* (2012). Among other awards, he is the recipient of a Guggenheim Fellowship (2002) and an Honorary Doctorate from the École normale supérieure de Cachan (2013).

Holger Strassheim is visiting professor at the University of Bremen and assistant professor at the Humboldt University of Berlin. During the spring term 2014 he was visiting research fellow at the STS Program, Harvard Kennedy School of Government. His publications include works on the changing role of expertise in welfare and risk regulation, regulatory science, multi-level governance, networks and public management collaborations. He is the author of *Netzwerkpolitik* (Nomos, 2011) and co-editor of *Governance als Prozess* (Nomos, 2009). He has recently published in *Evidence & Policy, Innovation: The European Journal of Social Sciences, Leviathan, der moderne staat (dms,)* and *Politische Vierteljahresschrift.*

Anne-Sophie Trébuchet-Breitwiller is professor at the Institut Français de la Mode (Paris) and an associate researcher at CSI of Mines-ParisTech. In 2011 she defended her PhD titled *The Making of Things Precious. Economic Anthropology of Luxury Products Drawing on the Examples of Perfume and Wine*, and she has published in the *Journal of Cultural Economy*. Her current research interests involve economic sociology, luxury, marketing and the building of markets.

Olav Velthuis is Associate Professor at the Department of Sociology and Anthropology of the University of Amsterdam. He is currently studying the emergence and development of art markets in the BRIC countries (Brazil, Russia, India and China). His books include *Imaginary Economics* (NAi Publishers, 2005) and *Talking Prices. Symbolic Meanings of Prices on the Market for Contemporary Art* (Princeton University Press, 2005), which received the Viviana Zelizer Distinguished Book Award of the American Sociological Association in 2006. He co-edited with M. Lind *Contemporary Art and Its Commercial Markets. A Report on Current Conditions and Future Scenarios* (Sternberg Press, 2012). His journalistic writings on the art market have appeared in *Artforum*, the *Art Newspaper* and the *Financial Times.*

1

Pragmatist Perspectives on Valuation: An Introduction

Michael Hutter and David Stark

A guest takes a sip of wine and puckers his lips as he puts the glass back down. An art expert claims in court that a recently discovered oil painting was not by Leonardo da Vinci. An international board of accountants suggests entering assets at their market value in the balance sheets of companies. All three cases involve acts of valuation. The wine taster, amateur or professional, signals the discovery of a surprisingly good wine, connoisseurs perform as arbiters of attribution to the work of a famous painter, and auditors use discretion in deciding which technique of estimating the present value of a firm's assets will be applied. All three cases involve uncertainty: something new is entering the world, and someone, or some group in society has to determine its worth, its dangers, and its potential. The new vintage has its distinct taste; the attribution of the discovered painting to Leonardo's hand remains speculative; the estimation of an asset's "fair value" accounts for unforeseen changes of worth since its acquisition.

Is there comparability in these cases of valuation? And if so, what do we discover about the failure and the success of innovations?

As modern society transforms itself into a society of continuous self-change, the scope for innovation widens to all processes that introduce something new. A very broad definition of innovation is needed to capture cases as diverse as the shapes of specific synthesizer sounds to new labor market policies, or from a new fashionable style of painting to the invention of a mathematical proof. New products, styles, and practices come to be selected and positioned as valuable in communities, organizations, and markets. This process is by no means trivial. In order to enter a social world of meaning, the unknown item must be recognized as new (Hutter et al. 2010). The quality of newness moves into the focus of research: How is it that an established understanding of that which is common can be interrupted by something, which then is recognized and evaluated as new by some group of arbiters? The challenge is to understand in greater detail the dynamics of social innovation, in their trajectories from situations of initial dissonance to the moments of valuation during which the new is confronted with the value scales of the established world.

In the social sciences, valuation made its first major appearance through John Dewey who demonstrated that value is a quality that has to be performed. The concrete action and practice of art criticism, for example, is a form of valuation (Dewey 1934: 309–25). Both artists and beholders are able to abstract that which is significant (Dewey 1934: 54; see also Muniesa 2011). He generalized these insights in later works (Dewey 1939, 1943). Since then, systematic studies have developed in very diverse empirical fields. For example, policy makers and institutions in the field of education seek measures with which to evaluate performance in teaching and research, and in the financial sector, objective performance indicators are complemented by judgments of future performance. Results have been published in a number of recent volumes (Beckert and Aspers 2011; Beckert and Musselin 2013). As more studies proliferate a "sociology of valuation and evaluation" has been declared (Lamont 2012) and a journal has been launched (Muniesa and Helgesson 2013; Vatin 2013).

Valuation has also become a central topic in economic sociology and in economics. There is inevitable uncertainty about the performance of products in markets for singularities, like restaurant food or novels. A broad range of devices and institutional regimes exists to help consumers make judgments about unknown qualities (Karpik 2010). Pure numerical price evaluation is an exceptional situation. In most other cases, the valorization through judgments of others provides the necessary information and reduces the risk of disappointing purchases (Hutter and Throsby 2008; Callon and Muniesa 2005; Fourcade 2011).

This volume opens a wide spectrum of social change for observation, ranging from microscopic change through the appearance of new entertainment goods to macroscopic change through new technologies. The authors seek out situations that bring an entire episode, from dissonance to valuation, into focus. They consider moments of valuation, captured in sites of dissonance. Implicitly, and in some cases explicitly, they work with a pragmatist perspective.

Taste, Test, Contest

De gustibus non est disputandum. We begin, indisputably, with taste. Then we immediately make it disputatious because we challenge the dominant view in cultural sociology that taste is *something one has*. Taste, in that view, is primarily symbolic because it is used for social purposes to mark distinction (Bourdieu 1984). By contrast the authors in this volume treat taste not as a noun but as a verb. It is *something one does*. It is a social process, to be sure; but if symbolic, it is also emphatically material.

Valuation involves a tasting, a testing. The studies here reject the claim that things just "have" some value, that their taste is intrinsic to them, and that tests reveal this natural value. At the same time they also deny the claim that the taste of things is something merely "attributed" to them, and that tests then do nothing more than reveal this value. We reject the dichotomy between natural objects (for which there is nothing to do but exploit the properties of things) and socially constructed objects (for which it is enough to show their arbitrary character as the stakes in social games).

We consider that the things to be tested and tasted are not given but made, and they are transformed in the very process of testing. Furthermore, tasting them or becoming attached to them is not like choosing some gratuitous label to enter a social logic of identity and difference; rather, identities are made and transformed by them. The chapters in this volume treat people's relationship with things as reciprocal interactions: making things and making us. As Antoine Hennion writes in his chapter: "So conceived, both operations (tasting and testing) are productive, open, and they remain tightly connected, referring less to an absolute divide between objectivity and subjectivity than to a continuous co-production of stabilization and transformation, both in the things and in our capacity to feel what they do" (Hennion in this volume).

To the two verbs—to taste and to test—we now add another: to contest. The adage *de gustibus non est disputandum* is misleading, for in matters of taste there can be disputes. Taste, whether the noun is understood as the quality of a thing or as the quality of a person, can be put to tests. And these tests are themselves contested. As the chapters here detail, the trials of valuation frequently involve disputes among different measures of worth, orthogonal principles of evaluation, and contending tests of value.

Situations—Situated Critical Moments

If our first steps were to consider *value* as a verb and, hence, valuation as a contested process, our next steps are to consider the contexts in which valuation occurs. Valuation, as Dewey explained, takes place in situations (Muniesa 2011). The situations of valuation are felt by the participants, and they are recognizable to observers. Rejecting a methodological individualism and a methodological institutionalism, we adopt a methodological situationalism (Suchman 2007; Latour and Woolgar 1986; Esser 1996; Knorr-Cetina

1981; Stark 2009: 31–3 and 185). Valuation is spatially and temporally localized. The authors of the chapters in this book are attentive to valuation taking place at a given site in a given moment.

The sites of valuation are often spatially marked: the dining hall, the court room, and spaces for professional meetings all have their material characteristics. They are sites for collective gatherings. Each one is equipped with certain technologies and devices. Valuation sites in which creators work are typically studios—for artists, architects, or musicians—or laboratories for scientists. These sites are carefully arranged and equipped with tools and materials that might be needed for experimentation and production. Sites for consumers are either private homes, or spaces that are equipped for a favorable experience, as is the case in concert halls, theaters, cinemas, and art galleries, or in luxury stores. Sites of valuation can also be marked through communication media platforms, from print magazines to TV channels and websites.

Moments of valuation are also temporally marked. In most of the case studies they have a recognizable beginning and end. They are identified as sessions, or trials, or experiments. They could consist of a sequence of meetings or of an era of changing evaluation standards, only recognized in hindsight. The duration of such moments might last for minutes, for hours, for months or even for several years. Unavoidably, the recognition of longer moments is fuzzier, both among participants and among observers. But it is still possible to observe the temporal extension and the internal timing of valuations.

Situations are characterized by the particular social assemblage of persons and things that are in place and in motion during a span of time. We prefer the language of situation over setting for two reasons. First, the authors of the cases in this volume tend to be attentive to how the situation is *set up*, indeed how it might be *a setup*, that is, increasing the applicability of some tests and precluding or posing obstacles to others. Second, the term setting is less evocative of the possibility of dispute, of things puzzling, of "Uh-oh, we have a situation on our hands here." Situations of valuation need not necessarily be troubling (and the more powerfully and gracefully they are set up, the less troubling they will seem to the participants). But they are, at least potentially, sites and moments of dispute and contention.

Newness Spans Dissonance and Valuation

In contemporary society innovation, for technological objects as well as for organizational routines, has become the norm of entrepreneurial strategy. Audiences in the public sphere, in entertainment, as well as in the sciences, expect the continuous emergence of new ideas, projects, and products. As cycles of variation accelerate, there must be procedures in place to determine whether the phenomena that disturb the present order are a mistake or an opportunity. Newness—the property of belonging to the set of items that find attention—is not inherent in a phenomenon. It is designated in situations that are, in hindsight, called "creative," but whose outcome had been highly uncertain (Hutter 2011).

Ask anyone to say what comes to mind when they think of the moment of creativity. Chances are that they will, at some point, mention a flash of inspiration when it clicked and they just got it. It's that "a-ha!" moment. The chapters in this book depart from that way of thinking in two respects. First, as social scientists, we think about creativity along less individualistic lines. For us, creativity is the property or outcome of some collectivity. In that sense, we are more interested in something like a *collective* "a-ha!" But that is not enough. The second differentiation characterizing our approach is that we are actually interested in the moments just *before* and *after* the "a-ha."

The chapters in this volume suggest that *these* moments are often characterized by *dissonance*. To grasp the notion of dissonance and its relevance for innovation we need to think more about what is actually happening in that "a-ha." One might use the word *recognition* to refer to such an experience: "It just came to me, and all of a sudden I recognized the new problem." But a simple understanding of recognition could be misleading here. For the more one recognizes an already existing pattern, the less new it is. Innovations disrupt old patterns and existing categories. Even when they take an already recognized pattern and adopt it (usually with modification) to a new category of problem, they are still breaking with old patterns of thought. For this reason, we think about innovation as a process of re-cognizing an aspect of the world (Stark 2009).

Recall now a recent "a-ha" moment and then think about the moment just before it. That was not likely to have been a settled moment, things were not all worked out. In fact, you might have experienced it as puzzled, or scrambled, perhaps even troubled. Although the process we are considering is not something in the mind of an individual, the analogy is helpful because dissonance refers to a similar process—one taking place in social situations. A situation is dissonant when there is more than one framework for assessing it, more than one value system for measuring worth.

The tension between two value systems can be intense.[1] The sovereignty of interpretation is contested. In the majority of cases, traditional values

[1] The sociological literature offers various models of a society with differing value systems, notably worlds of worth (Boltanski and Thévenot 2006), functional subsystems (Luhmann 2013) and institutional logics (Friedland and Alford 1991).

prevail: the new idea, proposal or product is overlooked or rejected. The moment in which the unknown item finds re-cognition as being new is often connected with expressions of surprise. The shift is experienced as a jump, as a losing of ground before finding new bearings.

Dissonance disrupts. It decreases the likelihood that things will flow on a smooth course towards conformity. For us, collaboration for creativity is not via the standard sociological account of shared understandings. Creativity is not smooth. It does not require that everything fit neatly together. The social settings for innovation are characterized by differences being held in tension (Stark 2009; de Vaan, Vedres and Stark 2014).

Dissonance is a musical metaphor. Some sound or sound constellation violates the aesthetic expectations of the listeners. Within the musical field, it connotes a temporary state of misunderstanding. This state ends in one of three ways: the dissonant sound is resolved back into the old resonant order, the dissonance persists, unrecognized, as a form of noise, or the aesthetic expectations adjust to a new resolution. The new resolution may be the result of initial intentions by the producers or the outcome of a misunderstanding that has found a successful interpretation—in either case, it becomes part of social change.

Newness, the hallmark of innovation, is a fleeting quality: since when and for how long is something called "new?" Its fleetingness makes newness difficult to observe. Using our distinctions, we can take this temporal instability into account. In order to become something new, for a while, a variegated process of social interaction takes place: the deviant, the tenuous, the rough, is recognized in specific situations, and it is estimated in the terms of established value scales.

The distinctions provide a heuristic for empirical investigation. Since newness spans dissonance and valuation in situations, the general pattern of comparison will rely on these dimensions: the terms of value attributed, the controversies and interpretations expressed, and the sites on which the historical, inevitably singular events took place, and continue to take place.

Kinds of Moments

This general pattern helps to organize the contributions from authors who were intentionally chosen to represent a variety of sub-disciplines and national backgrounds in the social sciences. Although the authors in this volume share an interest in social valuation, their chapters are positioned in diverse literatures. They are anthropologists, economic or cultural sociologists, organization researchers, historians or political scientists. The differences between their empirical evidence reflect the wide variety of appearances that innovation, or social change, might take. Four groups of chapters, characterized by similarities in their research objects, are distinguished.

A first section gathers *varieties of aesthetic valuation*. Aesthetic judgments are a particularly rich case because of the constant stream of novelties clamoring for the attention of their respective audiences. Aesthetic impressions impact on the human senses, they trigger mental experiences. The sensations fade with length and repetition, and are therefore replaced by new impressions. Trevor Pinch follows a rather long-term change in the sound material used in contemporary music: the genesis of a few successful synthesizer sounds, from their accidental invention to pre-fabricated sounds for the Mini-Moog took several decades. The sites were music sessions, fairs, and concerts. The dissonances were literally frictions of sound. The valuation by musicians exposed to the (not yet new) sounds, began with precise formulations of what it is that they recognized in the new acoustic signal, and it continued with the construction of keyboard-synthesizers.

Antoine Hennion observes equally minute expressions in a food sector where a new vintage has to be appraised every year. The sites are ubiquitous—dinner gatherings all over the world in which the wine being drunk is given social attention. The dissonance experienced in this case is even more physical than sounds: it comes with the sensual taste of a sip, from a specialized glass, in a symbolically charged social context. The contest of taste—as an acquired skill, social as well as individual, valorizing as well as evaluative—is carefully dissected by Hennion.

Michael Hutter is interested in the effect of artistically valuable visual forms on successful entertainment products. His sites are the studios of artists, film-makers, and photographers, where documented artistic inventions inspired or prefigured the symbolic content of some creative industries' products with unusually high margins of profit, and the show-rooms, cinemas, and magazines in which the products were received by their users or, rather, interpreters. Neither authoring nor interpreting are smooth processes, they are both driven and even motivated by the appearance of dissonances and gaps of understanding that are small enough to be overcome, and to be enjoyed.

In John Brewer's court case, the site and the situation are particularly explicit. Brewer uses the evidence generated during a trial to explore a major shift in the U.S. practice of appointing expert witnesses. By provoking the case of a painting falsely attributed to Leonardo da Vinci, the art dealer and expert, Joseph Duveen, expected the opinion of art connoisseurs to be recognized and respected in legal proceedings. But connoisseur judgments lost out to scientific evidence based on material properties, and were henceforth discredited as a device in legal contests.

The story told by Svetlana Kharchenkova and Olav Velthuis deals with the current contemporary art scene, where personal judgments of art experts still carry authority. The sites are art galleries, auction houses, and art fairs in

China and Europe, the valuations concern two new styles of painting introduced by a few Chinese artists. The styles are named "Cynical Realism" and "Political Pop," and a protracted debate has erupted around the claim that the name reflects a misunderstanding by Western art protagonists of those painting in their original Chinese context. It is this misunderstanding, the authors argue, which fuelled the success of the new styles in the foreign setting. Their moment of valuation, then, took much longer than the short, intense event of the Duveen case reported by Brewer.

The second section assembles *devices valorizing uncertain aesthetic experiences*. The chapters still address products for aesthetic experiences, aimed at various senses. The focus is on particular practices or institutions of valuation that have developed in order to channel attention to specific new products. A classic version of such practices is the literary review. Phillipa Chong is interested in a specific aspect of review writing. Her sites are the literary magazines where a moderately sized number of book authors and reviewers keep switching roles. Her interviews demonstrate how the critics' evaluative practices carry consequences not only for the books they review, but also for the evaluators themselves. Given the switch-role structure of the contemporary fiction market, she reveals how critics' social position in the field shapes what they put in their reviews: the social interdependencies that result from these entanglements between peers.

A more recent genre is restaurant reviews, a subset of which is examined by Sophie Mützel. The sites are again magazines, this time entertainment magazines in Berlin, during the years from 1995 to 2012. The new products to be critiqued are restaurant performances. The texts have no literary ambition; they are intended to inform and to attract. As she investigates these texts with topic modeling software, Mützel finds trajectories of taste, reflected in the rise and decline of specific food cultures within the city's culinary field.

Anne-Sophie Trébuchet-Breitwiller's sites are unique and they are meticulously designed. With ethnographic methods she enters the stores of a highly successful "alternative" perfume maker. The dissonance lies in the ruptures of traditional perfume-making practices, but also—and here is the focus—on the surprises encountered by a prospective buyer who enters the carefully scripted interaction with sales personnel and samples from the "edition" of perfumes offered. Valorization, in this business, has been developed into a creative skill.

The cases presented in the other sections of the book are less immediate in their sensory appeal. They demonstrate that the innovation pattern of situated dissonances and valuations works as well in an expanded social universe. The third section collects studies on *valuation in fields of practice*. Claude Rosental's field is artificial intelligence and one of its sub-disciplines called fuzzy logic. The sites are scholarly journals, conferences, and a newsgroup on the Internet, the moment of valuation started with a paper containing a mathematical proof that fuzzy logic can be reduced to binary logic. As the debate on the validity of the deviant claim developed, the contributions, conveniently assembled on the website, became ever more reflective of the valuation process itself. The valorizing nature of the contributions to a scientific and cultural dispute was partially recognized. About a year after the original challenge, a set of texts by renowned researchers in artificial intelligence, issued in a reputed journal, positioned the argument in a wider context, thus deflecting it. The moment ended, the debate ran out of steam.

Andrea Mennicken and Michael Power observe the field of financial accounting practice. They identify four moments during which unsuccessful and successful attempts were made to change fundamental features of that practice. Just as other valuations, accounting is a constrained way of producing fictions. The fictions, in turn, react to misinterpretations of monetary asset values in increasingly volatile markets. The sites in this chapter are highly dispersed. The dissonance in the historical situations erupted whenever new rule interpretations by accountants were challenged by a regulatory agency.

Liliana Doganova and Peter Karnøe investigate attempts to improve valuation procedures in the agricultural field. The sites are Danish livestock farms and the conflict is triggered through the environmental damage caused by slurry. Under the pressure of environmentalists and EU norms, state agencies promote more efficient technologies, and encourage farmers to invest in the necessary equipment. The Danish agency has developed a tableau of quantifiable indicators that aids farmers in determining their "optimal" choice. As in the accounting case, the impulse to innovate is not part of the professional regime under study. It is a "boundary conflict," a change in the surrounding economy or the political field that forces farmers into choices that are new and highly uncertain for them.

Holger Strassheim, Arlena Jung, and Rebecca-Lea Korinek also deal with valuative discourses in public policy. Their sites are labor market and food safety agencies in Great Britain and in Germany. While studying changes in expert cultures, they discovered the spectacular rise of behavioral interventions, so-called "nudging practices." Their comparative approach revealed that experimental evidence and concomitant assumptions on human behavior easily entered the British policy regime, but has received considerably less attention in German agencies. The reason, they argue, lies in different ways knowledge claims are validated and valued: Britain's communitarian culture of expertise is more open to the publicly demonstrable evidence, whereas expert credibility in Germany draws on institutional affiliation as testimony of sound judgment.

The fourth section contains studies of *valuation within organizations*. These processes are particularly difficult to research because they take place in private sites. Organizations, be they for-profit or not-for-profit, have little interest in revealing how they deal with their own creative procedures and with their techniques while introducing innovations from outside. The studies, therefore, rely on the intensive engagement of researchers in the respective organizations, and they are in themselves innovative in detecting and expressing moments of valuation. Ignacio Farías spent several months in the studios of internationally successful architectural firms. Here, newness means the recognition of a proposed project by an external jury. In the studios, sets of projects are developed simultaneously, and the team members serve each other as judges of quality. They do so in fluid sequences of interaction, without the need for formal review meetings. Their routine even has the reflective quality of changing its own criteria, if that seems opportune.

Ariane Berthoin Antal has studied organizations in the private as well as the public sector that invest in bringing people, practices, and products from the world of the arts together in the hopes of stimulating a "creative clash" from which new skills or ideas might emerge. These artistic interventions occur in clearly defined sites—projects, art installations, workshops—which she characterizes as interspaces in which the artists temporarily challenge or suspend the dominant organizational norms while experimenting with new approaches. As her interviews and surveys show, the artists engage with the emotional and physical responses of employees as resources for, rather than as barriers to, generating newness.

The management consultancy company in which Kimberly Chong was a participant observer for 16 months follows an entirely different approach to shaping the valuation frame of their employees. Added shareholder value, in terms of reduced cost or increased revenue, constitutes the core of the comprehensive systems of accountability which the consultants sell to clients. The company applies the same valuation frame to itself. Internal workshops are the sites where the components of the system are conveyed. Participants who misunderstand are corrected, and old Chinese traditions are invoked to strengthen the persuasion.

* * *

So, we now invite you to a tasting. The menu in our table of contents has 15 courses. We have asked our chefs to keep the portions small. No single essay will be filling in itself, but we are confident that the ensemble will be satisfying. It is, of course, a testing too—an opportunity for you to test pragmatist perspectives on the study of valuation. And we expect that there will be dissonance—dissonance in departures from currently prevailing ideas as well as dissonance in contestation among the positions argued by our authors. We hope these dissonances will lead to innovation. Whether and how they are recognized is up to you.

References

Beckert, J. and Aspers, P. (eds.) (2011). *The Worth of Goods: Valuation and Pricing in the Economy.* Oxford: Oxford University Press.

Beckert, J. and Musselin, C. (eds.) (2013). *Constructing Quality. The Classification of Goods in Markets.* Oxford: Oxford University Press.

Boltanski, L. and Thévenot, L. (2006). *On Justification: Economies of Worth.* Princeton, NJ: Princeton University Press.

Bourdieu, P. (1984). *Distinction: A Social Critique of the Judgement of Taste.* Cambridge, MA: Harvard University Press.

Callon, M. and Muniesa, F. (2005). Peripheral Vision: Economic Markets as Calculative Collective Devices. *Organization Studies* 26(8): 1229–50. doi: 10.1177/0170840605056393

de Vaan, M., Vedres, B., and Stark, D. (2014) Game Changer: The Topology of Creativity. *American Journal of Sociology* 120(3): in press.

Dewey, J. (1934). *Art as Experience.* New York, NY: Minton, Balch & Co.

Dewey, J. (1939). *Theory of Valuation* (Vol. II, No. 4). Chicago, IL: University of Chicago Press.

Dewey, J. (1943). Valuation Judgments and Immediate Quality. *Journal of Philosophy* 40(12): 309–17.

Esser, H. (1996). Die Definition der Situation. [The Definition of the Situation] *Kölner Zeitschrift für Soziologie und Sozialpsychologie* 48: 1–34.

Fourcade, M. (2011). Cents and Sensibility: Economic Valuation and the Nature of "Nature." *American Journal of Sociology* 116(6): 1721–77.

Friedland, R. and Alford, R. R. (1991). Bringing Society Back in: Symbols, Practices, and Institutional Contradictions. In W. W. Powell and P. J. DiMaggio (eds.), *The New Institutionalism in Organizational Analysis.* Chicago, IL: University of Chicago Press, 232–263.

Hutter, M. (2011). Infinite Surprises: On the Stabilization of Value in the Creative Industries. In J. Beckert and P. Aspers (eds.), *The Worth of Goods. Valuation and Pricing in the Economy.* London: Oxford University Press, 201–22.

Hutter, M., Berthoin Antal, A., Farias, I., Marz, L., Merkel, J., Mützel, S., Oppen, M., Schulte-Römer, N., and Strassheim, H. (2010). *Research Program of the Research Unit*

"Cultural Sources of Newness." Discussion paper SP III 2010-405, WZB Berlin Social Science Center.
Hutter, M. and Throsby, D. (2008). *Beyond Price: Value in Culture, Economics, and the Arts.* Cambridge : Cambridge University Press.
Karpik, L. (2010). *Valuing the Unique: The Economics of Singularities.* Princeton, NJ: Princeton University Press.
Knorr-Cetina, K. (1981). Introduction: The Micro-Sociological Challenge of Macro-Sociology: Towards a Reconstruction of Social Theory and Methodology. In K. Knorr-Cetina and A. V. Cicourel (eds.), *Advances in Social Theory and Methodology: Toward an Integration of Micro- and Macro-Sociologies.* London: Routledge & Kegan Paul, 1–47.
Lamont, M. (2012). Toward a Comparative Sociology of Valuation and Evaluation. *Annual Review of Sociology* 38(1): 201–21. doi:10.1146/annurev-soc-070308-120022
Latour, B. and Woolgar, J. (1986). *Laboratory Life: The Construction of Scientific Facts.* Princeton, NJ: Princeton University Press.
Luhmann, N. (2013). *Theory of Society* (Vol. 2). Palo Alto, CA: Stanford University Press.
Muniesa, F. (2011). A Flank Movement in the Understanding of Valuation. *The Sociological Review* 59(S2): 24–38.
Muniesa, F. and Helgesson, C.-F. (2013). Valuation Studies and the Spectacle of Valuation. *Valuation Studies* 1(2): 119–23.
Stark, D. (2009). *The Sense of Dissonance. Accounts of Worth in Economic Life.* Princeton, NJ: Princeton University Press.
Suchman, L. A. (2007). *Human-Machine Reconfigurations. Plans and Situated Actions* (2nd ed.). Cambridge: Cambridge University Press.
Vatin, F. (2013). Valuation as Evaluating and Valorizing. *Valuation Studies* 1(1): 31–50.

Part I

Varieties of Aesthetic Valuation

2

Moments in the Valuation of Sound: The Early History of Synthesizers

Trevor Pinch

Weird Shit!

In a small upstate New York town, Trumansburg (outside Ithaca), Robert Moog, an engineer trained at Columbia and doing a PhD at Cornell University, was acutely aware of the weirdness of electronic sound to the populace at large. He had built Theremins[1] as a teenager and now the musician who sold small numbers of his Theremins to schools—an experimental composer, Herb Deutsch—was working with him to create one of the first electronic music synthesizers. This synthesizer would be smaller, portable, cheaper and more versatile than existing equipment for making electronic music such as stand-alone oscillators or the massive RCA Mark II synthesizer housed at the Columbia-Princeton electronic music studio. Here is how Robert Moog described the first sounds of that first synthesizer when played by Deutsch:

> He [Deutsch] took this and he went down in the basement where we had a little table set up and he started putting music together...I still remember, the door was open, we didn't have air conditioning or anything like that, it was late spring and people would walk by, you know, if they would hear something, they would stand there, they'd listen and they'd shake their heads. You know they'd listen again—what is this weird shit coming out of the basement?[2]

In this chapter I will ask how and by what means sound is valued. It is at moments of dissonance or disturbance like the one just described when the

[1] Popular in movies for producing other worldly or scary sounds (used by Hitchcock in *Spellbound* (1945)), see Glinsky (2000).
[2] Interview with Robert Moog, June 5, 1996.

process of valuation becomes most manifest. In this case it is the entry of a new type of sound into our soundscape—electronic sound—which produces the dissonance.

Before the mid-1960s most people had not been exposed to electronic sound; they may have heard an esoteric composition by composers such as Karlheinz Stockhausen or John Cage or visited the Philips Pavilion at the 1958 World Fair in Brussels where Edgar Varèse's composition, Poème électronique, was heard by many visitors (Holmes 2012). But in popular culture the only electronic sounds likely to be heard were those emanating from marginal electronic instruments such as the Theremin, the trautonium,[3] or the Ondes Martenot.[4] Science fiction movies, in particular, used the Theremin for special effects or the one-off cybernetic circuits designed by Louis and Bebe Barron to make the eerie electronic sound track to *Forbidden Planet* (1956). The genre of "exotica," popular with the new fad of hi-fi in the 1940s and 1950s (Taylor 2012), also used electronic instruments such as the Theremin.[5] With the invention in 1964 of commercial electronic music synthesizers by Robert Moog and Donald Buchla (Theberge 1997; Pinch and Trocco 2002; Holmes 2012) this was all to change. From being marginal, electronic sound became part of the fabric of our soundscape. Today we are saturated with electronic sound from cell phones, computers, TV shows, movies, cars, numerous technological appliances and, of course, in many genres of music.

In the introductory quotation, Moog mentions four of the key characteristics necessary for a sound to be valued. First is *the place*: in this case the sound is heard, not within the usual circumscribed space of a studio or workshop, but has leaked out onto the sidewalk via an open door. This makes the sound available to non-specialized listeners—passers-by who encounter the sound for the first time. The second way he situates the sound is with an *active listener* who in this case is a casual listener unfamiliar with the sound of synthesizers. Thirdly, he describes how people listen to this sound by detailing their response—they shake their heads and listen again. The shake of the head allow us to infer that the listeners pay (a nice economic metaphor) attention (Hennion this volume)—what is this strange unfamiliar sound? Where does it come from? What is producing it? Attention is rare in the world of listening—unlike with the wine world studied by Hennion, we are living in a world where sound is ever present. We can say we are saturated by sound but alas not by wine. Thirdly *words must be used to describe particular sounds*—Moog invokes the words "weird shit" to describe these sounds which have not yet acquired specific meanings at the dawn of electronic music synthesis.

[3] Used by Hitchcock in *The Birds* (1963).
[4] Familiar mainly to French moviegoers.
[5] See, for example, Harry Ravel's *Music out of the Moon* conducted by Les Baxter (1947).

Placing a value on sound is a much newer activity than, say, valuing wine (Shapin 2011, 2012; Fourcarde 2012; Hennion in this volume). The media to give sonic objects a degree of permanence such that they can be subject to valuation techniques, categories, and procedures has existed for a much shorter period of time. It is with the rise of the phonograph in 1877 that sonic experiences, and of course mainly music, for the first time came to be captured and stored (Katz 2010). This enabled sound to be reproduced and hence travel in a way hereto impossible. In earlier periods the means of reproducing sounds were either locally grounded in specific material artefacts such as bells (Corbin 1998), or needed skilled practitioners—namely musicians—in order for sounds to travel. Musicians have stabilized ways of reproducing sounds via increasingly mass-produced instruments and a culture for passing on skills and musical signification (by way of standardized scores, and institutions such as conservatories which allowed training). Sonic objects such as bells acquired value but sound itself was much harder to value until it could be captured and reproduced independently from the objects and skilled practitioners who produced it.

For the sort of analysis I want to propose here, materiality matters but in a slightly different way than Hennion stresses in his chapter. The story of sound I want to tell is how, through technological mediation, sound has increasingly become embedded in material artefacts. Hennion describes how the very matter of wine matters in how it is tasted. It is an obvious point but a deep one: just as the taste of a wine must be capable of a surprise so too must sound. Part of the power of the new electronic sound came from the fact that sound was produced by electronic means and sounded different to anything experienced before. Sound, like wine can be a source of surprise. This means some sort of agency must be given the material world in how a sound is experienced. Materiality matters even if it is simply in resisting our ambitions for it (Pickering 1995). A Moog synthesizer, however well played, cannot easily make the sound of the Niagara Falls in full flight anymore than cheap plonk can easily be made to taste like a top bottle of Latour Burgundy. Note the weasel word "easily." Of course, these are tricky issues because the person doing the valuing is an integral part of the valuation put upon wine or sounds. To an uneducated palette a plonk-wine may taste the same as a fancy Latour, just as to those people who hate electronic music with a passion or who have never been exposed to it, debate whether it should even be called music. For them the sounds have no value and hence there can be no surprise.

Hennion instructively compares musicologists listening to snippets of music to the wine tasters taking their dainty sips. Electronic sound, in this respect, is more like wine than a symphony: it is the raw element of the sound—the weird shit—to which the listener pays attention when first hearing electronic sound. But then electronic sound, unlike the taste of wine, can be directly materially captured, transmitted and reproduced for multiple listeners, first with tape recorders and vinyl LPs, and later with CDs and digital sound files. Electronic sounds thus become subject over time to a process of cultural evaluation and refinement as listeners with the aid of technology more easily reproduce sounds and acquire tastes as to what sounds they like or desire. This is of course to an extent true of wine tasting where tasting sessions, with the aid of magazines and now carried out online, promote certain notions of taste and rankings of wines. But as with bells and musical instruments before phonographs and synthesizers, the wine itself in its full material richness must be reproduced for tasting to occur. With storage media and a way of reproducing sound independent from its recording, the material production of the sound is changed. A CD player can now replace a Moog synthesizer such that a thirty-dollar machine can replace a ten-thousand dollar machine. The sound may stay the same (but this in itself is a matter of judgment with audiophiles notoriously challenging different modes of sound reproduction (Perlman 2004)) but its possibility for dissemination is dramatically changed by its reproduction with other material means. With electronic sound, materiality thus figures more prominently in the story of how value gets stabilized because, unlike with wine tasting, sounds can be stored and reproduced. There is not yet a phonograph for taste or digital taste files to be sent over the internet.

It is through the broader engagement with science and technology in particular that sound becomes more material, particularly in the twentieth century (Jackson 2006; Pinch and Bijsterveld 2012). Sound over time becomes more thing-like—a commodity to be bought and sold on iTunes, a thing to be worn, as with personal stereos. Sound itself in the twentieth century becomes a new way to sell and market goods (Taylor 2012). Sound becomes not only listened to but also measured, regulated, and controlled (Thompson 2002; Bijsterveld 2008).

If we take one metric of value, money, it is startling to realize how valuable electronic sound and the means of producing it have become. For instance, Stanford University's second biggest earning patent is for a sound technology known as Frequency Modulated (FM) synthesis. This technology, developed in 1985 by John Chowning, was at the core of new polyphonic digital electronic music synthesizers such as the Yamaha DX7 and became the basis for a new technological means to produce sound in computers (e.g. Sound Blaster). Fast forward twenty years, and recall that the technology which set Apple on its route to becoming one of the most valuable companies (in terms of stock market price) in the world was another sonic technology, the i-Pod. Neither the i-Pod, with its capacity to play MP3 files of digital music (Sterne 2012), or FM synthesis, would be possible if it was not for the earlier generation of sonic technologies made by Moog and Buchla which permitted sounds to be produced, controlled, stored and reproduced by electronic means.

How then did the "weird shit" described by Moog turn into something of value? In order to tell this story we need to go back to the period 1964–71 when these synthesizers were first made and their sounds started to be valued in different domains.

Into the Crystalline

The Moog modular synthesizer (see Figure 2.1) was an expensive instrument in the rubric of another common way of valuing things (house prices), in 1964 it cost the price of a small house (this is Moog's own way of conveying how expensive the instrument was back then). Who would buy and use such an expensive instrument?

Moog expected the initial users to be academic music studios, such as the Columbia-Princeton studio and individually wealthy composers. He was surprised to discover that his second-ever customer was Eric Siday, a pioneer in a fast-growing specialist new industry where sound was highly

Figure 2.1. The Moog Synthesizer (photograph taken in Factory Studio, Robert Moog is left at front wearing pen protector)

valued—advertisements and jingles for radio and TV (Taylor 2012). Siday's work in this area was noteworthy enough for *Time* magazine to write about him (*Time* 1966). One famous sound he made was for an early Maxwell House coffee advertisement which featured a burping coffee percolator sound. *Time* magazine estimated that Siday had earned $40,000 for a few seconds of sound—a staggering amount of money for 1966. The entry of sound into the soundscape via its use in advertisements and jingles heard on radio and TV shows was an important way for electronic sounds to be domesticated into the 1960s American household (Taylor 2012). The monetary gain to be had from working in this new world of electronic sound for advertizing was not lost on two musicians who eventually became Moog's West Coast sales team. "That's for me" said Bernie Krause after reading about Siday in *Time* magazine.[6] Krause, a San Francisco based former folk guitarist with the Weavers, immediately boarded a plane to New York to visit Siday. He and his collaborator, Paul Beaver, sunk their life savings into purchasing a synthesizer, which quickly became their new instrument. Not only did they become Moog's West Coast sales team, famously selling synthesizers from a tent at the Monterey Pop Festival (see below), they also recorded their own highly regarded electronic music albums, made numerous ads, performed concerts, and became in demand in recording studios and on movie sound stages for sessions where Moog sounds were sought after.

The Modular Moog synthesizer was a complex instrument to control or "program" (early LPs refer to the operators as "programmers"), because different modules could be "patched" together in different configurations, and numerous knobs could be turned to also control the sort of sound made. Musicians could also interact with the instrument via a monophonic keyboard (one note a time) or via a newly designed ribbon controller which allowed for sweeps of sound. In the hands of a skillful synthesist, such as Paul Beaver, the instrument could generate a range of different sounds. But having this new sonic technology in combination with someone skilled enough to make sounds was not enough for particular sounds to acquire value. The sounds of value must be reproducible. This means first that they must be named or signified in such a way that the operator and listener can identify the relevant sound. Second the operator must have a way of returning to that sound. If the Moog sounds were to acquire value, the "weird shit" which Moog first referred to must somehow become differentiated enough, recognizable enough, and reproducible enough, for other musicians, record producers and the like to make use of the new sounds. The problem initially

[6] Interview with Bernie Krause, October 24, 1998.

faced by musicians and producers is nicely captured by the Doors keyboard player Ray Manzarek at an early recording session where Paul Beaver is playing the Moog synthesizer. The Doors had just broken through as a psychedelic rock band and were about to make their second studio album, *Strange Days*. Manzarek:

> Paul Beaver began plugging in a bewildering array of patch cords. He'd hit the keyboard and bizarre, Karlheinz Stockhausen-like sound would emerge. "Actually that sound you had about three sounds back was very usable. Could you go back to that?"... "That Crystalline sound" Jim Morrison joined in. "I liked the sound of broken glass falling from the void into creation." Which sound was that?" said Paul Beaver. (Manzarek 1998: 256)

The identification and naming of early Moog sounds had not at this point stabilized. There was what might be described as an "overflow" of sounds (Callon 1998). Even if a musician identified a sound he or she liked they were often unable to describe it in such a way that the Moog operator knew what they meant. What could be the sound "of broken glass falling from the void into creation" which Jim Morrison heard? We will probably never know! Paul Beaver failed to be able to return to the sound. Manzarek:

> "Which sound was that?" said Paul Beaver. "A couple back from where you are now," Rothchild [the producer] said. "It reminded me of the Kabbalah," said Jim. "*Kether*, the I AM, creating duality out of the one. All crystalline... and pure. You know. That *sound*." "Did I make a sound like that?" "Sure," Jim said, "A couple back." "Just go back to where you were," said Rothchild. And Paul Beaver began to unplug and replug patch cords, and twist little knobs, and strike the keyboard, which emitted strange and arcane and utterly unearthly tones that sounded nothing like Kabbalah or Kether, the crown of the Sefiroth. None of the sounds he was creating were pure and crystalline. And then we realized... he couldn't get back. (Manzarek 1998: 257)

The Doors eventually used some Moog sounds on their album *Strange Days*. The spacey sound of the Moog filter warbling can be heard echoing Morrison's voice as he announces on the title track that "Strange days have found us."

The Stabilization of Moog Sounds

A *sine qua non* for sounds to acquire value is they must be recognizable and reproducible. But a prior question has to be asked—what even constitutes a sound? If we mean by sound a vibration in the air detectable by the human ear, then the Moog synthesizer alone produces no sound. What it produces is a range of continuous and discontinuous electromagnetic wave forms which must be fed into a pair of head phones or an amplifier and loud speakers to be audible. That listening in the twentieth century via devices such as head phones and loud speakers is a form of practice which hardly needs description is a remarkable achievement of several centuries of auditory culture as Sterne (2006) has powerfully argued. Other technologies, such as the telegraph, stethoscope, telephone, and radio helped prepare for the sonic soundscape and listening practices of modernity.

What sort of sounds would listeners to the Moog synthesizer circa 1964–70 have experienced? To understand this we need to delve a little more into the technology. As with all musical instruments, the synthesizer has something vibrating (in this case electrical oscillators) as its source of sound. The oscillators produce different electromagnetic wave forms—typically, sine, triangle, saw tooth, and square—which have different overtones and musical properties.These wave forms can be mixed together and can also be frequency and/or amplitude modulated. This analog transistor technology involved enables a crucial form of control known as "voltage control." This means a musical property of interest, say the pitch of an oscillator, can be varied or controlled by simple changes in DC voltages.

The genius of the Moog synthesizer is the flexible patching arrangement. The output of a module, say a low-frequency oscillator producing a sinusoidal varying voltage, can be fed into another oscillator as an input and will frequency modulate the audio frequency of that oscillator according to the value of the voltage input. In this way a key musical property which is pleasing to the ear—vibrato—can be produced. Other modules on the synthesizer enabled even more complex changes to wave forms. For example an important module was the low-pass filter (the only module Moog patented) which removed higher over tones from complex waveforms. A monophonic keyboard (one note at a time) can be further used to control parameters such as the pitch of an oscillator. The sonic character of acoustic instruments depends a lot on how fast a sound grows and decays. For instance if you strike an individual piano key the sound builds up quickly (fast attack) and decays slowly. Moog developed an envelope generator which enabled sounds to be controlled from the keyboard in terms of the amount of their attack, sustain, decay and release (ASDR). Further sources of sound such as white noise could also be patched in. By feeding back outputs into inputs it was possible to assemble very complex sounds that bore little or no relationship to acoustic instruments.

When people refer to "getting sounds" or "getting particular sounds" on the Moog synthesizer they do not mean simply producing any sound. As long as it is set up correctly the synthesizer will make sound—indeed usually there is an overflow of sounds and the skill is to winnow the useful sound from the sonic overflow. Implicitly when people ask a synthesist such as Paul Beaver to go back to "a sound" or "the sound he made three back" they mean a sound that in some way has become recognizable and distinguishable to the ear. In other words a sound which is starting to acquire value (usually, in the context of recording studios, musical value), as opposed to the vast array of sounds that are physically capable of being produced by the instrument. A sound of value is not simply a particular pitch or wave form or envelope shape, it is made up of some combination of these parameters but is too complex to be described exactly in this way.[7]

The sounds which get valued in the early development of this sonic culture are crucially dependent upon the listening context. Wind, surf, explosions, thunderclaps, siren and other ear-catching sounds are good for demonstration purposes. But it is questionable whether these sorts of sound have any lasting value in musical compositions.[8] The novelty of a sound is itself a complex matter. A familiar sound, such as the wind, can *sound novel* when produced in a new way such as by a synthesizer. Also the pharmacological state of the listener is important. Hearing a thunderclap sound when tripping on LSD may itself produce a novel sonic experience (see below). A novel sound, upon repeated exposure to listeners, can also start to lose its attraction and quickly become stale or even kitsch. This is a problem for modern recording engineers who report that they can no longer listen to music as all they hear are "production values."[9] In other words the valuation of novel sounds depends in part on the sonic skills of listeners and their exposure, which in turn depends on how familiar the sound is from recordings, radio or live performance. In short the valuation of sound is itself part of complicated issues of political economy, such as the musical recording industry.

Real Head Music

Other complicating factors in the valuation of sound are the genre of music and the association of the experience of a particular genre with special pharmacological states of the listener. The Moog synthesizer arrived just at the moment when psychedelic or acid rock as a genre was taking off. The genre

[7] Only very occasionally does the actual shape of the waveform allow it to be named. For instance a triangle wave sitting on top of a square wave was called the "building the house" sound by Moog's studio musician, David Borden (because it looks like a house on an oscilloscope) when he wanted that sound produced by a Moog synthesizer in his ensemble, Mother Mallard (the first ever live synthesizer ensemble).

[8] But as Myles Jackson (2006) points out, the siren, when first developed, was used for several musical compositions. And all these sorts of sounds were eventually incorporated into psychedelic music—listen to for example the track "On the Run" on Pink Floyd's album *Dark Side of the Moon*.

[9] This point was made to me by Owen Marshall, a Cornell graduate student who is currently carrying out ethnographic research in LA pop recording studio.

was of course built upon the driving beat of rock 'n roll, but added strange washes of sound and unusual timbres (hence the popularity of the sitar). This music was experienced by listeners (and often performers) as they took psychedelic drugs such as marijuana and LSD. A key moment was the 1967 Monterey Pop festival which is generally regarded as the moment when the recording industry discovered the San Francisco sound. Of course, if the sound of an instrument is complicated, so too is the sound of a place. The festivals in this period, such as the famous Trips Festival, where the Buchla synthesizer first made its appearance, and the Monterey Pop Festival are important places for technologies to be discovered, experimented with, and valued (Schulte-Römer 2012).

It is at the Monterey Pop festival that the San Francisco sound first emerged on the wider stage, echoing the popularity of Bay area psychedelic rock bands such as the Grateful Dead and the Jefferson Airplane. The music recording industry was wealthy in this period and major recording labels offered large record advances to promising bands who conveyed the new psychedelic rock sound. Beaver and Krause were set up at the Monterey Pop festival in a booth with a Moog synthesizer and headphones. Bernie Krause describes what happened next:

> Little by little, as people got more stoned, they came skulking into the booth. And we'd set things up like thunderclaps and Hammond organs and people were really impressed...and they came in all guises...Like representatives of the Byrds, a couple of folk from the Beatles were there, and the Stones and lots of different groups. And ultimately many of them, because of the large advances that the record companies were offering, ended up buying the synthesizers from us right there on the spot. And I think we probably sold six or seven synthesizers at $15,000 a crack at that concert alone in maybe one afternoon. I mean it was unbelievable.[10]

The value of those particular thunderclap sounds is rendered directly into economic value by Krause in terms of sales of the synthesizer. It is easy to underestimate and difficult to conceptualize how the pharmacological state of the musicians and listeners at the time played into the value of sound. But it was clear that to the participants there was a direct correspondence. The newness of the mental space which the drugs made available seemed to take on a resonance with the newness of the Moog sounds.

The Moog synthesizer was used for the first time ever at a recording session on the West Coast for the album *Zodiac Cosmic Sounds*. Moog was himself invited to attend the session. He was famously a nineteen fifties-style

[10] Interview with Bernie Krause October 24, 1998.

engineer who always wore a pen protector in his shirt pocket. He captures the ambience of the recording session in his usual irreverent manner:

> All these people were going around with their shirts like this [unbuttons his shirt]. With a big, mother-fucking amulet right here [points to his belly] and their pot bellies sticking out. And they're talking to each other how great it is to be on dope...this kind of LSD over that kind, and the experiences they had...and they got through a couple of takes and they listened to it and it sounded pretty good—these big speakers and what not. "Oh man!" they said, "this is real head music. Every head in the country is gonna have to have this."[11]

There were many different sorts of sounds that made up "real head music." It is often hard for the participants from this particular era to describe or recall particular sounds.[12]

Only a few of the sounds of the Moog, however, were valued in the long run. Beaver and Krause noticed over time a subtle change in the sorts of sound demanded of them. They were often requested to make sounds that they had made for other musicians on earlier recordings, "Could you get us the sound you used on a Stevie Wonder date?" or "The Sound with the Byrds."[13] After a time Beaver and Krause found themselves using "a limited repertoire of 20 or 30 sounds that we got, that were very easy to patch and do. Finally, it just got to the point where it was becoming so simple and ridiculous that we were able to replicate those sounds even on a Minimoog or a Model 10 [smaller more portable synthesizers] and we didn't even bring the big synthesizer with us. Nobody wanted to explore."[14]

What Beaver and Krause were witnessing was the stabilization of sound. Instead of the endless possibilities which the Doors heard when Beaver first entered the recording studio with the novel instrument, about twenty to thirty sounds had become desirable enough to be recognized, described and which were "easy to patch" (in other words reproducible). It is hard to capture now, fifty years later, what exactly those particular sounds were. But I want to examine as a site of dissonance one specific sort of sound made by the Moog which has endured right into contemporary genres of electronic music. This is the legendary "baruuump" bass sound—a sound reproducible by the Minimoog and often taken as part of the defining genre of hip hop and much electronic dance music.

[11] Interview with Robert Moog June 6, 1996.

[12] A feel for the range and timbres of the Moog sound and the dominance of the "fat sound" produced by the Moog low pass filter in the bass range can be experienced by listening to classic Moog psychedelic records from the period such as TONTO's Expanding Head Band's 1971 release, *Zero Time*.

[13] Interview with Bernie Krause October 24, 1998.

[14] Interview with Bernie Krause October 24, 1998.

He Turned as White as a Sheet

The Moog modules needed to produce the bass sound are oscillators, a keyboard, the low-pass filter, and envelope generator. When a saw-tooth bass waveform is filtered and shaped by the envelope generator the sound is astonishing. It has the qualities of a plucked sound, the resonances and timbres of a bass guitar, but is a much fuller, flatulent sound with much more sonic energy and an extended sustain. Again it is notable how hard it is to describe this sound.

One of the first ever recordings on the East Coast to use a Moog synthesizer was a track by Simon and Garfunkel called "Save the Life of a Child" on their 1968 album *Bookends*. Simon and Garfunkel had been involved with the organization of the Monterey Pop Festival and, although folk artists, were heavily involved in experimenting with psychedelic sound. Art Garfunkel:

> Robert Moog had invented the synthesizer, and was having courses in New York, for those who wanted to learn about this. And we would go to these classes to learn about the instrument. And we arranged to bring the whole thing into the studio, it's an enormous contraption. We didn't know how to use it. We knew it could make these wonderful buzzy saw-tooth sounds. It could do so many different things, but we said let's use it as a bass instrument, let it be a pretend electric bass, and we tried to exploit the buzzy sound at the low end. So that's the first time it had ever been used.[15]

Moog himself described how this sound was made:

> Paul Simon was doing "Bookends" and I brought it [the synthesizer] in—you know they paid me well for this—they just wanted it there for a couple of days . . . I sat around watching the session . . . at that point I could get sounds pretty fast. I knew the equipment. But one sound I remember distinctly was a plucked string, like a bass sound. Bump, bump, bump. How did this go Baruuump. Then it would slide down—it was something you could not do on an acoustic bass or electric bass and John Simon [the producer] finally did it with a pedal. And I can remember, while John was fooling around and getting this sound and sort of playing it, a couple of session musicians came through. One guy was playing a bass and he stops and he listens and he listens. He turned as white as a sheet.[16]

Moog points to several interesting things in terms of the valuation process of this bass sound. First off he notes that his services do not come gratis, "they paid me well for this." Being able to produce sounds is a source of value to session musicians (as Beaver and Krause also found). Moog by this point "knew the equipment" and more to the point "Could get sounds

[15] Art Garfunkel interview at http://www.artgarfunkel.com/articles/bookends.html (accessed February 12, 2014).

[16] Interview with Robert Moog June 5, 1996.

pretty fast." Moog describes a particular sound made in the session as a "plucked string." Then by using onomatopoeia he phonetically reproduces the sound (something he was particularly skilled at doing) and points to it as a sound which an acoustic or electric bass cannot make. The particular value of this sound is pointed to by Moog by drawing in yet other listeners—in this case a passing session musician who played bass and "who turns as white as a sheet." This image wonderfully conveys the potential value of this sound as that of putting the bass session musician out of work. In short he can see his skill capital will be devalued if this major upheaval in this moment of dissonance prevails.

So here sound is valued in terms of displaced labor. This nicely demonstrates the double meaning of valuing sound: the sound is appreciated (valorized) by a population of listeners, and the money paid by these consumers for the new sound results in less sales with old sounds measured by the personal expected loss of income for the musician. The point here is that the baruuump bass sound builds upon a familiar old bass sound and transforms that sound and the way of producing it. Such a sound, because it taps into a sonic palette already available and goes beyond that palette, is likely to be recognizable to a knowledgeable practitioner as being a moment of major dissonance. For instance, it is less likely that a session musician would turn white if he or she had strolled through the Doors session and had heard the same crystalline sound at the edge of creation that Jim Morrison heard.

The baruuump bass sound can be heard at the start of the Simon and Garfunkel track, "Save the Life of a Child." Malcolm Cecil, another prominent Moog musician who made most of Stevie Wonder's early Moog sounds, used the ribbon controller through the low-pass filter to get a similar "baruuump" bass sound on the track "Boogie on Reggae Woman."

The baruuump bass sound went on to become a valued part of the Moog sound and is still desired today. Obviously the full story of the value of the sound here would have to delve into several genres of music, such as Hip Hop and Electronic Dance Music where the rhythmic sound of bass is important. Listening in these genres of music, as it is for tribal drumming and centuries of organ music, is not only done through the ear but also through the body.

The Yawling Sound of the Minimoog

Another highly recognizable sound is what is known as the monophonic yawling sound produced in particular by the Minimoog. This sound first arose to prominence in a genre known as "progressive rock" in the mid 1970s. It was a period when synthesizer soloists such as Keith Emerson and Rick Wakemen became rock stars and the sound is often associated with them as performers.

Figure 2.2. The Minimoog Synthesizer (musician is Chris Swansen)

The Minimoog (developed in1970) is a portable keyboard instrument which was built specifically for live performance (see Figure 2.2). Whilst the Modular Moog was largely designed by Moog alone, the Minimoog was produced by a small team of engineers. The Minimoog originated as a lunch-break project by one engineer who wanted to assemble modules in a fixed combination which corresponded to the sorts of configurations routinely used in demonstrating the Modular Moog. This would allow the main sounds of the instrument to be more conveniently demonstrated.

How do the engineers determine which sounds are most useful? The answer is remarkably straightforward—the sounds that they repeatedly set up to demonstrate to visiting musicians. It is exactly the same thing that Beaver and Krause noticed whereby they were only asked to produce a limited repertoire of about twenty to thirty sounds. The engineers could learn the values of different sounds because of a unique space that Moog had built—a factory studio. The origins of this studio can be traced to a two-week summer electronic music seminar Moog organized early on in his manufacturing career. He invited prominent and up-and-coming electronic music composers to use his rudimentary equipment at his factory. The benefits were two-fold—the composers got to use and hear about the latest equipment and he got to observe and listen to how they used it. It is from this symbiotic relationship that the engineers started to learn which sounds were valuable to composers. The factory studio was set up for studio musicians to use late at night (to explore the capability of the instrument and also for idiot-proofing gear) and for demonstrating and instruction to customers who visited during the day (customers were given more instruction depending on the amount of equipment they purchased). Moog and the engineers who worked at the factory started to recognize which sounds worked best in demonstration. It was the next step to build a smaller Moog which contained "hardwired" versions of these same desired sounds. This saved the time-consuming process of patching up each new sound. Now the sounds could be reproduced more easily by using switches, turning knobs, and hitting notes on the built-in keyboard.

We can think of the factory studio as another place, like the Monterey Pop festival where value can be assessed or tested. It enabled a huge variety of musicians to interact with the Moog and for the engineers to participate in and learn from that interaction. Value was in a very real sense made in the course of these interactions. It led directly to the Minimoog synthesizer which is widely recognized as the template for all future synthesizers and the invention of which has been recognized with numerous awards.[17]

The Minimoog has its own properties as an instrument and its own distinctive sounds. In particular new controllers, known as "pitch and modulations wheels" are located at the left-hand end of the keyboard and became part of the standard architecture of keyboard synthesizers—and can still be found on keyboard synthesizers to this day. The Minimoog as an instrument is much simpler to operate and much cheaper to purchase than the Modular Moog. In another interesting form of valuation it was described in 1970 as costing the same as a rock group's van. In other words, although sold at only $1,195 at the time it was still an expensive instrument.

Sound becomes stabilized to a degree by hard wiring because it assumes a particular patching configuration in making the sounds. Not every sound that could be made by the Modular Moog could be made by the new instrument. And not everyone saw the value immediately in the new instrument. Moog himself regarded it at first as little more than a toy and was initially resistant to mass producing the instrument.

The Minimoog has three oscillators with a variety of wave forms which can be routed through the filter and envelope generator. The filter and envelope generator are triggered by pressing monophonic keys which also control the

[17] For example, the Polar Prize of Music in 2001, a technical Grammy in 2002 (shared with Steve Jobs), and Moog's induction into the Inventors Hall of Fame in 2013.

pitch of the oscillators. Keyboard controls such as modulation, pitch bending and portamento (ability to glide between notes) means that the instrument is capable of producing fast piercing monophonic solos akin to a lead guitar but with a much richer and more varied sonic energy and range of tones and timbres.

One of the key features of the sound is that the monophonic sonic energy enabled the keyboardist in rock groups to join the lead guitarist and perform solos which cut through the drum, bass and rhythm guitar accompaniment. This was noted by Minimoog salesmen who had to persuade musicians to buy the instrument (the Minimoog was the first ever synthesizer to be sold in retail music stores and part of the story is how it gets to be sold there, Pinch 2003). The keyboardist's role up until that point had mainly been in the background doing "fills." The soloing capability of the Minimoog enabled the keyboardist to move center stage and become "the" soloist on a par with the electric guitarist who could produce pure electronic sound by means of feedback (Hendrix is of course the most famous example).Sometimes guitarists even swapped their instrument for a Minimoog just to perform a solo.

The use of keyboard controls such as portamento over several octaves and the pitch bending of individual notes adds to the sound's effectiveness. The success of this sound—how it became highly valued—is more complex than for the baruuump bass sound documented above because it also involves performance. This use of the Minimoog was part of a new genre—progressive rock—which included visual aesthetics (style of clothing, hair and so on) and more crucially demanded virtuosity in live performance. Groups which formed part of a second "British invasion" such as Emerson, Lake and Palmer and Yes played to stadiums throughout the US. The Minimoog as an instrument helped enable the virtuosity of performers such as Rick Wakeman and Keith Emerson—it enabled the audience to see with their own eyes and hear with their own ears virtuosity being performed. For example, a virtuoso performer such as Keith Emerson kitted out in all his pomp rock finery in live concert often exaggerated his pitch bending gestures visually with raised arm motions enabling the audience to see how bodily gesture and sonic effect were aligned. Changing the resonance of the filter during a solo by manipulating the "Frequency Cut Off" control on the Minimoog also enriched the sonic quality of the sound. This is a sound which thus becomes valuable not only within a particular genre of music but also within a particular style of performance. The value to the Moog company was noted by the sales force as sales of Minimoogs always followed in towns where Emerson, Lake and Palmer had just performed.

In this case the value of the sound is also grounded in the aesthetics and architecture of rock stadium performances, genres of music, as well as the recording industry. Furthermore the sound is also part of gendered performances with the main performers of progressive rock being male (although see the example of Heart fronted by two women below). For a while the keyboard became equivalent to the "technophallus" electric guitar (Waksman 2001). Keyboards, as they got lighter, started to be worn around the chest as guitars and eventually special guitar-style synthesizers were designed such as the Moog Liberator (see Figure 2.3).

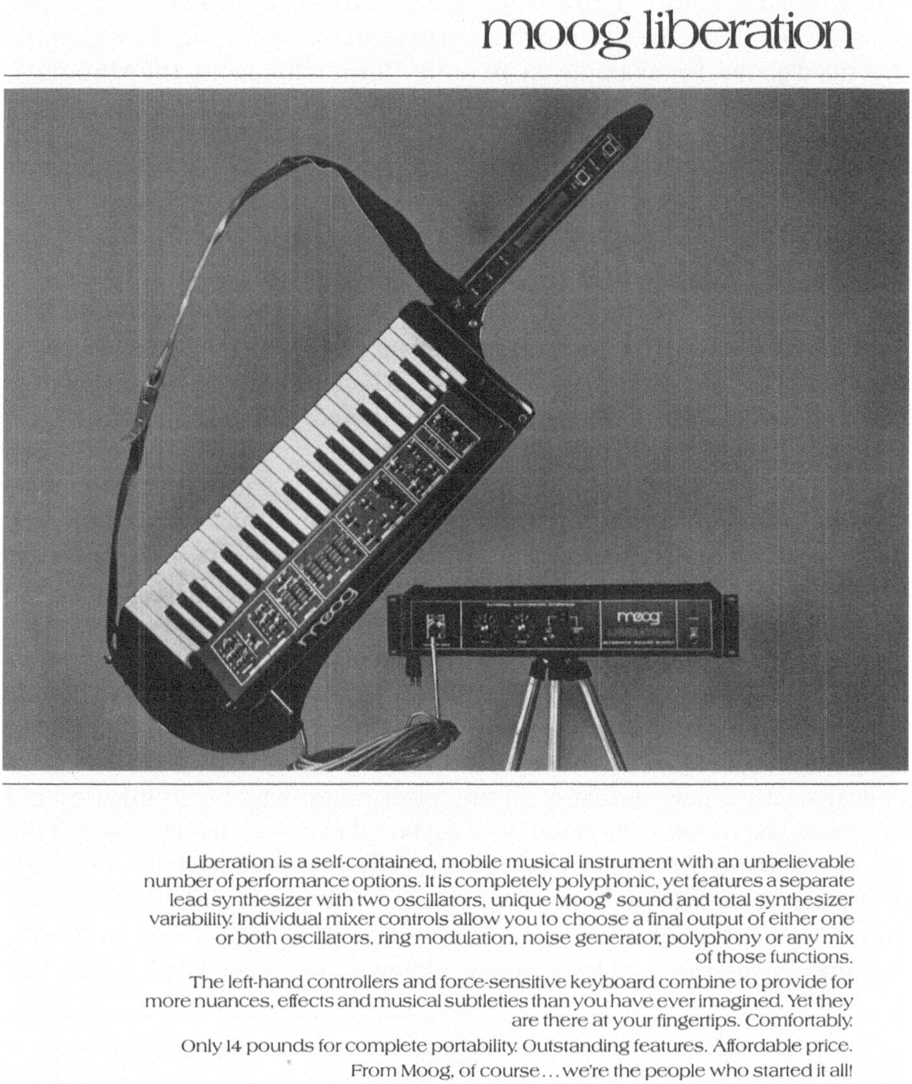

Figure 2.3. The Moog Liberator (original advertisement)

The type of yawling Minimoog solo described here is to be found on Emerson Lake and Palmer's successful 1970 single "Lucky Man" (towards the end of the recording).[18] A nice visual example is the Minimoog solo performed by a guitarist in the band Heart on their 1976 song "Magic Man." Within the genre of early progressive rock this is a typical generic type of Minimoog sound and performance practice and the musician can be clearly seen to adjust the filter "Cut Off" control as he plays the solo (note that the guitarist has his guitar casually slung over his back as he plays the Minimoog solo).[19] The Minimoog is still highly valued as an instrument today because it can make the baruuump bass sound referred to earlier and the type of lead sounds described above. For example, in an online forum discussing the Minimoog a musician writes, "Dance and Techno musicians like the MINI especially for its capability to make "boomin' basses" and "in yer' face" leads and of course the strange noises!!!"[20]

Buchla—Dissidence and the Politics of Valuation

Lastly in concluding this essay on the valuation of electronic sound I want to turn to the Buchla synthesizer (Figure 2.4) which produced a different range of sounds which became valued for rather different genres of music.

Donald Buchla, like Robert Moog, received his training at an elite university (UC Berkeley) and he is recognized as having come up with the modular voltage controlled synthesizer idea at almost the same time. Buchla, however, had a very different artistic vision for the instrument and had a different configuration of modules and controllers. Buchla worked at the Tape Center in Haight Ashbury in San Francisco and this brought him into contact with avant garde and experimental composers such as John Cage and later Steve Reich. Buchla was also part of the Bay area psychedelic scene being friends with Ken Kesey and the early Grateful Dead (for whom he made equipment). Buchla rejected the standard keyboard as a controller. His vision was more radical—with a new source of sound, electronics, why be stymied by old interfaces taken over from traditional keyboard instruments? He rejected the use of keyboards and preferred arrays of plates that produced different voltages when pushed. He designed a way of controlling his synthesizer that was an electronic form of music concrète. It enabled the synthesizer to be programmed by repeated cycles of voltages known as a "sequencer." Buchla had

[18] The solo was actually performed in the studio on a Modular Moog as the Minimoog was not yet available.

[19] Video clip of Heart performing "Magic Man." Minimoog solo begins at 4:34 and can be seen at 4:56. www.youtube.com/watch?v=0uvr3dmptvg (accessed February 12, 2014).

[20] www.sonicstate.com/synth/moogmin/ (accessed February 12, 2014).

Figure 2.4. Buchla 100 Synthesizer with composer Mort Subotnik

a radical aesthetic. For example, the module known on the Moog as "white noise" for the hissing ambient sound it produced, was labeled by Buchla as "The Source of Uncertainty."

Buchla's synthesizer, like Moog's, produced a bewildering array of sounds. Many of these sounds were also suitable for explorations of psychedelic sound and indeed Buchla discovered this before Moog. Buchla's synthesizer (operated by Buchla himself) was on stage at the famous San Francisco Trips festival (Turner 2006) where bands such as the early Grateful Dead and Big Brother and the Holding Company (before Janis Joplin joined), performed. A Buchla synthesizer was even installed in Ken Kesey and the Merry Prankster's bus "Further" for their famous trip to try and turn America on to LSD. But particular sounds never standardized as "the" Buchla sound.

Part of the reason for this is that Buchla rejected the idea that music should be about keyboards and standard octaves. Moog standardized his synthesizer around a technical standard called the volt-per-octave. This meant that a one volt change of input into a module, such as an oscillator, produced an octave change in pitch. Other early commercial synthesizer companies such as ARP in Boston, and EMS in London used a version of the volt-per-octave standard. This meant that in fact part of musical culture—that music is about octaves—was being hard wired or embedded into synthesizers and the sorts of sounds they could produce. But musical culture, as Buchla and experimental musicians such as John Cage pointed out, did not have to be about conventional octaves. Buchla's synthesizer, because it never used a standard keyboard, had no standard equivalent to a volt-per-octave. Indeed Buchla rejected the whole idea of standardization regarding himself as being more akin to a violin manufacturer where each instrument should have its own quirks and special resonances. He valued the synthesizer as an instrument, not as a machine.

Given the importance of conventional melody as embodied in the twelve semitones of the octave it meant that it was much harder for particular Buchla sounds to acquire musical value. Musicians and engineers would point to particular sonic qualities of Buchla-sounding instruments such as the purity of his sine tones, the dramatic arpeggios of sound which the sequencer produced and so on—they could tell the "sound" of a Buchla instrument overall, but could not point to any one particular sound it made. A good listening point here is Mort Subotnik's 1968 album, *Silver Apples of the Moon,* made on a Buchla synthesizer.

Paths Taken and Paths not Taken?

The Buchla story provides a counterpoint to the story of Moog. It is about a path not taken and a whole different way of valuing sound. The Buchla sound in a way has resisted commodification. It reveals some of the politics underlying metrics of valuation. Buchla rejected mass production and hence tried to resist the logic of capitalist production values. The Moog synthesizer, via the path of the Minimoog, became the standard design for keyboard synthesizers mass produced today by Japanese multinational corporations such as Korg, Roland, Yamaha, and Casio. Buchla has survived with a much smaller niche market. Moog has been widely lauded as the inventor of the synthesizer (even recently honored with a Google Doodle) and Buchla has largely been neglected. Specific Moog sounds can be recognized, identified and reproduced, unlike Buchla sounds. Later generations of synthesizers have presets and software pull-down menus which reproduce generic Moog sounds. Moog sounds today have become part of our mass culture: they have acquired value.

The process whereby electronic sounds acquire value is complicated. It involves special places such as studios and factories, it involves technologies, and it involves musicians, record producers, and ordinary listeners. The purveyors of mass culture such as the recording industry, movies, radio and TV also play a role. Taste and value, as we have seen, do not arise naturally.

They are not exclusively properties of people either, although when taste and value have stabilized they may appear to be that way. Value is as much a part of a technology and the material world as it is a property of people or social classes (think of the musical value of octaves embedded within the Moog synthesizer). The recognizability, stability, technological means of rendering and controlling, practices (including gendered practices and performances), and situated locations (such as the Moog factory studio or festivals), all set within a specific cultural context (notably the psychedelic sixties), are important ingredients in understanding how in this case electronic sound began to acquire value.

References

Bijsterveld, K. (2008). *Mechanical Sound: Technology, Culture and Public Problems of Noise in the Twentieth Century.* Cambridge, MA: MIT Press.

Callon, M. (1998). *The Laws of the Markets.* London: Blackwell.

Corbin, A. (1998). *Village Bells: The Culture of the Senses in the Nineteenth-Century French Countryside.* New York, NY: Columbia University Press.

Fourcarde, M. (2012). The Vile and The Noble: On the Relationship Between Natural and Social Classification in the French Wine World. *The Sociological Quarterly* 53: 524–545.

Glinsky, A. (2000). *Theremin: Ether Music and Espionage.* Urbana, IL: University of Illinois Press.

Holmes, T. (2012). *Electronic and Experimental Music: Technology, Music, and Culture.* 4th ed. New York, NY: Routledge.

Jackson, M. (2006). *Harmonius Triads: Physicists, Musicians, and Instrument Makers in Nineteenth-Century Germany.* Cambridge, MA: MIT Press.

Katz, M. (2004/2010). *Capturing Sound: How Technology Has Changed Music.* Berkeley, CA: University of California Press.

Manzarek, R. (1998). *Light My Fire: My Life with the Doors.* New York, NY: Putnam.

Perlman, M. (2004). Golden Ears and Meter Readers: The Contest for Epistemic Authority in Audiophilia. *Social Studies of Science* 34: 783–807.

Pickering, A. (1995).*The Mangle of Practice.* Chicago, IL: The University of Chicago Press.

Pinch, T. (2003). Giving Birth to New Users: How the Minimoog Was Sold to Rock and Roll. In N. Oudshoorn and T. Pinch (eds.), *How Users Matter: The Co-Construction of Users and Technologies.* Cambridge, MA: MIT Press, 247–270.

Pinch, T. and Bijsterveld, K. (eds.) (2012). *The Oxford Handbook of Sound Studies.* New York, NY and Oxford: Oxford University Press.

Pinch, T. and Trocco, F. (2002). *Analog Days: The Invention and Impact of the Moog Synthesizer.* Cambridge, MA: Harvard University Press.

Schulte-Römer, N. (2013). Fair Framings: Art and Cultural Festival as Sites For Technological Innovations. *Mind and Society* 12(1): 151–165.

Shapin, S. (2011). Changing Tastes: How Things Tasted in the Early Modern Period and How They Taste Now. The Hans Rausing Lecture 2011, Uppsala Universitet, *Salvia Småskrifter* 14: 7–47.
Shapin, S. (2012). The Tastes of Wine: Towards a Cultural History. *Rivista de Estetica* 52(51): 49–94.
Sterne, J. (2006). *The Audible Past*. Durham, NC: Duke University Press.
Sterne, J. (2012). *MP3: The Meaning of a Format*, Durham, NC: Duke University Press.
Taylor, T. (2012). The Avant-Garde in the Family Room: American Advertizing and the Domestication of Electronic Music in the 1960s and 1970s. In T. Pinch and K. Bijsterveld (eds.), *The Oxford Handbook of Sound Studies*. New York, NY and Oxford: Oxford University Press, 388–407.
Théberge, P. (1997). *Any Sound You Can Imagine: Making Music/Consuming Technology*. Hannover and London: Wesleyan University Press.
Thompson, E. (2002). *Soundscapes of Modernity*. Cambridge, MA: MIT Press.
Time (1966). Composers, Swurpledeewurpledeezeech! Time, November 4.
Turner, F. (2006). *From Counterculture to Cyberculture*. Chicago, IL: The University of Chicago Press.
Waksman, S. (2001). *Instruments of Desire*. Cambridge, MA: Harvard University Press.

3

Paying Attention: What is Tasting Wine About?

Antoine Hennion

What is a good wine? Do you drink a wine, or a label—and a price? In this chapter, I will focus on wine tasting settings. I will address the minute relationships that have to be set up between bodies, small groups of people, and specific *agencements* (Callon 2013) in order to get an agreement. Not on wines themselves—agreeing on ways of disagreeing is sufficient—but on acceptable ways of proceeding in order to get divergent valuations discussed and circulated.[1] Wine is a good case to scrutinize how sensations connect to markets (Chauvin 2010; Muniesa and Trébuchet-Breitwiller 2010). In a pragmatist approach, I will delegate the issue to the actors themselves.[2] Wine producers, critics and amateurs do not only invent procedures to taste wine, they do not only dispute about which is best: through those debates, they also gradually produce definitions of quality and make explicit the criteria necessary to measure them (Shapin 2011, 2012). In so doing, they closely connect issues concerning the taste of wine on the palate, appropriate ways of making it, and conditions that permit its appreciation. The intermediary concept of attention, that has been scrutinized by experimental psychologists (Grabenhorst and Rolls 2008; Veldhuizen and Small 2011), and that organization studies have also underscored (Ocasio 2011), is curiously left aside by sociological approaches in sensory studies. It will appear to be central here: organizing

[1] From diverse perspectives, valuation studies are growing fast, and have become a specific field of inquiry (Stark 2009, 2011; Vatin 2009; Hutter 2011; Muniesa 2011; Fourcade 2011; Lamont 2012; Mallard 2012; Helgesson and Muniesa 2013).

[2] The empirical data originate from a long-term research project in which Geneviève Teil and I have compared various forms of attachment (Hennion 2004, 2007, 2010), in this case music and wine (Hennion and Teil 2004). Our project has shown taste not as a passive, over-determined social game (Bourdieu 1984), but as a reflexive and equipped activity.

tasting amounts to defining a whole pragmatics of attention, by connecting bodies, values and settings.

A Second of Attention

Scene one: The dinner is well under way, everyone is merrier, chatting and interrupting one another. A guest serves wine to his neighbor, who reaches for his glass, takes a sip, and puts it down again while still in conversation. He goes back to his meal, looks over, talks to another guest.

Cut to *scene two*: The same setting: same guests, same atmosphere, same gestures. The man takes his glass and starts to drink. At that point, he pauses, gives it two brief sniffs, drinks again, then puckers his lips as he puts the glass back down, before proceeding to continue the discussion where he had left it.

Very little sets these two scenes apart. A second of attention "lent," as we say in French, to what is being drunk. It is not necessary for the guest to make a comment or congratulate the host. Neither is it imperative that the person drinking be fully aware of a change of state, of intentionally shifting during the dinner from one type of presence (through conversational exchange) to another (that of his palate to the wine he is drinking, and of himself in his act of tasting), and back again. That would be a very heavy descriptive apparatus indeed, weighing down this instant with too burdensome a charge, the drinker with an overly defined intention, and the course of action with a rupture that no one has felt. What happens here is far more delicate, smooth and natural. It is an ordinary displacement of the sort we operate in all situations without being aware of it. No need to isolate two contrasting courses of action, as though in reality we were not permanently caught within a web of states, of modes of presence to ourselves, to others and to objects that interlace, overlap, and encase themselves like parentheses. We sweep crumbs off the table, we discreetly scratch ourselves, we laugh at someone's slip of the tongue, we apologize after too brusque a gesture, we interrupt ourselves for a moment to follow a conversation within earshot. We are constantly performing these minute controls: What is so-and-so doing? What are they laughing about out there? It's late! I wouldn't mind another helping of that dish that is being taken away. The ordinary state is this spontaneous management of a plurality of relations to one's body, to others, to things and events—and not some univocal positioning of oneself towards a definite object of attention. Although often taken for granted when considering taste, it is this latter disposition which in fact constitutes the exception: it requires an explicit effort, aimed at a precise object, it presumes intentionality and will, a material and temporal frame, training, time and favorable conditions. The (however slightly) more attentive act of tasting performed by the second guest compared to the first one still doesn't call for this whole arrangement. It is enough to open one of these many enfolded parentheses without disturbing the course of events more than necessary.[3]

Still, even reduced in this way to a nuance—a pause marked by a movement of the lips—this small gesture introduces an important distinction. However briefly expressed, this gesture signals another disposition of the self, the act of savoring moving to another level. Less an intention than a more focused attention, together with a stronger presence of the tasted object—each reinforcing the other without a primary cause. He has not only drunk, he has drunk *this* wine. There is no taste without this minimal ordering that makes an experience appear itself to itself, this slight shifting of the self with oneself. However faintly these elements come into view, without effort or motive, a parenthesis is opened within the course of what is happening, inflecting, directing and putting it into a frame. I drink and "I drink," I feel the effects and I pause to consider what "it" makes me feel. On the other hand, there is no taste either without this intensification on the part of the object which itself responds or provokes: it, too, adjusts itself, moves forward, unfolds, taking its time, slowly revealing all its flavors: A more salient contact and a heightened presence. Before any verbalizing has occurred, and still before the need to know more about the wine or who tastes it, I would like to focus our attention precisely on this moment of attention. As does the wine-taster, I want to reinstate, for our analysis of taste, the centrality of this open and inquisitive moment that marks itself through the sudden appearance of an intensified exposure, leading to a decentring both of the self towards the object and of the object towards the self. The minimal term of "attention" conveys this idea particularly well, as it brings together, in the same light stroke as that of the taster, the two displacements reaching for contact: to grant attention to, and to capture the attention of.[4]

Yet, this minimum already implies quite a lot: some reflexivity on the side of the taster[5] ("Hey! This wine's not bad . . . ": to whom other than oneself as the yet undefined subject of experience is this curious interjection addressed, this "hey!" by which one summons the required attention?). And on the part of the object, it implies a capacity to interrupt, to surprise or to respond. This right to advance is their own form of reflexivity, their power to make themselves more present. Objects are never already there, inert and available,

[3] These issues had been reformulated by Thévenot (1990).

[4] See Merleau-Ponty's invaluable analyses, especially (1964).

[5] The reflexive character of an activity (here presented in its most basic form through the simple fact of entering into a self-identified disposition) does not automatically mean that there is reflection on the part of the actor—this would involve a much higher level of calculation and consciousness of one's action, as well as a shift from a simple variation in one's modes of presence in situation, to the register of deliberate action.

at our service. They reveal, hide, or impose themselves. Luckily, language is less strict than the philosophies of the subject or of intentionality, since it unscrupulously grants objects this use of the reflexive form (as we would say of a wine that needs time to "express itself") that amateurs know only too well: beautiful things only offer themselves to those who offer themselves to beautiful things.

Tasting Without Seeing

Change of set, *new scene*: Again, a group of people are drinking around a table. Except that now they are paid to do it. Food critics, oenologists, retailers, experts in winemaking, and distinguished sommeliers are sitting facing a row of glasses quarter-full. The bottles are placed in covers, their label replaced by anonymous numbers. Their tasting is concentrated, silent, regulated by a succession of gestures and moments, and recorded as they happen in fast and precise notes.

Here it is, the constrained situation that we evoked above—and, by contrast, we can see the whole *agencement* that it requires. These moments of total inclination towards their objects alternate with equally regulated moments of relaxation, pauses during which all this efficient but exhausting attentiveness loosens up a little: the body relaxes, some give a little cough, others get up, talk to each other, a hint of irony in their voice, then, suddenly, everyone returns to the task at hand.

Later, faced with the bottles at last taken out of their cases and with the tasting session now officially over, the participants' tongues will loosen on another mode, either to talk about something else, or to reminisce enthusiastically about the competition and the surprises it revealed. Another interesting phase: the work of professionals goes on. The vocabulary is elaborate, the factors taken into account are analytical and varied, the argumentations precise and characteristic. These provide the connoisseur with factual or technical information, but also with positions and options chosen between different ways of evaluating the session, of assessing the wines themselves, of debating the evolution of such and such a winemaking technique or the work of critics (Teil 2001). During these open exchanges, everyone records everyone's opinion, confirms or rectifies their own choice, realizes that next time more attention needs to be paid to certain impressions, or strengthens their convictions about what really matters. After having gathered for the duration of the competition by the formalism of the exercise and the concentration required for its achievement, the diversities of professional aims eventually resurface: some think about their article, or the rising prices, others about their next *menu dégustation* or the future trends devised by influential critics.

A Matter of Expertise?

Is scene number three, of a blind wine-tasting competition, merely a wide-angled version of scene number two, in which we could notice our drinker marking a brief freeze-frame? Let us first try to examine things in this linear way, by placing both scenes on either end of a scale going from the degree zero of tasting to its most elaborate form; from a slight digression caught in the flux of other everyday interactions (so slight, in fact, that it might not even be clear whether it occurred at all, as is expressed in the passage from scene one to scene two), to the formalized and controlled setting up of a contact aimed at producing reasoned assessments (scene three). Incidentally, we can note that, moving from the dinner tasting to the professional blind testing competition, a whole range of previously indefinite variables has been systematically extracted and isolated. Their analysis and elaboration make the formal setting of the competition possible, but also help to distinguish between the amateur and the professional wine taster: a delineated frame, objects registered and put forward so as to allow for their systematic comparison, a set of mastered skills, competencies acknowledged by and representative of a variety of positions within the wine-making industry, as well as, for some participants closer to the art critic's position, the proficiency and right to defend a certain aesthetics, or a personal conception of their taste.

And we could push the analysis further in that direction, since behind those personal aspects regarding participants themselves lies the long and winding pathway that had to be mapped out in the *longue durée* for this confrontation to take place, here and now, between a collection of classified objects and their assigned judges: methods of production, oenology and sensory analysis, tasting techniques and professional formations, ranking and standardization of quality, marketing, development of a more autonomous and expert group of critics, and so on. That is to say, expertise: this way of representing the difference between the two scenes amounts to arranging them along a line that goes from the simple, more or less occasional amateur, to the confirmed expert. If that is the case, the professional organization of the act of tasting would only reveal (in objects, in bodies) and express (in a particular discourse, through professions and eventually on the market) the kind of analytical and historically produced variations that the sporadic amateur can only perceive in a flawed, unstable, and inarticulate way. Taste formation would therefore amount to a gradual move away from fleeting impressions and towards objectivity, mastery, the certainty of one's own judgment, the knowledge of the products and their tastes. Experts have learned to distinguish and assess the qualities of objects (they have all the necessary "equipment" at their disposal, from the *dispositif* set up to formulate comparisons to the training of their palate, from the vocabulary that acts as a resource for isolating what they feel, to the knowledge about wine-making processes that helps in sorting out causes and origins), whereas the indiscriminate amateurs could only access these qualities through the confusing prism of their moods, penchants and habits.

There is something photographic about this particularly objectivist representation: expertise is akin to the "developer" of properties contained within the objects. The same qualities act upon both amateurs and experts, but only the latter's experience and skills actualize what, to the former, remains virtual.

A White for a Red

Yet, experts do get it wrong and quite often, too. Not only on those fine distinctions that cause the delight of any specialist—to the contrary, our exceptional virtuosos excel precisely at this sort of challenge, like when they have to distinguish between two good years in the same *cru*, both with an equally rich tannin. No, these proficient tasters, who are able to spot the different chateaus of Margaux, can also make the most blatant mistakes. In the milieu, everyone has dozens of anecdotes to tell on the subject: a white wine mistaken for a red, a "Vieux Papes" likened to a Latour, by one of the most reputable palates. What's going on? Are wines simply indistinguishable without their labels? Could their actual effects be so far removed from the differences that have been officially tagged onto them? If so, what about taste and quality? Are the certainties of objects really that fragile? Is it not just an indication that progress still needs to be made, for even more analytical and better trained professionals, even more typified, controlled and stabilized products, and for the professionals of taste to incorporate more scientific techniques in their training? Is it not time to develop the measuring instruments that will at last enable to control perfectly the influences that tend to disturb the experts' evaluations?[6]

Of course, there is no lack of good reasons if the point is simply to let decent professionals off the hook. Experts themselves are the first to analyze the imperfections of tasting and to think up ways to refine their procedures: the temperature, the erosion of sensibility due to repetition, the unpredictable effects of certain associations of flavors, the weariness that comes with this succession. Some put things in perspective: after all, when you discard the few big mistakes and other false values, the majority of assessments remain

[6] For a critique of the influence of fashion and cultural biases on tasting, see Renvoisé (1996), Teil and Hennion (2004).

remarkably consistent. And what's more, isn't it a big part of wine's charm precisely not to let its effects be reduced to analysis so easily? "There's no accounting for taste ...": making a virtue of necessity, one can now relish the impossibility of reducing the world to objective criteria and praise wine for magnifying in such unexpected ways the joyful perseverance of Nature's fantasy.

But if our project is to understand taste, we cannot leave it at that. Because, in this perspective, the whole model of expertise as the embodied and objectivized "developer" of the object's qualities is rendered null and void.[7] Moreover, if we now have a look on the side of chemistry and visit INRA[8] laboratories, we can witness the same disjuncture at work between, on the one hand, the precision with which analyses of wine's components are carried out, and on the other, the difficulty to translate these quantitative properties in the realm of a perceived "taste." In an attempt to stabilize descriptions, experts have ended up splitting taste in two distinct registers, one taking charge of the variations located in the object, the other those related to the taster's subjective evaluation. The model has indeed changed: experts are not super-amateurs anymore, their acquired competencies exist on another plane, that of an objective grasp on "the" taste of things—meanwhile, on the side of people's tastes, the variety of forms and modalities of appreciation proliferate ad infinitum. The object characterized with increasing accuracy by the techniques of the expert-taster is no longer the same object that led our earlier drinker to pause; and on the taste of this amateur, nothing is certain anymore. The scoring cards used in wine tasting may now contain a descriptive section to separate objective grade from personal judgment, but the experiment proves difficult, especially when any identification of common criteria can be so swiftly denounced as a result of trends or worse, of social interests.

In reaction to the tendency for objectivist expertise, some tasters decided to push their own tastes forward and, following the model provided by film critics, have adopted a "subjectivist" approach. In fact, scientific studies and the sort of systematic training that facilitates a breaking down and identification of the wine's components reconstruct another object. But, above all, by isolating the wine from its taster, they establish a very peculiar relationship between the two. We have now returned to our initial question. What differentiates the amateur from the expert is not a matter of increase in knowledge,

[7] Bessy and Chateauraynaud (1995) have developed a very interesting theory of expertise, proposing a sociological reframing of Merleau-Ponty's phenomenological analyses, and conceptualizing judgment as an uncertain trial which destabilizes both the criteria of the judge and the folds of the object.

[8] The Institut National de la Recherche Agronomique is the French public research body on agriculture and food.

but of two divergent modalities for framing the relationship between the wine and its tasters: first, the judgment of taste relying on the taster and what he likes; second, the description of the taste inscribed in the wine itself and its properties, as uncovered by the technical equipment, the expert's blind testing/tasting, a strict procedure, the collectively approved list of all the relevant and superfluous factors of assessment. If there is no reason to think (as a spontaneist theory would have it) that knowing better should prevent from tasting better, as the former can often support and expand upon the latter, knowing is not simply the more objective prolongation of tasting either. Although not incompatible, knowledge prompts a kind of inner negotiation with oneself. And while it does not affect certain professionals, it can become an obsession for others: "If I keep analyzing, will I still be able to appreciate?"[9] The expert's act of tasting, geared towards a different object than that of the amateur, thus preserves intact the mystery of taste: why do we like certain things and not others?

The Temptation of Sociology

Having arrived at this point, the stubborn explorer is in great danger of falling into sociology: if objects do not govern tastes, shouldn't we look for the logic of taste in the system of classification rather than in the objects they serve to categorize? Doesn't taste tell us more about the dispositions of the person tasting than about the properties of the thing tasted? Now, the sprinkler is sprinkled, and the blind test is reversed: could the whole montage, which appeared at first to closely align taste and judgment, be a mere illusion? Instead of the wines being evaluated by the experts, is it not the wine's indiscipline that questions expertise itself—not as a competency but as the self-appointed representative on matters of taste. Suddenly, the fancy rhetoric and pseudo-scientific criteria seem destined first and foremost to mask the deeply arbitrary nature of tastes. By covering this arbitrariness behind the natural character of the objects they are supposed to be based on, the montage serves more to provide buyers with all the required signs, rather than to attend to the quality of wine as such. After all, the economic implications of such a substitution on a market would be far from negligible. This very hypothesis seemed plausible enough for the INRA to swiftly appoint economists and a couple of sociologists in the 1990s to work together with chemists and oenologists.

[9] This formula sums up the terms of a constant debate amongst amateurs, between those who promote verbalization and discussion as means of accessing ever-deeper strata of knowledge about an object, and those who consider this discursive operation as inevitably reductive and having disruptive effects on perception.

Without having to go back to Bourdieu, it is enough to listen to the expert oenologists. Faced with this question, they have established a radical break between the taste *of* wine and the taste *for* wine—a distinction only overlooked because of the approximations of language. On one side, the objective: the wine's physico-chemical properties on which experts and laboratory assistants can continue to work. On the other side, the subjective: the infinite variety and cultural enigma of "tastes," those that make people favor such and such a product. The objective measurement only serves to control specific features, sanitary conditions, standards of origins and components. As for the subjective tastes, ask the psychologist, the artist, the economist, or even the sociologist, because I have nothing left to say. Something has been ruptured compared with the clarity of the objectivist model. And it is tempting to go on from there and, considering that if the logic of the object cannot explain everything then it explains nothing at all, to reject the object altogether. Critical strands of sociology have supported this radical denunciation—taste is arbitrary, there is no link between the properties of products and the kind of attachments they bring about. These experimentalists, who used to swear only by the virtues of organoleptic enquiry and dreamed of purifying wine production from its useless folklore by taking it back to the laws of chemistry, have been turned by sociology into manufacturers of social artefacts, that is, of tastes radically autonomous from the objects that carry them. Whether they obey some elusive coherence of the social order, reflect cultural systems, or vary endlessly according to the unbounded proliferation of subjectivities, tastes have lost their anchor and now sail away with the wind. Once under the custody of critical sociology, tastes have become social necessities, but at the price of being contingencies of their object—whereas their experts wished to make taste a necessity of the object by getting rid of their part of social contingency.

A radical critique for sure, but with limited gain. In front of us, the dualist wall between objectivity and subjectivity seems to have never been so high. But there is another lesson to be learned from the sometimes spectacular failures of blind testing exercises. No, it is not to be found in the characteristic back-and-forth between an unshakable trust in the object and its sudden evaporation in the arbitrariness of taste, the logic of which is to be found elsewhere (in a person's tendency, within history and culture, in the psycho-cognitive variety of individuals, in the hidden game of social distinctions). We can learn by listening to the experts themselves as they compose the sociology of their genuine competency. Some had not been waiting for sociologists to carry out a direct critique of the blind testing technique (Fribourg and Sarfati 1989; Peynaud and Blouin 1996). The montage seems flawless: if there is anything objective about the taste of wine, then please taste just what is in front of you, with no additional information. Cultural determinisms, conformism towards oneself or others, effect of reputations—all possible biases have been eliminated and only the product now remains.

False evidence: At this point, a comparison with music can be telling.

The Object-Wine or the Object of Wine?

Scene four: Bring together a group of musicologists, musicians, critics, and other great amateurs to perform a blind test. All in line, each with a pair of headphones, oblivious to what they are about to listen to. Twelve quartets per series, ten seconds each. Have you taken notes? Good: And now, to the symphonies.

What do we get from this? Everything but music: musicology. That is, not just anything, of course, but a form of expertise, in the strictest sense that we have just questioned: instruments, phrasing, writing of the different voices, rhythms, style, interpretations. There are more than enough variables to be perceived in these few seconds of recording. The majority of these avid listeners would have had the time to identify the period, the composer, often too the name of the work. In music, the problem is not there: any respectable amateur won't make any more blatant mistake on the repertoire they know best, except maybe in case of tricks (paradox, imitations, etc.), but then even those tend to confirm a capacity to navigate amidst the forest of history and genres.

However, difficulties emerge if they are being asked, not to know and identify "what it is," but rather to appreciate, to describe "how it is"—and what it does: how beautiful it is, how enjoyable, what it evokes, whether we like it (or feel we might like it) and why. And for this very reason, the incongruity of the blind test is revealed. Listening to a symphony is not the same as comparing twelve short samples. Music, as a particular "matter" with its own analytical character, is not as difficult to identify as the fleeting and condensed aromas of wine. It is precisely because music mobilizes so much "cultural material," as one would say that it presents itself with such a wide range of noticeable characteristics. But what about *music,* in the sense of an aesthetic effect, of the emotion or delight that it conveys or not? Not a single participant to this little game would imagine being able to "get into it," to let oneself be taken by it, to feel its quality, within the limits of this curious set up, without being given the time to even appreciate whether one likes it or not. Musical extracts don't make a work, not even a piece: this simple difference illustrates very well the distinctiveness of wine, which, on the other hand, is so concentrated that it seems to offer itself completely in the moment of its tasting. So much so that it is hard to distinguish between the object-wine and the object *of* wine; that is, between the analytic features that a trained practice of tasting can detect and what would be the equivalent of the "work" of wine: what it does, what it affords, the pleasure and sensations it brings about, the ways in which it grasps, seduces, moves, beginning with the object-wine. But the wine tasting exercise never departs from this "beginning." To make sure that only the wine speaks (of its origins, its making, its components), it silences wine instead of allowing it express itself.[10]

Reformulated in this way, the paradox of the blind test does not lead us back to this series of clichéd and false oppositions: between objective and subjective, knowing and tasting. Nor does it lead to the thesis of a cultural arbitrariness of taste unwittingly determining the real causes of our attachments—which is only a more sophisticated version of the same duality, pulling us further away from the possibility of thinking up any link between the two facets of taste, and leaving us only with a physico-chemical description of products on one side, and with the history and sociology of fashions and practices on the other. It is not the taster's own *tastes* that disrupt (or, conversely, guarantee) the true appreciation of the *taste* of wine, as we would tick a box in a questionnaire. For there to be taste, and for wine to produce its effects, as in music, there must be a passing, temporality, exposure. All things suppressed in the blind test.

One could wonder whether the persistent mistakes of reputed tasters are not due to some mischievousness on the part of taste, as it protests against its containment during the test: its unexpected effect of surprise and pleasure confuses things; all of a sudden, the expert thus affected loses track and becomes an amateur again. The point is not to grant less power to the object (in a theory of taste), but to give it even more (in the act of tasting). In the complex move from taste to tastes, the matter is more than ever one of contact with the object—but of an object unfolding, becoming plural. Between a string of musical notes and a work of art, we proceed layer by layer, through a chain of mediations, and not by dramatically falling over on the other side of a binary line of fracture. The question of tastes invites us not to turn our back on the object and to look elsewhere for their real causes, but to reconsider the object here, in front of us, in situation—as appearing, unfolding, as trial and temptation, not as a sum of components. The same goes for the action of wine.

In other words, what is repressed in this debate between those who think it is all in the wine (provided you learn to appreciate it better) and those who oppose this naïve naturalism to the cultural arbitrariness of taste or the diversity of subjectivities is precisely the essential, the only thing that makes tasting worthwhile: this moment of attention and appropriation to

[10] Cf. the aesthetics of things discussed by É. Souriau (1956) in terms of an open-ended unfolding of their potentialities.

which I gave emphasis in the introduction. This moment of a problematic, uncertain formation of associations between features and qualities, effects to be measured and effects felt. A zone of contact within which the object and its effects, as well as the amateur and his or her pleasure are given time to unfurl: associating both—or, rather, not dissociating them—is exactly what Dewey (1939) was after as he used the word "valuation," against the dualist one of evaluation.

A Pragmatic Tasting

A number of wine specialists have by themselves rejected the possibility to even improve the objectivity of the blind test, without feeling the need to compare, as we just have, their art to that of musicians, and even less so to ground their argument on the sociology of taste (luckily, as with the "critical" state it is in, they might have quickly abandoned their quest). They have duly noted that there is no more love of wine in trying ten in a row, taking notes before spitting them out, than there is love of music in listening to a succession of short extracts. And, more essentially, they have taken on board the fact that, since it is impossible to bracket out the tasters from their act of tasting, they might as well fully assume their participation. Exactly in the spirit of what has just been said about the listening of a symphony, these specialists now promote what they call "hedonistic" tasting. Even the act of comparison, whose benefits are clear for analyzing and perceiving components, turns into an obstacle when it comes to letting oneself be engulfed in the whole experience. Taking the opposite view to that of the blind test, our brave oenologists looking for the object of wine (and not the object-wine) have endeavored to let proliferate all that can make a wine *do* something. I am using these rather heavy-handed phrases on purpose, as we cannot replace them by others, as we so often do mechanically, without directly invoking the tastes of the taster, that one likes a wine, finds it pleasurable. This would be to remain in the duality of taste and tastes.

There are two aspects to this proliferation. First, the preference for tasting "in clear," i.e., knowing what we drink, the objective being to give a wine more space to reveal what it is, or pretends to be—no object goes without a project. Then, a shift of focus in judging towards the taster, with a *dégustation d'auteur*, a form of tasting that makes no secret of its arbitrary preferences, but exposes them through reasoned choices, whose principles can then be debated by the other participants. The point here is not to add to the objective qualities of the wine any cultural counterpart, selective grid, cognitive frame or screen of reception. Rather, it is to measure the importance of the pragmatic dimension of taste (as momentum, action, unfolding of presence) which, if met with success, links a now multiple product to a moved amateur, making them present to one another.

Scene five: A good meal, light yet delicate and varied; a carefully chosen place, and some involved guests—not necessarily wine specialists; often, a theme or a special occasion; through various modalities, the host's effort will revolve around the idea of assembling all the favorable conditions to give a certain wine all the attention it deserves.

From then on, our hedonistic tasters, good empiricists that they are, do the job of analysis for us. They must put together a full-blown sociology of taste—but not a sociology of "tastes" that would treat them as mere illusions and call for underlying causes outside of them. Who other than a sociologist would think of taking everything into consideration when talking about the taste of wine—the look of it, the labels, the rituals, the whole game of interactions among the guests, the snobbery, the prices: everything *except* what's in the glass itself! Just because they have rejected the objectivist reduction of the blind test doesn't mean the hedonists have had to throw the object with the wine in their glass. To the contrary, they give it all the more importance by choosing not to reduce it to an immediate addition of components, equivalent to their measurement, as is the case in blind testing. They give it sufficient time and space to develop. They are pragmatists without knowing it, and all we have to do is follow them.

A Theory of Attachments

Looking at the montage of the wine-tasting dinner, we can recognize a sort of minimal framework that supports any kind of attachment. It requires objects, a collective, equipment and conditions, and the commitment of bodies and souls (Hennion 2010). The point is not to assign a particular role, or predominance, to any of these elements, but to outline something like an initial scaffold, reconfigured for each specific attachment.

In any attachment, no element is ever "given": they are produced, tentatively and without guarantees, they come forward and reveal themselves in the course of production. This goes for the tasted objects as well as for the collective of amateurs, the body being put to the test, the techniques and materials mobilized—but above all, for the montage that will make them all define each other. The terms of attachment are never already there, but appear and redefine each other in situation, in a reflexive and problematic way—this is exactly what the performance, the tasting, and the pleasure are for. That implies bringing all of this together, training faculties and perceptions, learning skills and know-how, a repertoire, systems of classification, and techniques in order to make all the differences in the object speak, being aware of one's body as it becomes sensitive to these differences and not only trains itself, but truly invents and shapes itself in the process. Nothing is given, which is why taste is always an attention: not to taste from what one knows, but to discover oneself sensing through the trained or repeated contact with what was previously unperceived. How interesting would wine be if it were nothing more than what we want to make of it? What would the point of music be if it only ever delivered what we put into it?

Which brings us to our final theme, one that is omnipresent for amateurs: that of the *sur-prise,* that which adds up to the bliss of tasting. Why the hyphen? This is not just a play on a word: all the terms matter in the issue of taste and the keyword of surprise can summarize it effectively. First, there is *prise* (meaning "grasp" or "hold" in French), in the dual, active and passive sense of the word, making it perfectly adapted to talk about taste: the hold of the object, refined in the act of grasping, but also, on the amateur's side, the act of holding and allowing oneself to be held, each calling for, and not conflicting with, the other. Then, there is *sur* (in French, the prefix "over"): because the object adds something—that's all its charm!—as much on the side of tastes, savors and effects than on the side of bodies, sensations, what is felt. The object of attachment holds and it *sur-prises,* that is, it *over-holds* us more than we can ever expect. Finally, there is "surprise" in the usual sense, the one that gives the final touch. Indeed, this is one of the essential criteria of any good object, provided of course that we have identified its paradoxical nature—the constructed, rigorous, controlled, attentive and trained infrastructure from which this surprise can "suddenly" appear. The surprise comes from what is being added, above and beyond what is done and expected, and it cannot be perceived against a neutral background, undifferentiated and unincorporated. Under these conditions, it cannot emerge, since it has nothing with which to contrast—yes, it's good, but it could be something else.

Taste is best described as a machine for generating differences—but not in the sense of a reduction to a predictable mechanics, an available supply of differences of another (social or ritual) order to be projected onto the fictitious screen of objects. Rather, in the sense that objects cannot be grasped without procedures and that they are not given, but must appear; they can only be held, and felt, through a series of trials, and through a body that has itself been put to the test, produced and trained by its tasting activity: A dedicated place and time; the presence of objects; a meeting of guests; an assortment of dishes and wines. Besides *all* these or, more accurately, *from* these, come a shared sense of anticipation, a good disposition (which the idea of the themed dinner helps to produce). So, what have these "hedonists" and their empirical montages for magnifying the appreciation of wine taught us? That everything counts when it comes to taste, not as independent variables to add up in order to secure an outcome, but as uncertain mediations that support each other to bring about states, transform beings, make moments hold together. We only had to liberally rationalize the methods they have developed by themselves. Let us dream: what if this other sociology of taste, the program of which we have just outlined under their control, stopped its conflict with the imaginary hold that objects have on us? Why not, instead, fully acknowledge this very hold, so that sociology could better describe this art of a strong and reflexive relationship that through taste, humans establish towards things, others and themselves?

"Yes, Matter Does Matter!"

In conclusion, I would like to follow up this query, and to address in a straightforward way a central issue in the present book: what may lead sociology to engage into valuation studies? For me, it's the need to pose an urgent question on the value of what we do[11]: do things matter or not? Can they be only taken for granted and estimated from the outside by experts—sociology coming then only *before* or *after* evaluations to analyze uses, biases, power games, cultural criteria, social determinations, unknown logics of markets, and social appropriation? To study evaluation means to answer with a vigorous "NO!". Valuation studies require a threefold methodology: first, to take evaluation as an activity, not as a passive assessment of realities "already there"—this means that we pay the sharpest, minute attention to devices, tools, competences, debates involved (Helgesson and Muniesa 2013). But this is not enough. It implies that to value things is not only an object of investigation, but is necessarily our own activity as sociologists studying it. Finally, it also involves that we grant a certain degree of agency to things: they react, resist, make us do things. It has a lot of consequences: one does not listen to interviews the same way, ask the same questions, pay the same attention to hesitations, to wanderings between criteria, to reappraisals of advices, if one looks at how actors concretely proceed to make decisions in uncertain situations (Callon et al. 2009), or if one is on the lookout for evidence of tacit norms and rules applied without knowing.

Things are concerns, emerging in concrete situations, from unpredictable uses and effects, and through valuation debates. They are not inert things, prisoners of a fixed materiality (Ingold 2007, 2011). In that case, evaluation would be an issue but for technicians and experts. Hence the necessity of going back and forth along the spirals that the rich series of words derived

[11] As Stark (2009) among other authors reminds us, it goes back to the "Parson's pact" separating social values and economic value, and distributing each to watertight disciplines.

from value indicates so well: valuation (or, even more directly, "valuing"), evaluation, valorization (Dewey 1939). Such a shift is a crucial issue for sociology today.[12] Those operations are all ways of re-opening the question of the value of things, once one no longer considers criteria as already being available (evaluation produces them while using them), and properties of things as their own (valorization makes them emerge, become consistent, resist, be recognized, as much as it exploits them). Valuation is between the two, as a co-production of attachment to objects and of objects of attachment—being attached/attach oneself to; it is a moment when a contact or an experience of things makes both the characters of the thing appreciated and a sensitiveness to those differentiated characters appear: "it's good/I like it," in the case of taste, a case in which those two aspects are the closest. The "tasting/testing" oscillation is not a simple play on words to express our debates, it is central.

Finally, I will formulate the issue in more personal terms, leaning on an explicitly pragmatist stance. To me, the interest in valuation studies is that they reject the dualist stance in which sociology has left the question of value. They refuse both the naturalistic claim that things just "have" some value, and the sociologist claim that value is but "attributed" to them, quite independently of their nature, after social or cultural logics. Valuing things is no issue if one maintains such a divide between natural things taken as being determined by nature, and cultural objects seen as "socially constructed," which are more or less arbitrary (Latour 2013; Hennion 2013). In both cases, determined or arbitrary, the only thing sociology can do with things is to exploit their properties, if considered physical, or to show that they are only stakes of social games, if considered cultural. Taking valuation seriously is precisely a claim against such a double impotence. Yes, matters matter! On one side, things are not given, but emerge from interaction, lines of action, relations, and this has to be empirically documented. On the other, tasting them or being attached to them is not like choosing some gratuitous label to enter a social logic of identity and difference: it implies being made by them, transformed, possessed. Things are not tokens, nor totems. So there is in fact nothing like the "on one side, on the other" I just used: experience, testing/tasting, being possessed/possessing: our relationships with things are reciprocal transactions, making things and making us.

Hence a constitutive ambiguity, with regards to the taste of things, between testing them (trying to make properties of things appear, be stabilized, to putting them into words and figures, etc.), and tasting them (trying to "be" the thing, to make oneself incorporate what it makes or produces,

[12] It could be referred to as a "pragmatic turn"—in so far as one is not too dizzy after so many recent turns!

to create an in-between state where both the taste *of* the thing and the taste *for* the thing may deploy). So conceived, both operations (tasting and testing) are productive, open, and remain tightly connected, referring less to an absolute divide between objectivity and subjectivity than to a continuous co-production of stabilization and transformation, both in the things and in our capacity to feel what they do. Maybe the word "dissonance" sounds so expressive to our ears precisely because it nicely underlines, with a perfume of mystery, this profound indeterminacy. Things are not inert, fixed. Things happen, surprise, they are relational and, depending on situations, they are retaken through action, controversies, inventions, etc. And such clinamens do not arrive to order, at the place and time we expect. Speaking of dissonance involves a sort of renunciation to control and mastery. In that sense too it is very pragmatic: it's not (or not only) about better supporting and encouraging innovation, and making its organizational forms take more adequately into account the irruption of unexpected matters. A more radical understanding of dissonance insists on the fact that this constant irruption of things cannot and must not be reduced through a more clever management of newness, but puts into question control, mastery, management itself. Uncertainty is a constitutive feature of judgments, not an artificial bias due to implicitness: this requires at a much larger level new forms of attention to the world, to what happens, to the unpredictable emergence of common realities. Such a politics of attention can only be collective, debated, progressively realized—it cannot be "managed," as shown by the example of organic wine: it has been a social movement related to ecology, nature, new forms of markets, considerations on the consumer as being active and committed, on the winemaker as being attached to his *terroir*, among a thousand other relations. It was not just a capacity of winemakers to adapt themselves to new market sensitivities (Teil et al. 2011).

Valuation studies do not all agree—in fact they disagree on many points—but on the whole, their posture entails a pragmatic shift: they take concerns seriously (e.g.,: fairly assessing quality, even if in a disputable way); they estimate that, even indefinite, qualities are not arbitrary, and that tools and procedures, even if fragile, imperfect means, do express something real, even if plural—the quality of things, however defined; and they grant actors the competences to make evaluations, by confronting and discussing this uneasy, controversial quality. Last but not least, to do so situates the sociologist in close continuity to actors, who in a reflexive way deal with the same issue: what is going on?—then sociologists too have to be sensitive to uncertain events, whose meaning and interpretation are not clear from the beginning. To me, this openness, this uncertainty, is what makes it so interesting for sociology to consider our relationships to things in new terms, by taking them as *pragmata*, things as they are not given, "in their plurality," in a world "still in process of making" (James 1909a: 210; 226),[13] and symmetrically, by paying minute attention to what happens in the intercourse with them through various devices and material supports, along the complex series yet discussed (valuing-evaluating-valorizing). At least this is the way I would phrase the program of valuation studies. Of course, there are many ways of addressing it, and each would stress one aspect or another more, maybe quite differently from the way I have done, in a pronounced, provocative pragmatist tone. So much the better: it will lead to stimulating dissonances, and help enrich the most promising valuation studies.

References

Bessy, C. and Chateauraynaud, F. (1995). *Experts et faussaires*. [Experts and Forgers]. Paris: Métailié.

Bourdieu, P. (1984). Distinction: A Social Critique of the Judgment of Taste. London: Routledge.

Callon, M. (2013). Qu'est-ce qu'un agencement marchand? [What is a Market *Agencement*?] In M. Callon et al. (collectif) (eds.), *Sociologie des agencements marchands*. Paris: Presses des Mines: 325–440.

Callon, M., Lascoumes, P., and Barthe, Y. (2009). *Acting in an Uncertain World*. Cambridge, MA: MIT Press.

Chauvin, P.-M. (2010). *Le Marché des réputations. Une sociologie du monde des vins de Bordeaux*. [The reputation market. A sociology of the World of Bordeaux Wines]. Bordeaux: Féret.

Dewey, J. (1939). Theory of Valuation. In O. Neurath, R. Carnap, and C. W. Morris (eds.), *International Encyclopedia of Unified Science*. Chicago, IL: University of Chicago Press, 2(4): 1–67.

Fourcade, M. (2011). Cents and Sensibility: Economic Valuation and the Nature of "Nature." *American Journal of Sociology* 116(6): 1721–1777.

Fribourg, G. and Sarfati, C. (1989). *La dégustation. Connaître et comprendre le vin*. [Tasting. Knowing and Understanding Wine]. Suze-La-Rousse: Edisud.

Grabenhorst, F. and Rolls, E. T. (2008). Selective Attention to Affective Value Alters How the Brain Processes Taste Stimuli. *European Journal of Neuroscience* 27(3): 723–729.

Helgesson, C.-F. and Muniesa, F. (2013). For What It's Worth: An Introduction to Valuation Studies. *Valuation Studies* 1(1): 1–10.

Hennion, A. (2004). Pragmatics of Taste. In M. Jacobs, and N. Hanrahan (eds.), *The Blackwell Companion to the Sociology of Culture*. Oxford/Malden, MA: Blackwell, 131–144.

[13] James writes also "What really *exists* is not things made, but things in the making" (2009b: 263).

Hennion, A. (2007). Those Things That Hold Us Together. *Cultural Sociology* 1(1): 97–114.
Hennion, A. (2010). Vous avez dit attachements?... [You said attachments?]. In M. Akrich, Y. Barthe, F. Muniesa, and P. Mustar (eds.), *Débordements. Mélanges offerts à Michel Callon*. Paris: Presses de l'École des Mines, 179–190.
Hennion, A. (2013). D'une sociologie de la médiation à une pragmatique des attachements. [From a Sociology of Mediation to a Pragmatic of Attachments]. *SociologieS*. http://sociologies.revues.org/4353 (accessed December 10, 2013).
Hennion, A. and Teil, G. (2004). Le goût du vin. Pour une sociologie de l'attention. [The Taste of Wine. For a Sociology of Attention]. In V. Nahoum-Grappe and O. Vincent (eds.), *Le goût des belles choses*. Paris: Éditions de la MSH, 111–126.
Hutter, M. (2011). Infinite Surprises: On the Stabilization of Value in the Creative Industries. In J. Beckert and P. Aspers (eds.), *The Worth of Goods: Valuation and Pricing in the Economy*. Oxford, New York: Oxford University Press, 201–220.
Ingold, T. (2007). *Lines: A Brief History*. Oxon, UK: Routledge.
Ingold, T. (2011). *Redrawing Anthropology: Materials, Movements, Lines*. Aldershot: Ashgate.
James, W. (1909a). *The Meaning of Truth*. New York, NY: Longmans, Green & Co.
James, W. (1909b). *A Pluralistic Universe*. New York, NY: Longmans, Green & Co.
Lamont, M. (2012). Towards a Comparative Sociology of Valuation and Evaluation. *Annual Review of Sociology* 38(21):1–21.
Latour, B. (2013). *An Inquiry into Modes of Existence*. Cambridge, MA: Harvard University Press.
Mallard, A. (2012). Developing Uses, Qualifying Goods: On the Construction of Market Exchange. *Consumption, Markets and Culture* 15(2): 191–211.
Merleau-Ponty, M. (1964). *Le visible et l'invisible*. [The Visible and the Invisible]. Paris: Gallimard.
Muniesa, F. (2011). A Flank Movement in the Understanding of Valuation. *The Sociological Review* 59: 24–38.
Muniesa, F. and Trébuchet-Breitwiller, A.-S. (2010). Becoming a Measuring Instrument: An Ethnography of Perfume Consumer Testing. *Journal of Cultural Economy* 3(3): 321–337.
Ocasio, W. (2011). Attention to Attention. *Organization Science* 22(5): 1286–1296.
Peynaud, É. and Blouin, J. (1996). *Le Goût du vin*. [The Taste of Wine]. Paris: Dunod.
Renvoisé, G. (1996). *Le Monde du vin: art ou bluff?* [The World of Wine: Art or Bluff?]. Rodez: Éditions du Rouergue.
Shapin, S. (2011). Changing Tastes: How Things Tasted in the Early Modern Period and How They Taste Now. The Hans Rausing Lecture 2011, Uppsala Universitet, *Salvia Småskrifter* 14: 7–47.
Shapin, S. (2012). The Tastes of Wine: Towards a Cultural History. *Rivista de Estetica* 52(51): 49–94.
Souriau, É. (1956). L'œuvre à faire. [The Work to Be Done]. *Bulletin de la société française de philosophie* 25 février 1956: 4–44.
Stark, D. (2009). *The Sense of Dissonance: Accounts of Worth in Economic Life*. Princeton, NJ: Princeton University Press.

Stark, D. (2011). What's Valuable? In J. Beckert and P. Aspers (eds.) *The Worth of Goods: Valuation and Pricing in the Economy.* Princeton, NJ: Princeton University, 319–338.

Teil, G. (2001). La production du jugement esthétique sur les vins par la critique vinicole. [The Production of Aesthetic Judgments of Wine by Wine Critics]. *Revue de Sociologie du Travail* 43: 67–89.

Teil, G. and Hennion, A. (2004). Discovering Quality or Performing Taste? A Sociology of the Amateur. In M. Harvey, A. Mcmeekin, and A. Warde (eds.), *Theoretical Approaches to Food Quality.* Manchester: Manchester University Press, 19–37.

Teil, G., Hennion, A., Barrey, S., and Floux, P. (2011). *Le vin et l'environnement.* [Wine and the Environment]. Paris: PEM.

Thévenot, L. (1990). L'action qui convient. [The Appropriate Action]. In P. Pharo and L. Quéré (eds.), *Les Formes de l'action,* Paris: Éditions de l'École des Hautes Études en Sciences Sociales (Raisons pratiques 1), 39–69.

Vatin, F. (ed.) (2009). *Évaluer et valoriser. Une sociologie économique de la mesure.* [To Evaluate and Valorise: An Economic Sociology of Measurement]. Toulouse: Presses universitaires du Mirail.

Veldhuizen, M. G. and Small, D. M. (2011). Modality-Specific Neural Effects of Selective Attention to Taste and Odor. *Chemical Senses* 36(8): 747–760. [doi:10.1093/chemse/bjr043]. www.ncbi.nlm.nih.gov/pmc/articles/PMC3175104/ (accessed December 12, 2013).

4

Dissonant Translations: Artistic Sources of Innovation in Creative Industries

Michael Hutter

One of the perfectly groomed hands of a woman rests on a silver mirror, the other one, embellished with a jewel bracelet and a ring, touches a bottle of Jergens Lotion. The image in Figure 4.1 was published in Ladies' Home Journal in 1928. Its creator was Edward Steichen, at the time one of the most prominent photographers. The hands of the mostly female readers rarely looked like those in the advertisement, and they were surrounded by more mundane objects. But they could buy a bottle of the very same lotion, be touched by its content and engage in a fantasy of luxury. The attributes of worth, assembled and compressed into one image by Steichen and his studio, were translated into the value attributes of a hand cream.

Introduction

The photograph of the hands touching a bottle of lotion (Figure 4.1) was clearly invented and produced for a commercial purpose. However, its effectiveness as an image, reproduced in the small space of a journal page, depended on a number of artistic features. Those features were characteristic of Steichen's style of photography. Steichen was known to draw on a vast repertoire of artworks and of artistic practices. In this particular image, he cites an arrangement of hands that can be found in many portraits by Ingres, created 70 years earlier (Figure 4.20). Steichen's skills as an artistic photographer seem to be directly related to his successful inventions as a commercial photographer.

Figure 4.1. Walter Thompson Agency: Advertisement for Jergens Lotion, with Edward Steichen photograph, Ladies' Home Journal, Jan. 1928, in Johnston 1997: 138.

Yet, it would be too simple to attribute the transfer of value from an artistic source to an industrially reproducible commodity to an opaque mental process within an exceptionally talented individual. The process is, in large parts, social and thus empirically observable: there exists a product that is made primarily of symbols. The symbols carry meaning, and the assembled meaning contained in a photograph, or a novel, or a movie is able to generate an experience for a person who exposes her or his senses to the product.[1]

The production of symbolic goods differs markedly from material production. Rather than transforming raw materials into a sequence of identical forms, like cars, a "master copy" is invented, produced and then copied through a medium of reproduction, like print. Reproduction in electronic charges, or digital copying, increases in effectiveness while decreasing in cost. In consequence, a growing share of total economic value is generated by sectors that generate such symbol-based experiences.

Recently, these sectors have been categorized as creative industries. They encompass the traditional cultural sector (e.g., operas and museums), analogue and digital media (e.g., TV and videogames), and various services provided around original symbolic forms (e.g., design, fashion, architecture and advertising) (DCMS, 1998; UNCTAD, 2008). In all these sectors, master copies are generated and then repeated, either in live performances or in activating the copies drafted from the master copy. The cost for a single master copy can exceed hundreds of millions of Euros, while the cost of further copies might be infinitesimally small.

In all of these branches, artistic sources are at the initial end. The finished product, like the journal ad with Steichen's photo, reveals connections to concrete, assignable artistic works of the past. It seems to draw part of its sensory and emotional impact from such sources.

But how is this "transfer of value" from artistic sources to symbolic products to be modelled? The use of quotation marks indicates that the notion of "transfer" is only roughly indicative, without any specification of what is transferred, and how it is transferred. By means of three historical cases, selected from different branches of the creative industries, this question will be pursued.

In a first section, I will discuss attempts to employ the notion of translation as terminological option for the transfer of value. I will also argue that the trajectory necessarily includes phases of dissonance. The following sections apply the concept of "dissonant translations" to three cases for which the social process of creation is well documented: a successful Disney movie, Steichen's advertising photos, and Vuitton handbags with a pattern designed

[1] Scitovsky (1976) suggested that there is already a class of products demanded for their mental stimulus rather than their physical comfort.

by Murakami. In a last section, I will argue that these trajectories, in which artistic value is translated into values purchased in markets, should be recognized as moments of valuation.

Value Transfer Through Dissonant Translations

The search for experiences is strongly driven by a desire for the unknown, for the surprising (Hutter 2011). Symbolic products, e.g. novels or vintage wines, carry meaning, and meaning is of interest as new meaning. Once the novel is read, the meaning is known, and the product's value decreases. In a few instances, new interpretations continue to be drawn from the artefact, as is the case in Shakespeare's sonnets, but also with old bottles of wine. Producers in the creative industries are focused on offering novel products at fairly regular intervals. Because the symbolic content of their products is novel, nobody knows how successful a new offer will be. Most of the products will have a very short life span, while some of them become canonical and continue to generate new value.

Creators, producers and providers can only guess what will "work for" potential users—what will generate affective value in their minds. In most instances, the creators do not invent the features that generate these values from scratch. They work along a familiar, recognizable basic form, and add or detract some of the features. In doing so, they employ symbolic forms that have proven their impact on users—by being praised, conserved and re-performed (Dewey 1939). They integrate these older works, or a few of their strands of meaning, into their own symbolic forms, and thus make them more likely to become valuable. As users experience the content of the novel good—of a concert, of a garment worn—they make it their own and are willing to pay for future variations of it.

How is the value of an original invention infused into novel products—into new kinds of movies, or into new photos in ladies' magazines, or into new fashion accessories? I will discuss the explanatory reach of the concept of translation. It is an attempt to grasp the paradox of a form of meaning that is continued and newly invented at the same time.

The notion of "translation" is frequent in inquiries that investigate certain kinds of value transfer. The common English term indicates a change of signifier for an unchanged signified content. One assumes that something expressed in the idiom of the original artistic source is translated into another idiom. The assumption implies a high degree of regularity for such activity. Translators have small degrees of freedom in their semantic choices. The variability of translation in creating new symbolic products is much larger.

Recent contributions to cultural sociology operate with a different meaning, a meaning that stems from the French term "translation" (see Callon 1986 and Latour 2005).[2] In this interpretation, translation refers to a displacement[3] or delegation of power within networks, and networks are characterized by "the ability of each actor to make other actors do unexpected things" (Latour 1993: 129). Such a notion of networks is helpful in linking the capacities of original sources with the affective impact of symbolic products. Whether the delegation of political power that occurs between the phases of problematization and mobilization (Callon 1986) has similarity with the selection of artistic features by creators and producers, and the subsequent effect of creative products on their users, is yet to be shown. In organization theory, there has been an attempt to employ the concept outside of the political field, for the chain of activities leading from ideas to the appearance of commercial objects (Czarniawska and Joerges 1995). In this model, ideas "travel" to their materialized destinations within Latour-type networks. Would it make sense to employ a similar notion of translation to explain the particular path from an initial artistic form to its later appearance in the form of a commercially valued creative product?

I suggest two specifications to the translation process: it is assumed to be affected by dissonance, and it is assumed to go through two complementary phases: One of the phases takes place among those actors who create, produce and reproduce the novel product, the other one takes place among those who interpret and experience the product, or a copy of its symbolic form. I will call the first one the "authoring phase," the second one the "reading phase."

In both phases, value change is perceived as an instance of dissonance. Dissonance is the term suggested by Stark to indicate what happens when "alternative principles of the valuable" encounter each other (Stark 2009: 211). Dissonance indicates a change between worlds of understanding, a movement through a period of indecision, confusion, even panic (Stark 2009 and Chapter 1 of this volume). Dissonance is more than just rupture; it is a particular kind of constructive change, just as a change in harmonic structures might for a while be heard as noise, and only after that moment as meaningful, emotionally charged music.

Stark investigates the innovative effect of dissonance within organizations, but the phenomenon is just as observable within less firmly coordinated fields of interaction.[4] The process has a mental as well as a social dimension. Moments of mental dissonance are observed through introspection,

[2] In French language, *translation* differs from *traduction*—a distinction easily overlooked in English discussion.

[3] The common meaning of the French term refers to the displacement of vehicles along rails or other paths.

[4] I use the terms "field," "play of meaning," and "social system" synonymously.

social dissonance is observed in public debates amongst the two groups of agents noted above—those who might collectively be called authors of creative products, and those who can read or otherwise perceive the meaning of the product and then report about their experience.[5]

In the authoring phase, artistic value principles clash with commercial cost-revenue-principles. The selection of features of form, the ability to be "inspired" by some major artwork or by some lesser form of aesthetic expression, comes into contrast with calculations of revenue, based on future demand for copies of the master copy. The sales are expected to repay and reward the enterprise of production. These expectations, inevitably, are governed by observations of the complementary phase of dissonance among readers.

In the reading phase, the copies are used to create experiences. The dissonance is now between expectations concerning the further trajectory of the experience—the suspense of a novel or a trapeze act, the surprise of seeing a bike in a design studio, or witnessing a murder in a movie. The dissonance contains a stimulus that can result in disappointment or in pleasure. The value of the stimulus is not only mental, but also influenced by the observation of disappointment or joy in others, and by the possibility to establish links between members of an audience by sharing judgments about the experience's qualities.

Both types of dissonant translation are driven by dissonance—the translation from the initial artistic spark to the product template, and the translation from the acquired copy to the critical judgments of those who experienced the product. Both phases are phases of valuation.

Can dissonant translations be detected in the material documenting the genesis and the market success of creative industries products? This chapter explores three cases with conditions favorable to a positive answer: The creative products were highly successful at their time of market introduction, and their "making of" is fairly well documented. If these cases do not show indications of dissonant translations, then the validity of the model for less outstanding products is questionable.

The authors of the three products employed different kinds of reproduction media: repeat movie theater performances, mass-printed advertisements and custom-printed surfaces of fashion accessories. The kinds of symbolic content translated are also diverse: particular graphic outlines, complex studio arrangements to generate painterly effects, and visual marks distinctive enough to be recognized as trademarks.

[5] Note that an individual human actor can operate as an agent in both phases.

Three Cases

Louis Vuitton's Murakami *Handbags*

In luxury goods production, the seller's historical pedigree "rubs off" on the buyer or user whose family may never have owned objects that are older than a decade. *Louis Vuitton* has accumulated such a pedigree because its founder cornered the luggage market since 1853 with the triple innovation of a flat-bottom canvas trunk, a secure lock and a virtually indestructible canvas. The company grew into a global market leader for luxury leather goods. In 1987, family control ended when the major investor, Bernard Arnault, bought out the company and merged it into a larger conglomerate of luxury goods production, *Louis Vuitton Moët Hennessey* (LVMH).

The major steps toward a world-wide operating enterprise are credited to Georges Vuitton, the son of the founder. By the end of the 19th century, the company was so successful that it had to protect its products from counterfeiting. In consequence, Georges had the "Monogram Canvas Pattern" designed, and registered it as a trademark in 1896. Still today, the company profits from the instant recognizability of the Monogram Canvas.

The prestige of the brand is maintained at high current cost: frequent photo campaigns in print media, spectacular sponsoring events, and exclusive shop presence in key centers of wealth concentration are all part of the strategy. In 2001, art director and fashion designer Marc Jacobs initiated a consistent practice of temporary collaborations with leading artists. After cooperating with Stephen Sprouse on a graffiti pattern and Robert Wilson on a window scenography, Jacobs signed up Takashi Murakami in 2002.

Murakami was born in Tokyo in 1963. His early training was in animation drawing for *manga* cartoons and *anime* films. In addition, he studied *nihonga*, a specific school of traditional Japanese drawing and painting techniques. He developed his own cartoon figures, connected into a cartoon world with dozens of characters, surrounded by detailed narratives as to their origins and relationships, and supported by video sequences set to electronic music. Central characters are "DOB," who faintly resembles "Mickey Mouse," and the couple "Kaikai" and "Kiki." After a stay in New York in 1994, Murakami gained increasing recognition and reputation in Western art circles. His popularity among a young audience in Japan was greatly increased by his 7-year-long cooperation with *Louis Vuitton*. Within those circles who represent traditional versions of Japanese art, he is seen as someone who has turned venerable cultural values into cheap merchandise, as well as someone who has tricked Western art collectors into paying high prices for artworks that use popular Japanese means of expression.[6]

[6] On the history of the Japanese art scene whose most prominent member is Murakami, see Favell (2011). See also the bibliography in MOCA (2007).

Figure 4.2. Takashi Murakami: Monogram Multicolore, 2003 Kaikai Kiki Co., Ltd.

Murakami designed patterns that were used on scarves, rugs, shoes, umbrellas and stores of the company. Most of the attention concentrated on patterns designed for handbags that were "issued" in limited editions of a few thousand objects.[7] To get a sense of the kind of variation invented by Murakami and his studio, three particularly attractive designs will be described in greater detail. They were all introduced in 2003.

Monogram Multicolore (Figure 4.2) leaves the four components of the Monogram pattern and their arrangement intact, but changes the color scheme. The background is either white or black, and the signs are brought into a scheme of light colors that is adapted from Murakami's works.

Cherry Blossom sprinkles Murakami-style pink blossoms sparsely over the original Monogram Canvas. A few years later, *Cerises* was issued which substituted pairs of cherries for the blossoms.

[7] The exact size of the editions is not revealed by the company.

Figure 4.3. Takashi Murakami: Eye Love Superflat, 2003 Kaikai Kiki Co., Ltd.

The most successful of the patterns seems to have been *Eye Love Superflat* (Figure 4.3). In this pattern, the round flower sign is substituted with a sign of particular relevance in the Murakami universe, the "jellyfish eyes." Jellyfish eyes have become Murakami's trademark.[8] Both trademarks, that of the image maker and that of the luggage maker, are woven into a seamless pattern, within the same color scheme.[9]

The market launch of the Murakami designs on Vuitton products was accompanied by an animated film, titled *Superflat Monogram*. The heroine, a small girl, falls through a long tube into a world inhabited by Murakami cartoon characters, interspersed are Vuitton logos. In 2009, a second video titled *Superflat First Love*, was released. Purportedly commemorating six years of

[8] In April 2013, Murakami's first feature-length film, titled *Jellyfish Eyes*, premiered.

[9] Noteworthy is a late addition in 2008, *Monogramouflage*, which shows the monogram pattern in black on a background of abstract shapes in beige against camouflage green.

Vuitton/Murakami cooperation, it featured the same girl as a teenager, again falling into the Murakami world, where she meets a young luggage maker.

Murakami has also created works for the art market that use his Vuitton designs. Already in 2003, he used acrylic paint for a version of *Eye Love Superflat*, measuring 180 x 180 cm. The canvas was sold in 2010 for $700,000. In 2004, he added silkscreen editions of the same pattern. Since 2007, pieces of canvas from the handbag production with Murakami patterns have been stretched on regular picture frames, signed by Murakami and sold in editions of 100 copies.

From 2007 to 2009, Murakami's major works were shown in a large retrospective exhibition. With variations due to the physical location and local opportunities, the exhibition was installed at the Museum for Contemporary Art in Los Angeles (2007), the Brooklyn Art Museum (2008), the Museum für Moderne Kunst Frankfurt (2008), and the Guggenheim Bilbao (2009). Handbags from the cooperation with *Vuitton* were exhibited in all four of the venues (Figure 4.4).

The translation by the authors—in this case a network connecting the artist Murakami, the members of his workshop and employees of Vuitton—is

Figure 4.4. Vuitton bags with Eye Love Superflat Décor, exhibited at © Murakami, Museum für Moderne Kunst Frankfurt, 2008. Photo by the author

clearly recognizable: Patterns created and performed in the art world are translated into patterns suitable for handbags and other fashion accessories, under the stipulation that the original brand appearance remains recognizable. The company sets constraints through the shapes of its signs and objects, the vinylized canvas material and other criteria that ensure cost effectiveness. Some of these features function, at the same time, as indispensable carriers of value in their own right.

Murakami's three variations can be read as paradigmatic specimens of translation: (1) substitution of the color range he has developed as an artist, (2) addition of one of his artistic symbols to the original pattern, (3) substitution of color range, and substitution of one of the four components of the Vuitton pattern for a symbol central to Murakami's oeuvre. The originals of these variations gained their meaning and their first valorization in the world of the visual arts, with its web of cultural references, aesthetic codes and performance practices. The variations and their framing as Vuitton handbags translated them into fashion items. They gained new meaning for those who acquired the Murakami-styled handbags and employed them for their own social statements.

Murakami and Jacobs actively generated dissonance between an artistic logic of value and a commercial logic of value. In the two American venues of his retrospective exhibition, Murakami created a store-like structure. In the store, two kinds of products were on display and sold by sales personnel: Vuitton handbags and trunks with conventional and with Murakami design, and the editions of multiples mentioned above (Figure 4.5).[10] The multiples alone are reported to have generated revenues of $4 million (Boehm 2009: April 23).

Vuitton Shop was framed by the powerful artistic context of the exhibition. Bags and canvases were assembled in the particular arrangement of a "live" shop that, in turn, was part of a "live" museum exhibition of major works by a "live" artist. Commercial transactions took place, and the conversations with the sales girls and the exchange of money units for a bag or a canvas were turned into part of the artistic performance. The observer could experience the shift between economic and aesthetic valuation, always feeling the other valuation as a contextual presence.

In Murakami's own opinion, *Vuitton Shop* was the "heart of the exhibition,"[11] and he called it "my *Fountain*" (Thornton 2008: 212).[12] Thus, he attributed to

[10] The multiples were sold in five patterns, each in an edition of 100. The first fifty items were sold at $6,000, the second half of the edition sold at $10,000 per frame. The sales brochure stated that the canvases were "revisited" by Murakami, and that "the original canvasses have been hand mounted on a 16 x 16" chassis signed by the artist."

[11] Quoted in a press release of the Brooklyn Museum of Art, published in the Art Newspaper, March 27, 2008.

[12] Marcel Duchamp's *Fountain* was a urinal whose shape, tipped sideways, vaguely resembled a traditional mother and child figure. It was rejected from the art show at which it was entered in 1917.

Figure 4.5. Takashi Murakami: Vuitton Shop, exhibited at © Murakami, Museum of Contemporary Art Los Angeles, 2007. Kaikai Kiki Co., Ltd.

the bags and trunks a status comparable to Duchamp's seminal work. This strategy is pointed out by the chief curator of the Los Angeles Museum of Contemporary Art, Paul Schimmel: "Takashi found exactly the point that would irritate both me and Louis Vuitton. He took the materials that he had printed for various [Vuitton] products—the white one, the cherry one, five different sorts of patterns—and he had them stretched like paintings and made into a very large but numbered edition. He is sort of selling this rather high-end multiple up in the Louis Vuitton boutique" (Wilton 2007: November 19).

The difference in logic is used to create a juxtaposition that brims with dissonance. For those consumers engaged in Murakami's creations, this juxtaposition generates fascination. Fascination is a particular, positively valued version of dissonance. In the present case, the mind is attracted by the ambivalence of artistic and commercial value, by the play of oscillations between them. The success of Murakami's work is triggered by a design so carefully balanced that it provokes the fascinated attention of those who seek that kind of experience.

The commercial dimension allows a particular attachment to the products: they can be bought at prices that maximize the seller's revenue for a given limited number of copies sold. The number of those consumers who are able and willing to pay that price is much smaller than the number of those (mostly) women who can "read" the bags and are fascinated by them as well. That difference becomes part of the experience of owning one of the bags: women carry them at events in their social circles, and they collect comments and gazes by non-owners, as the appreciation of the bags with their singular patterns "rubs off" onto their carriers. This appreciation motivates the monetary value attributed to them.

Disney's Snow White And The Seven Dwarfs

Snow White and the Seven Dwarfs premiered in 1937. With 80 minutes of playing time, it was the first full-length animation film in cinematic history. The film was hugely successful in movie theaters around the world. It generated revenues of more than $8 million on its initial international release, and provided the financial means for the expansion of Disney's entertainment company (Allan 1999). The complexity of creating and producing a feature film led to a series of supporting innovations, like the introduction of story-boards, the direct transfer of drawings to celluloid and a new charge structure for distributors and exhibitors. 1,200 people were involved in producing the 250,000 frames that made up the master copy.

The plan to produce a successful feature animation film was commonly called "Disney's Folly" by the cinema community of the 1930s. It was considered impossible to assemble an animated film that would hold the audience's attention for the length of a feature film, and that would engage them emotionally (Gabler 2006: 213f.).

Snow White was one of the fairy-tales in the collection of the Brothers Grimm. The collection, published in countless versions since 1812, inscribed this set of tales into the collective memory of Western civilization. In 1917, Walt Disney had seen a film based on the tale, and it had left a deep impression on him.[13] In the summer of 1935, Disney used an eleven-week vacation in Europe to search for books with potentially usable illustrations for an animated film version of the fairy-tale, which he then sent to the Disney Studio Library.[14] In addition to the books Disney collected, the library contained works by numerous European artists, like the Norwegian Theodor Kittelsen, the German illustrators Hermann Vogel and Ludwig Richter, the English Arthur Rackham and the French Gustave Doré.

Among the chief authors responsible for the visual style of the project in the Disney studio were a number of European or European-trained artists: the Swiss-born and trained Alfred Hurter, who had responsibility for the continuity of style and character; Ferdinand Horvath, trained in Hungary, and Gustav Tenggren, trained in Sweden. All three worked for the film in the newly invented position of "inspirational artists" producing a novel category of sketches in the

[13] The film had been lost by the time Disney began to create *Snow White*. Gabler (2006: 217).

[14] The Disney Studio Library records show entries for 15 titles from Italy, 81 from England, 90 from France, and 149 from Germany. Allan (1999: 31).

production process which delivered background contexts and determined the visual characteristics of the characters before the animation artists began to draw them in motion (Canemaker 1996; see also Allan 1999: 46–48).

The challenge was to present the figures in such a convincing way that the knowledge of their artificiality was suspended. I will use material on the creation of the two central locations in the film, and on a particularly exciting scene of the film, to demonstrate the two dissonant translations.[15]

At the outset, viewers see the castle where Snow White and her stepmother, the Queen, live (Figure 4.6). The shape of the castle can serve as an example for the complexity of the amalgamation process.

The oldest of the visual sources is a view of the Château de Saumur on the calendar leaf for September in the *Book of Hours* for the Duc de Berry, created by the Limbourg brothers around 1410 (Figure 4.7). The image, in its exaggeration of turrets and pinnacles, became a famous invention.

Figure 4.6. Snow White's castle, Disney Enterprises

[15] The findings are predominantly based on research by Robin Allan (1999). The material available to a broader public was greatly extended through the catalog of an exhibition conceived by Bruno Girveau and Roger Diederen (2008). Allan (2008) condenses his findings in one of the chapters of the catalog.

Figure 4.7. Limbourg brothers: Book of hours for the Duc de Berry, 1410, Victoria and Albert Museum. In: Allan 1999: 43

Figure 4.8. Alcázar of Segovia. Frank Kovalchek/flickr

The second source is also an invention, although it took the shape of spires on a real building (Figure 4.8). The Alcázar of Segovia was built around 1200, but gained its unique slate spires centuries later. Philip II had them added in 1570 as a reference to the castles in the homeland of his wife, Anna of Austria. They constitute a theatrical gesture rather than an architectural tradition.

The castles commissioned by Louis II of Bavaria were explicitly theatrical. The most successful of them, at least in terms of continuing tourist attraction since the early 20th century, is Schloss Neuschwanstein (Figure 4.9). Images of Neuschwanstein had become part of the collective canon by the 1930s.

The challenge lies in designing a castle that symbolizes political power and architectural craftsmanship, yet is light enough to be considered credible as

Figure 4.9. Schloss Neuschwanstein, colored photography, ca. 1900

the home of a young girl. The dissonance for the authors lies in this balance. The three visual sources just cited reveal that they use theatrical, already symbolic images and versions of real, material castles to come up with their solution. I will comment on the dissonance experienced by the viewers later.

Next, we take a closer look at the passage of Snow White through the forest. The Evil Queen orders her huntsman to kill Snow White in the woods, but he lets her escape. She flees into the dark forest and eventually falls into a deep hole (Figure 4.10).

Those who experience the movie look into a dark world composed of roots that turn into the jaws of an animal, before a background in which bundles of roots look like nerve cells. Such interpretations of the dark world were inspired by Romantic drawing traditions, most prominently by the work of Gustave Doré and Arthur Rackham (Figures 4.11 and 4.12).

For the particular frame that shows Snow White falling, it is possible to compare the earlier artist's sketch (Figure 4.13) with the final version (Figure 4.10). Note how Snow White's skirt became longer, and her legs more puppet-like.

Here, the moment of dissonance for the viewers can be pinpointed. The dissonance occurs in the mind of the viewer, most likely a child or a mother, as part of a larger movie audience. Accompanied by pulsating music, Snow White falls out of her familiar world and is on her way to another world,

Figure 4.10. Snow White falling. Disney Enterprises

Figure 4.11. Gustave Doré: illustration for Dante's "Inferno," 1861. In: Allan 1999: 50

Figure 4.12. Arthur Rackham: illustration for "A Midsummernight's Dream," 1908. In: Allan 1999: 17

the fairy-tale world. This is the moment where viewers cover or close their eyes, and where it is decided whether they tell others that the novelty is "too much," or "too little." Today, 75 years later, the thrill seems slight, because generations of viewers since then have been exposed to much rougher dissonances. It is also slight because the Disney technique has been, and still is, one that minimizes its extent, and thus increases the number of potential viewers.

After the fall, Snow White lands softly, sings the song *With a smile and a song,* and is led by the animals of the forests to a bucolic glade (Figure 4.14). The colors retain the menacing purple, but the sunrays break through, the brook glitters, white birds spread their wings, and, in the middle, there is the dwarfs' cottage, tiny in scale compared to the surrounding trees.

Figure 4.13. Snow White falling, artist's sketch, 1937. Pastel, 19.2 x 25cm. Walt Disney Animation Research Library, in Girveau and Diederen 2008: 163

The features of the cottage are, again, translations. The documentation from the Animation Research Library shows that the inspirational artists used templates like the fairy-tale illustrations by Ludwig Richter (Figure 4.15),[16] or a photograph of the house of the crazy inventor Rotwang in Fritz Lang's film *Metropolis* (Figure 4.16).

Similar translations took place in creating the protagonists of the play. I will focus on the dwarfs. They contain the wealth of Disney's own archive of animated figures. Each of them overemphasizes one particular trait—such as bashfulness, hay-fever grumpiness—and turns it into a comical form (Figure 4.17). Their rhetorical characteristics imitate contemporary vaudeville and radio comedians, recognizable to the American audience during the first release.

In the case of the dwarfs, the dissonance between the aesthetic value of the original sources and the "Disney version" is particularly strong. The dwarfs gain their emotional impact by being situated at the borderline between the

[16] The edition of illustrated stories was sold under Richter's name, indicating his reputation as an illustrator.

Figure 4.14. The dwarfs' cottage, Disney Enterprises

Figure 4.15. Ludwig Richter: illustration from "Buch für Kinder und Kinderfreunde," 1853, Girveau and Diederen 2008: 118

Figure 4.16. Fritz Lang: Metropolis—Dr. Rotwang's House, film photograph, 1926, Friedrich Wilhelm Murnau Stiftung. In: Girveau and Diederen 2008: 66

Figure 4.17. Six of the seven dwarfs, Cel set-up; 54.6 x 90.2cm. Walt Disney Animation Research Library, Girveau and Diederen 2008: 177

animal and human domain. They are miners, working in the bowels of the earth, but they spend their private time above ground, in human-like dwellings, and they behave like humans. In romantic stories and illustrations, their features oscillate between friendly and hostile. The version by Kittelsen, created at the beginning of the century, demonstrates that difference (Figure 4.18).

In the Disney film version, the initially dark and hostile figures are rendered harmless, with an emotional and endearing value. The audience gets

Figure 4.18. Theodor Kittelsen: The troll mines, 1907. 44.5 x 33.5 cm. Theodor Kittelsen Museum

engaged in the entertaining, slightly distorting, comically exaggerating motions of the seven dwarfs. Their actions make a much stronger emotional experience bearable: the dissonant drama of danger, suspense and deliverance, witnessed in colored animation, sound and song.

The technique of translating old artistic values, in literature as well as in music and the visual arts, became standard for the films that followed *Snow White*, like *Pinocchio* (1940), *Fantasia* (1940), *Bambi* (1942), *Alice in Wonderland* (1951) and *Sleeping Beauty* (1959). None of them surpassed the landmark success of *Snow White*, but all of them contributed to the growth of a new sector of the creative industries.

Edward Steichen's Commercial Photography

Advertising deals with a particular kind of dissonance—the dissonance of promise. An expectation is formed that links an emotionally desirable future state with a tool to attain that state. Imagining a future state of well-being, like having won a lottery or leading a life of affluence, is an attractive experience in itself. Although the experience is pleasant, it generates an emotionally charged difference between the present state of mind and the state once the product is attained. Pleasure is promised once the advertised product is purchased and thus enters the buyer's material world.

To make the promise convincing, producers hire creative agencies to invent visual compositions, to be used in mass-media advertisements. The results are symbolic products—drawings, photographs, short videos—just like those that are offered on entertainment markets. But their functionality is different: they do not contain, rather they promise a valuable experience that is connected to a particular product. The dissonance between artistic source and commercial commodity is therefore sharpened: the artistic features support the advertising image in its emotional effect on the viewer, but that aesthetic effect bears no relation to the biological effect of a hand lotion, to take the example of Steichen's photograph. The result is a continuous debate about seduction and truthfulness in advertising.

Probably the first artistic photographer contracted for advertisements was Edward Steichen.[17] The style of the photos, produced between 1923 and 1936 for a major advertising agency and for fashion magazines, was a major innovation in their field. The images changed advertising photography as well as high fashion photography.

Edward Steichen (1879–1973) had several consecutive artistic careers. He started as an avant-garde painter and photographer in the Pictorialist style. Pictorialists favored vague outlines, tonal contrasts and symbolic content. His first exhibition in London in 1900, at the age of 21, was called a "bombshell" (Haskell 2000: 21). After 1908, Steichen became successful as an artistic photographer in the modernist style, with clear tonal values and stark contrasts (Rexer 2001). Steichen's works were promoted in *Gallery 291* and published in the magazine *Camera Work*. Alfred Stieglitz called Steichen "the greatest photographer... that ever lived" (Becker 1982: 341). In 1923, Steichen started a further career when he signed a contract with *Condé Nast Publishing* to deliver, as "chief photographer," monthly fashion and socialite photographs to be published in *Vogue* and *Vanity Fair*. Also in 1923, he signed a contract with the *Walter Thompson Agency*, a major advertising firm (Simon 1998: 105). After establishing a further reputation as America's leading commercial

[17] Before Steichen, the usual technique for advertising images was sketches and drawings.

photographer, he terminated his contracts in 1936. From 1947 until 1962, he curated the photography section of the Museum of Modern Art in New York.

The images Steichen produced with his studio during his second career made use of emotional codes developed in symbolist painting. They introduced new techniques of lighting and they positioned the single frames of advertising photographs into fictional stories. The images chosen for demonstration are selected from an advertising campaign for Jergens Lotion.

The Thompson Agency handled the account for the Jergens Company since 1922. The lotion had been a locally successful brand and was to be launched into a nationally marketed brand (Johnston 1997). The agency was committed to modernist style, a look that stood for a practical, young, realistic attitude (Simon 1998: 106).[18] Steichen's style of "straight photography" was paradigmatic for the Thompson approach. In addition, the agency employed behavioral psychologists to "tap into women's supposedly instinctive motivations" (Simon 1998: 107). It wanted the most powerful visual statement for their product, and Steichen used his capacities and skills to produce such statements at the rate of dozens per month over a period of thirteen years.

The photograph in Figure 4.19 is part of a full-page advertisement for Jergens Lotion. It is positioned in the upper half of the page, framed with a passe-partout. The signature "Steichen" appears on the right side of the lower rim—a form of presentation common in high art. Contrast and lighting are employed to convey the message. While the image appears to be a straightforward representation of a mundane action, it contains an element of fiction—the hands kneading dough are flawless. The flawlessness seduces the spectator. The hands belong to another social world, to a lifestyle in which they are meant to lift opera-glasses rather than buckets. The dissonance generates "apprehensions and allurement" (Simon 1998). In the surrounding text, the beneficial effects of applying *Jergens Lotion* are explained. The feeling of tension can be relieved by buying and applying the lotion in the carefully designed bottle, shown in the lower right corner of the advertisement.

The Jergens campaign was continued until 1936, with a constant flow of new photographs. Later advertisements situate the hands in positions of luxury (Figure 4.1). In Steichen's photograph, the advertised product has crossed the border between the fictional and the material, between the fantastic and the real. The image tips the emotion from desire to (trans-)action. The soft hands in the dough and next to the bottle are at the epicenter of these highly dissonant compositions.

Steichen's fashion photos in *Vogue* and his portraits of celebrities in *Vanity Fair* are similar in construction and effect. The women or men are positioned

[18] "... advertising agencies came to perceive photography as a truly American medium in its modernity: It suggested newness, novelty, industry, efficiency."

Figure 4.19. Walter Thompson Agency: advertisement for *Jergens Lotion* with Edward Steichen: hands kneading dough, in: Ladies' Home Journal, Jan. 1925. In: Johnston 1997: 46

in ways that bring out the best in the surfaces photographed. There is no attempt to expose deeper layers of the subject. The goal is to sell a seasonal line of dresses, or to bolster the reputation of an artist or a socialite. The goal is commercial, which imposes constraints on the choice of subject matter and technique of execution, but the image is created according to Steichen's criteria of artistic validity (Haskell 2000).[19]

[19] Steichen was very vocal in advocating this argument. In exhibitions of his works, he insisted on showing the photographs from advertising campaigns together with works that were produced solely for appreciation in his art world.

Figure 4.20. Jean-Auguste-Dominique Ingres: Princesse de Broglie, 1853, Musée du Louvre, Paris

Steichen had lived in Paris for many years, and there he gained detailed knowledge of the European painting tradition. As a critic observes, Steichen's photographs "use the formal potential of great paintings... By consciously or unconsciously drawing on a vast repertoire of tried and tested portrait types—in his own inner Louvre, you might say—trained painter Steichen removed the existing constraints on the genre of 'beauty pictures'" (Bezzola 2008: 194).[20] For example, the skill of Ingres in painting hands was a model for Steichen's presentations of hands (Figure 4.20).

[20] The painters most often mentioned in the literature on Steichen are Parmigianino, Ingres, Whistler, Manet, and Degas.

Steichen's "inner Louvre" was not limited to Old Masters. Pictorialist photography, which he had practiced in his youth, shared the aesthetic goal of conveying meaning behind the image with the much wider symbolist movement. Symbolist painters extracted elements of mythology and of dream imagery as a visual language for the world of the soul. Such imagery found its way into compositions like the one in Figure 4.1.

Steichen's ingenuity of turning visual experiences into novel works did not stop with translations from the world of images around him. He ran a studio with dozens of assistants, and the range of experimentation in the studio extended to the complete sequence of practices in developing photographic prints. All material factors, from lenses to papers and from lighting sources to chemicals, were tested. Eventually, these practices became standard for commercial photographers. Steichen shaped commercial photography for decades, and inspired fashion photographers like Cecil Beaton, Richard Avedon, Helmut Newton and Guy Bourdin (Bezzola 2008).

In the art world, Steichen's works, including his commercial photographs, enjoy a solid and growing recognition. Originals achieve high prices at auctions, and they are present in all the major public collections of photographic art. Many of the works continue to be reproduced in catalogs and in books that assemble iconic photographs.

Linking Social Worlds Through Dissonant Translations

In the introduction to this chapter, I claimed that translations of value from already existing artistic sources to new creative products take place under conditions of dissonance, for the network of creators as well as for their audiences—networks of readers, viewers and listeners. Did the presentations of three products from different branches of the creative industries show traces of such dissonant translations?

The readers experience the products in the three cases mainly as viewers—they read, or decode, advertisements, of films and of handbags. Their viewing experience involved valuations—the expected benefits from displaying beautiful hands, the thrill of following a heroine through her adventures, the gratification from wearing an exclusively stylish handbag. What kinds of dissonance experienced by the consumers could be detected? In Steichen's photos, the readers feel the incongruity between their present physical state and the potential future state symbolized in the image. The visualized situations in Disney's films allow readers/viewers to suspend their disbelief that animated drawings could ever be truthful representations of persons with a rich inner life. The handbags branded by Murakami and Vuitton permit their owners to distinguish themselves from those without such trophies.

In sum, the translations of value in the three cases cross different boundaries: the boundaries between present and future, between reality and fantasy, and between privileged inside and unprivileged outside. The readers/users/viewers contribute to the values themselves as they imagine certain futures, or certain miracles, or certain kinds of prominence. For them, valorization and the reflection of that valorization in value judgments and purchases flow into one.

The networks of authors in the three cases are dominated by famous individuals, but a closer look reveals more complex arrangements. Steichen was surrounded by a professional studio, his photos were processed by a major advertising firm and distributed in a national journal's outlets. Disney depended on hundreds of collaborators, from the inspirational artists down to the colorists of the individual frames. He also depended on risk-tolerant banks and on a competent distribution network. Murakami employed up to a hundred young artists in the 2000s (Favell 2011: 54). The collaboration with Vuitton made the manufacturing and sales resources of a globally active enterprise accessible.

The translation from original artistic values into a novel product followed different strategies. Steichen taught his team how to display the surfaces of soaps, towels and socialites as symbolic surfaces, based on compositions invented by masterly painters mostly of the 19th century. Disney hired artists with European classical training, exposed them to images from the Romantic period and coached them in drawing new kinds of personages understood by American children. Murakami sold his artistic reputation, delivered a commissioned set of designs for accessories, and participated in the promotion campaigns for their sale.

In each of the cases, the strategies chosen were publicly contested. Steichen had to defend his commercial engagements, Disney became proverbial for the infantilization of cultural heritage, and Murakami is still not taken seriously in conservative art circles. The heated debates document that crossing the boundary from the artistic to the commercial world of worth inevitably generates dissonance. The creators valorized their new products by linking their symbolic content to the content of artistic sources, and the readers found that linkage meaningful, even exciting.

It seems, then, that the notion of dissonant translation does have empirical content. If that is the case, then it can be employed to broaden the scope of larger theories explaining social change. Such theories are basically structural. They model the social world in terms of fields of action, or plays of meaning, or systems of communication. Fields, plays and systems differ in their practices, their semantics and their grammars of worth. But how can processes of mutual influence between such worlds of worth be modeled if the actors involved in each single interaction cannot be taken into account?

One response to this question is to interpret the process as irritation: changes in the environment of a field of action are not understood in the way knowledge is transmitted, but interpreted as disturbances or attractions that make no sense in the other field, yet lead to new ideas and formulations. As in biological evolution, such irritations might trigger variations and mutations. Another response assumes that reactions of critique lead to attempts of change and compromise.[21]

The approach involving dissonant translations recognizes the structural differences between fields of action, and yet introduces processes between and within agents that transcend these differences: a symbolic form that has prior meaning in the artistic world passes through groups or teams of authors and gains meaning in the commercial world, through the demand of audiences of readers that are able to read and experience the form, and to pay its price. Actors are included in their capacity as communication-processing agents: first as code writers, or creators, then as code readers, or experiencers. Both modes of agency can be active in a single actor.

Dissonance seems to contradict translation because meaning is interrupted twice, by authors and by readers. But in fact, dissonance opens the wide scope of new variations and new interpretations that generate new value. The dissonant translations discussed in this chapter, from the grammar of an artistic expression to the commercial grammar of a product worth paying for, represent only one aspect of the much wider applicability of the concept to all the other boundaries between worlds of worth. But they are sufficient to explain how economic growth can be driven by largely immaterial symbolic value, contained in creative products.

References

Allan, R. (1999). *Walt Disney and Europe. European Influences on the Animated Feature Films of Walt Disney*. London: John Libbey & Company.

Allan, R. (2008). Disneys europäische Wurzeln. [Disney's European Roots] In B. Girveau and R. Diederen (eds.), *Walt Disneys wunderbare Welt und ihre Wurzeln in der europäischen Kunst*. [Walt Disney's Wonderful World and its Roots in European Art] Munich: Hirmer, 37–51.

Becker, H. S. (1982). *Art Worlds*. Berkeley, CA: University of California Press.

Bezzola, T. (2008). Lights Going All Over the Place. In W. A. Ewing and T. Brandow (eds.), *Edward Steichen: In High Fashion. The Condé Nast Years, 1923–1937*, Minneapolis, MN: FEP Editions, 187–197.

Boehm, M. (2009). Louis Vuitton Suit Adds Fraud Allegation. *Los Angeles Times*, April 23.

[21] See Luhmann (2013) for the first, Boltanski and Thévenot (2006) for the second approach.

Boltanski, L. and Thévenot, L. (2006). *On Justification: Economies of Worth*. Princeton, NJ: Princeton University Press.

Callon, M. (1986). Some Elements of a Sociology of Translation. Domestication of the Scallops and the Fishermen of St. Brieuc Bay. In J. Law (ed.), *Power, Action and Belief. A new Sociology of Knowledge?* Sociological Review Monograph 32, London: Routledge, 196–223.

Canemaker, J. (1996). *Before the Animation Begins. The Art and Lives of Disney Inspirational Sketch Artists*. New York, NY: Hyperion.

Czarniawska, B. and Joerges, B. (1995). Winds of Organization Change. How Ideas Translate into Objects and Actions. *Research in the Sociology of Organizations* 13: 171–209.

DCMS (1998). *Creative Industries Mapping Document*. London: Department for Culture, Media and Sport.

Dewey, J. (1939). *Theory of Valuation*. International Encyclopedia of Unified Science. Vol. II, No. 4. Chicago, IL: University of Chicago Press.

Favell, A. (2011). *Before and After Superflat. A Short History of Japanese Contemporary Art 1990–2011*. Hong Kong: Blue Kingfisher Limited.

Gabler, N. (2006). *Walt Disney. The Triumph of the American Imagination*. New York: Alfred A. Knopf.

Girveau, B. and Diederen, R. (2008). *Walt Disneys wunderbare Welt und ihre Wurzeln in der europäischen Kunst*. [Walt Disney's Wonderful World and its Roots in European Art] Munich: Hirmer.

Haskell, B. (2000). *Edward Steichen*. New York: Whitney Museum of Art.

Hutter, M. (2011). Experience Goods. In R. Towse (ed.), *Handbook of Cultural Economics*, Cheltenham: Edward Elgar, 211–215.

Hutter, M. (2011). Infinite Surprises: On the Stabilization of Value in the Creative Industries. In J. Beckert and P. Aspers (eds.), *The Worth of Goods. Valuation and Pricing in the Economy*, London: Oxford University Press, 201–222.

Johnston, P. (1997). *Real Fantasies. Edward Steichen's Advertising Photography*. Berkeley, CA: University of California Press.

Latour, B. (1993). *We Have Never Been Modern*. Cambridge, MA: Harvard University Press.

Latour, B. (2005). *Reassembling the Social: An Introduction to Actor-Network-Theory*. Oxford: Oxford University Press.

Luhmann, N. (2013). *Theory of Society* (Vol. 2). Palo Alto, CA: Stanford University Press.

MOCA (2007). *Murakami*. L. A. Museum of Contemporary Art. Los Angeles, Museum of Contemporary Art, Los Angeles and Takashi Murakami/Kaikai Kiki Co., Ltd.

Rexer, L. (2001). Steichen's Legacy. *Graphis: The International Journal of Design and Communication* 5: 30-31.

Scitovsky, T. (1976). *The Joyless Economy: An Inquiry into Human Satisfaction and Consumer Dissatisfaction*. New York, NY: Oxford University Press.

Simon, J. (1998). Review of Real Fantasies: Edward Steichen's Advertising Photography. *American Periodicals* 8 (Fall):105–107.

Stark, D. (2009). *The Sense of Dissonance. Accounts of Worth in Economic Life*. Princeton, NJ: Princeton University Press.

Thornton, S. (2008). *Seven Days in the Art World*. New York, NY: Norton.
UNCTAD (2008). *Creative Economy Report 2008*. New York, NY: United Nations Conference on Trade and Development.
Wilton, K. (2007). *Paul Schimmel on Selling Murakami*. November 19. http://de.blouinartinfo.com/news/story/269044/paul-schimmel-on-selling-murakami. (accessed December 12, 2013)

5

Evaluating Valuation: Connoisseurship, Technology, and Art Attribution in an American Court of Law

John Brewer

In the Spring of 1929, towards the end of a boom that was shortly to be followed by a spectacular bust, the sidewalks outside New York's State Supreme Court were crowded with people waiting to attend the trial of *Hahn versus Duveen*. The world press was also in attendance: the trial was to be reported throughout the United States and Europe—notably in Britain, France, Germany and Italy. Yet this was no sensational case of fraud, bank-robbery or serial killing, but a dispute over a single painting, a work that its owners, Harry and Andree Hahn, claimed to be a Leonardo da Vinci, and that the powerful Old Master art market dealer, Sir Joseph Duveen, dismissed as a copy of a picture in the Louvre known as *La Belle Ferroniere* (see Figure 5.1). Mrs. Hahn was suing Duveen in the New York courts for "slander of title" (disparaging her goods) claiming compensation to the sum of $250,000 because Duveen had dismissed her picture in a New York newspaper and therefore, she claimed, made it impossible to sell. Public fascination at the trial and the newspaper coverage that accompanied it were part of a long-standing preoccupation with the burgeoning Old Master art market as a symbol for the wealth generation, riches and power of the new American plutocracy; but the conflict itself was framed as "the battle of the experts," a struggle over the evaluation of a work of art, that would have profound effects on its market value. A court—a judge and a panel of jurors—were being asked to make a legal determination that depended upon an aesthetic judgment that, in turn, would have profound economic consequences. But, as all the litigants and the public knew, this was not merely a case about the authenticity of a particular work of art, but about different regimes of truth

Figure 5.1. After Leonardo da Vinci: La Belle Ferroniere, private collection

and value, sustained by different actors, values and institutions. What was at stake was not merely the $250,000 that would have set up the (relatively impecunious) Hahns for life, but the entire system of art attribution and valuation as it operated in the early twentieth century (Brewer 2009: 103–162; Decker 1985; Secrest 2004).

The Old Master Art World, 1880–1930

We cannot begin to understand what was at stake in 1929 without first examining the Old Master art world as it developed in the late nineteenth and early twentieth centuries. The years between the 1880s and the great crash of 1929 saw a boom in which Old Masters were shipped in unprecedented numbers out of Europe and into the United States. The arrival of extremely rich and aggressively acquisitive American collectors (Huntington, Mellon, Frick, Morgan and the like) transformed what had been a relatively stable market in which national museum directors competed with aristocratic and plutocratic Europeans for Old Masters, into a ferocious competition to acquire Europe's cultural heritage. These arriviste collectors wanted masterpieces, authenticated works by the greatest figures in European painting.[1]

The most obvious effect of this was a sharp increase in prices. Between 1870 and 1912 the cost of works by Fragonard increased between ten and twenty-fold. Titian's *Man with a Red Cap,* sold in 1876 for £94.10s. and reached a price of £50,000 in 1915. Rembrandt's *Nicholas Ruts* cost £375 in 1850 but £30,000 in 1904. The German art historian, museum director and expert on Netherlandish art, Max Friedländer, looked on in incredulity: "The quality of works of art which drift about in the market is declining. The number of dealers and agents who want to live by the sale in the art market grows continuously. The difference in value between a picture by Rembrandt and one by Ferdinand Bol is increasing. The hunt for valuable things becomes ever madder and more relentless" (Friedländer 1942: 181–2). As Friedländer noticed the spread of the market was growing wider: the prices of the best objects were rising at a much greater rate than those at the bottom or the middle. At the same time market pressures were such that the size of the recognized oeuvre of the most famous artists, after a long period of shrinkage, began to grow. As Gary Schwartz has shown in his examination of seventeen scholarly catalogs of the paintings by Rembrandt, the size of Rembrandt's oeuvre fell by an average of 4.5 pictures a year between 1836 and 1883; rose by 8.4 pictures per year between 1883 and 1923, and since then has been declining at a rate of 7.1 Rembrandts per annum (Schwartz 1993: 314–16).

At the same time, there were a number of features of this boom that increased opportunities for chancers and dodgers. First and foremost the Old Master art world was marked by asymmetries of information that were built into its structure and operations, and which many of its most important actors were eager to exploit. This was a new situation. Before the transformation of the 1880s, dealers, experts, and collectors shared a common interest and enthusiasm—indeed knowledge—of the art they traded, collected and wrote about. The Old Master world was relatively enclosed, and consisted of royal and patrician collectors, many of whom had inherited works of art, scholars and curators in museums—who were drawn from the educated elites of Europe, and dealers who had worked in the trade since the late eighteenth century. We should not idealize this snobbish and exclusive world, yet what held it together was not just an admiration for Old Master art, but a shared knowledge about it. The new very rich collectors, however, were clients not because of their knowledge, but because they were preposterously rich; they were largely ignorant of what they were buying. Most collectors, though intelligent and ruthless, had little or no education in the arts

[1] The best recent general account of this phenomenon is Gennari Santori 2003.

(JP Morgan and John G. Johnson were exceptions), and had no time to learn (a task they sometimes consigned to their wives). These collectors had a passion for "names"—Leonardo, Raphael, Titian, Giorgione, etc.—but no real ability to identify them and to distinguish them from their followers, pupils and copyists, much less a faker.

In fact what these collectors were bent of acquiring was not just an Old Master, but much more. They were after a chunk of European heritage, a manifestation of what Novalis described as "the memory of humanity." Old Masters were revered as signs of cultivation and culture, they helped define the old elites of both America and Europe—the Brahmins of Boston, the aristocracy of Europe—their ownership granted access to such circles, and gave butchers like the Wideners and brewers like the Fricks, a sort of social cache. Quite apart from their value as paintings, Old Masters had added value to the new collectors as the means to public respectability—think of Frick remembered as the cultivated art collector, not as the ruthless proponent of violent union smashing—and they enabled them to pose as trustees of culture and civilization rather than as "robber-barons." In this respect Old Master collecting was different from the collecting of French academic art (notably Bouguereau) and nineteenth-century landscapes (especially the hugely forged Corot and the Barbizon School) that catered to a private taste rather than something that could be portrayed as civic minded. No wonder the new collectors were willing to pay exorbitant prices.

Crucially, the body to which they were most concerned to represent themselves as collectors was not the old school of dealers, curators and collectors (though they were powerless without their expertise), but a modern mass public, the readers of the newspaper press. The American "squillionaires" opened up the art world to a mass public, but only in a vicarious way. Before 1914 there was extensive coverage of how millionaires had acquired Old Master art for America. But almost none of it was on public show; the public was given a vicarious role in its acquisition. And press coverage of the art market—essentially gossip on a new scale—was also deeply ill-informed about art. Remarkably little was said about works of art and their singular properties (except the occasional comment about rarity); commentary was almost entirely concerned with deal-making, sales and prices. Essentially the stories were just like all the other robber-baron stories—about gouging competitors and cutting deals—but in a good cause because it was about the patriotic acquisition of art. The radical commodification of Old Master art happened at exactly the moment when it became an object of widespread public interest and concern (Gennari Santori 2003: 27–70).

The aggressive, super-rich but ill-informed new collectors came to rely on a growing class of experts or connoisseurs. As the son of one of the most important dealers, Seligman, wrote in his memoir:

> The man who had no informed opinion of his own, yet could not bring himself, sometimes wisely, to rely entirely upon the word of the dealer, had recourse to the services of a third party–the professional expert, the art historian, or the consulting connoisseur. This was particularly true of the new collector of paintings to whom names were more important than the work of art and such attributions as "anonymous artist of the 15th century" or "school of" were anathema. Thus the consultant became indispensable to the dealer for he could usually supply a name as well as the detailed data which the client demanded. (Seligman 1962: 125)

This new class of experts did not consist, on the whole, of museum curators, who were often expressly forbidden from giving attributions, nor of academic art historians, who were few and far between, and also concerned with other issues than attribution. Rather, as discerning amateurs, the new connoisseurs grew out of an older tradition, combining collecting, small-time dealing and attribution as a means of pursuing their studies. They were not professionals, and they had almost no formal qualifications, just a knowledge and love of art. Many of them struggled over the tension between art and mammon—with what Olav Velthuis calls "the hostile worlds view" (Velthuis 2005)—that the values of commerce and art are necessarily different and normally in conflict. None was more conflicted than the greatest expert of the era, Bernard Berenson, but most, like Berenson himself, ended up making some sort of compact with the devil. The rewards and opportunities—not just to make money but also to study art—were simply too great.

The new market for Old Masters had a number of effects. Above all, it intensified the pressures to make decisive attributions. As Friedländer wryly commented, "Dealers and collectors are not served by suppositions; they demand a positive choice." (Friedländer 1942: 179). Such choices, as we have seen in the case of Rembrandt, expanded the number of marketable goods. In the world of museums and expert amateurs, complex judgments about workshop collaborations, about masters and apprentices, were often discussed as a scholarly and aesthetic issue; now they were likely to be swept aside. At the same time the market had a paradoxical effect on connoisseurship. On the one hand, experts were eager to claim their attributions as safe and "scientific" and were happy to focus on the art object itself (the commodity), rather than on a larger context. On the other, the rhetoric and language used about pictures was resolutely unscientific. Robber-baron collectors did not want to hear about technical details; they wanted to hear about the genius of the artist and in the work. The issue of quality was increasingly subordinate to the question of attribution; no Raphael was a weak painting, and a weak painting was never a Raphael. A sort of inversion occurred in which the attribution determined the quality, rather than the quality the attribution.

One other feature of this new Old Master art world needs to be emphasized. Though the drama of the huge movement of cultural goods from Europe to America was enacted, as Flaminia Gennari Santori has shown, on a large public stage and in the blazing light of the American and European press, the actual transactions were conducted among a relatively small number of highly important actors—the exceptionally wealthy collectors, a clique of international art dealers of whom Joseph Duveen was the most important and powerful, and a number of experts that could be counted on the fingers of two hands. In what was a relatively small and enclosed circle—though one, of course, that crossed national boundaries—attribution was in many ways a matter of personal negotiation, of face-to-face deliberation before an individual work of art. It took place privately—in the inner sanctum of the dealer's gallery, at the private residence of the owner or buyer, or in the study of the expert. Thus relations of trust that were fostered within the group were highly personal, and at the same time there was a collective interest in controlling the flow of information to outsiders.

Connoisseurship

So, when the client or a dealer turned to an expert, what did they expect? First and foremost they wanted a reliable and safe attribution of a creator to a work of art that would guarantee their investment. But how was this to be achieved? Where did certainty lie in a world of imponderables? The answer supposedly lay in the practice of connoisseurship, a process of valuation. Connoisseurship, according to the art historian and critic, Gary Schwartz, is the method of "close analysis and comparison as a means of judging the quality, authorship and authenticity of works of art" (Schwartz 1993: 313). Few art historians or connoisseurs would dissent from this view, though some might want to elaborate on it. But we should notice that, at the outset, there are two different tasks that are different in kind contained within this definition. One is aesthetic, making a qualitative judgment about the artwork. The second is empirical: did an individual artist named x paint/sculpt this work? Was this work produced in sixteenth or nineteenth-century Italy? Both of these rather different evaluative procedures contributed to the valuation of a work of art but there is an (unstable and potentially shifting) relationship between the two, as well as a question about which should have the higher priority. Historically, the operation or practice of connoisseurship and its priorities has varied, not just according to the sorts of technique and technology that were deployed, but also because of shifting notions of what connoisseurship ought to do, and what sorts of people/experts should practice it. The case of *Hahn versus Duveen* came at a critical moment in the history of connoisseurship, one in which a tradition of enlightened amateurism based on intuitive understanding came face to face with legal demands that attributions be based on "facts" established by the modern forensic techniques that had come to be used in courts of law.

In the nineteenth century, connoisseurship relied on two main methods—one visual, the other historical. The first depended upon thorough visual observation of the palpable characteristics of the picture using the "educated eye." This was usually characterized as an act of informed intuition, based on an accumulated visual acquaintance of works of art (in this way the eye became educated or cultured). It was, and indeed remains a cliché among connoisseurs of this sort that this process is intuitive, even sub-conscious and that, in the words of the great German expert Max Friedländer in his essay "On Intuition and the First Impression," that "the first impression is deeper than all subsequent ones, of different kind and of decisive importance" (Friedländer 1942: 17). This is the sort of process examined in Malcolm Gladwell's recent book, *Blink* (Gladwell 2006). The credibility of such judgments rested, in the final analysis, on how much trust was invested in the connoisseur evaluating the work of art, a trust that, in turn, depended upon the demonstrable familiarity the connoisseur had with original works of art. At the same time it was widely understood that some connoisseurs had a better "eye" than others. This was not simply a matter of experience—of having seen more pictures—but also of greater visual acuity, usually thought of as instinctive or "a gift," though obviously capable of cultivation. Intuition, familiarity, acquaintance—such language speaks to a sort of intimacy and affective relation that enables the connoisseur to identify not just the hand but the spirit of the artist.

The second method was one of historical analysis that used documents to trace the origins and subsequent history of a painting or art object. It was—and, again, remains—a matter of creating a paper trail that traces works of art through the documents they engender—commissioning contracts, inventories, collection, sale and auction catalogs, testamentary and other legal documents—all of which produce what is generally referred to as a picture's "Provenance." The skills here are those of the historian, familiar with and trained in the use of archives; its concerns are empirical rather than qualitative. The limits of such an approach lie less in its method than in the constraints imposed by the archive: huge numbers of pictures survive, but their documentary provenances do not.

Before the American boom, both of these methods were used—often together—in order to catalog museum collections and inventory national cultural treasure. As a systematic endeavor to order knowledge, it was quite often claimed that connoisseurship was "scientific" and that it had made significant discoveries, though few went as far as the famous Italian connoisseur,

Giovanni Morelli who, towards the end of his career, claimed to have developed a new kind of scientific evaluation of pictures that did not depend on a swift intuitive glance, nor on archival investigations, but rather on a system of identifying the morphological characteristics of different artists and schools. Famously his technique relied on making attributions by examining the formal execution of small details—ears, eyes, hands and feet—that identified the artist's hand. His conclusions, he claimed, were "based upon indisputable and practical facts, accessible to every observer, and are not merely subjective and aesthetic, dependent upon individual taste and impressions, as is usually the case in critical writings on art" (Morelli 1900: vol. 2). In fact Morelli's method was little more than a new and more rigorous way of looking that proceeded from the particular to the general, rather than vice versa. Morelli's claims were always controversial among art experts, and even his early followers, like Bernard Berenson, were eventually to conclude that his method did not answer the qualitative questions that, for them, were at the heart of art connoisseurship and criticism. But Morelli's claims, which were widely publicized in the art and literary journals of the late nineteenth century, helped foster a strong public belief in the possibility of rendering connoisseurship "scientific." They created high expectations, ones that connoisseurs were to find hard to fulfill (Brewer 2009: 42–70).

The rich collectors of Old Master art were, by and large, uninterested in the grounds on which a connoisseur evaluated a picture. They wanted to be reassured about a work's qualities, conveyed in suitably hyperbolic language and confirmed by an attribution to a famous name. Because they were buying not just the work but the cultural cache that surrounded it, they expected their experts not to be technicians but men of cultivation, lovers of art, amateurs in the old sense of the term. A successful connoisseur therefore needed to present himself as the guardian of civilization, at once protecting and affirming the value of its assets. His calling was a higher one—concerned, as we shall see, to connect itself with the spiritual not the material. Such experts self-consciously distanced themselves from any consideration of the material qualities of pictures—hence their tendency to downplay or ignore the *history* of an object—as detracting from an appreciation of the art-work as the expression of transcendent, universal values. This of course, was much like Duveen's ploy, in which he persuaded clients that they should pay very high prices for paintings because they were priceless.

It is therefore not surprising that the experts retained or employed by dealers and collectors never had recourse to what we today would understand as science. This was in part because before the 1930s there were few techniques that made possible the scientific investigation of paintings. This was not for want of trying, but the site at which these developments took place was in the museum rather than the art dealer's showroom, and its chief object was not attribution but conservation. The first scientific laboratory in a museum opened in Berlin in 1888, the second in Vienna seven years later; but it was only in 1926 that the Louvre acquired a lab, and not until 1934 that London's National Gallery followed suit. There were two areas in which art science was beginning to advance in the 1920s. The first was in the chemistry of pigments, a field in which a Scottish scientist, Professor A.P. Laurie, had established himself as a leader, developing micro-chemical techniques to determine the make-up of pigments and mediums in Old Master paintings, and using infra-red photography to reveal hidden layers of painting and earlier workings of pictures, most famously exposing a Rembrandt signature that enabled scholars to date a self-portrait. A friend of the British artist, William Holman-Hunt, Laurie was not only head of an important scientific institute in Edinburgh, but also Professor of Chemistry at the Royal Academy of Arts in London (Eastaugh 2009). He offered his services to Duveen over the Hahn affair, not because he was a supporter of the dealer's view of the Hahn painting, but because he was intrigued by the technical issues the case raised. Duveen reluctantly allowed Laurie to be involved—he was not a believer in the value of art science—not realizing, as we shall see, that Laurie's evidence was to have a decisive effect on the outcome of the trial.

A second area in which art science was beginning to develop was in the use of x-ray photography. The first "shadow graphs" of paintings were produced in Dresden and Frankfurt as early as the 1890s and by the 1920s German-speaking museums in Europe and Alan Burroughs of Harvard's Fogg Art Museum were developing the first archives of x-rayed art works (Bewer 2010; Padfield et al. 2002: 62–3). Burroughs, in particular, had travelled widely, assembling a broad range of x-rays of works in European and American museums, including the Louvre in Paris. He had x-rayed the Louvre *Belle Ferronniere*, and he was to be called to give evidence at the Duveen trial.

The culture of those developing scientific techniques to reveal the physical properties of pictures was very different from the individualistic world inhabited by the amateur connoisseur. The technical advances of the 1920s were consolidated in the establishment of the International Museums Office in 1926, the publication of a new periodical, *Mouseion* from 1926, and the organization of a huge international conference, held in Rome and sponsored by the League of Nations in 1930, devoted to "the Study of Scientific Methods for the Examination and Preservation of Works of Art." From the outset the aim was to use international cooperation to develop protocols and procedures to identify, analyze and treat works of art. Most of the discussion in Rome (where Laurie was a prominent participant) was about conservation, but the conference also examined what place "scientific" investigation could play in attribution. The conclusions they drew were modest and pragmatic, calling for greater collaboration between scientists and more traditional connoisseurs, committed to the importance of the educated eye (Brewer 2009: 152–4).

On the whole the intuitive and technical methods of attribution had occupied two rather different terrains, the former holding sway in the Old Master art market, the latter chiefly confined to museums, though, as the 1930 League of Nations conference made clear, there was growing professional pressure to deploy a combination of the two, a move resisted by dealers and experts in the art market. Yet it was hard to keep them apart, and they were to confront one another in spectacular fashion during the trial of *Hahn versus Duveen*.

The Trial

The case of *Hahn versus Duveen*, though it was fuelled by the ambitions of the Hahns, need never have come to trial. Duveen could probably have bought off the Hahns, but there is abundant evidence that he decided to take part in what he planned as a show trial, in which traditional connoisseurial expertise would be paraded before the public, just as sensational art purchases had been touted in the press. Clearly Duveen intended the trial as a means to educate the public in the ways of traditional connoisseurship, but he may also have felt that legal recognition of the legitimacy of "the educated eye" would give his experts and their methods more authority in the marketplace. And it is tempting to speculate that Duveen, who in his business practices exploited every new technique and technology, may have been aware of the challenge that recent scientific techniques posed to traditional connoisseurship. From a scholarly point of view they could only serve to enhance the security of attributions, but, as technical mechanisms for establishing facts, they also depleted the personal influence of the dealer and the experts he employed. It would have been quite typical of Duveen, who relished risk, aggressively to assert the legal validity of the method that he saw as central to his business.

Duveen's case—that he had every right as an expert to express his opinion about the Hahn picture and that his views were backed almost unanimously by testimony he had solicited from some of the art world's most distinguished connoisseurs—was opposed to the Hahn's, which claimed a conspiracy on the part of Duveen to exclude their painting from the canon, because he did not represent them in the market (Hahn 1946). They dismissed his so-called experts on the grounds that they were in various ways compromised by their relations with him, and, in any case, refused to act scientifically in their attributions, relying solely on intuition. From the very outset, then, the issue before the court was one of who was entitled to make evaluations, and on what grounds, though the jury, as part of their deliberations, also had to determine the authenticity of the Hahn painting, a task for which none of them had any expertise or experience.

From the very first days of the trial, when he took the witness stand, Duveen pushed his version of connoisseurship, one that rested entirely on the visual acuity of the tutored or educated eye. He explicitly rejected documentary evidence (even though it favored his cause) and the investigation of provenance. History, he said, was unimportant if the picture itself was "wrong." At the same time he and other expects (with the exception of the chemist, Laurie) disparaged the use of a new range of technologies that were just coming into use for the dating and identification of works of art such as x-rays, chemical analysis, and the use of ultra-violet light. Such techniques addressed empirical questions that had to do with the artwork as object, but they did not speak to the vital issue of discerning a picture's quality—its spirit or genius—associated with the work of a great master. Blithely indifferent to the question of the legal status of informed intuition as grounds for evidence, Duveen engaged in what many press commentaries referred to as a class in art appreciation in the courtroom (Brewer 2009: 124–6).

Duveen bolstered his intuitive rejection of the Hahn painting as a Leonardo with a hugely impressive dossier of art expert opinion. Back in 1923 he had brought the Hahn picture to Paris to be compared with the *Belle Ferronniere* in the Louvre, assembling a group of experts who then were deposed by lawyers. This was probably the most important single meeting of such connoisseurs since the conclave in Dresden in 1871 to compare the Darmstadt and Dresden versions of Holbein's Madonna. Bernard Berenson, Langton Douglas, a rival of Berenson and British expert, Roger Fry, the famous critic who had worked at the Metropolitan Museum in New York, Adolfo Venturi, an Italian expert on Leonardo, and the Directors of the National Gallery in London and the Rijksmuseum in Amsterdam, among others, concurred with Duveen that the Hahn picture was not a Leonardo, and that it was probably a Northern European copy of the seventeenth century. No major expert was prepared to defend the Hahn painting, a remarkable show of unanimity, given the long history of conflict among them.

The role that this testimony played in the New York trial of 1929 was rather unusual. Duveen was only able to secure the attendance of one of his experts, Robert Langton Douglas (Berenson refused outright), so that the expert testimony had to be read into the legal record during the trial, a circumstance that undoubtedly diminished its impact. Nevertheless, the firepower that Duveen was able to command was all the more impressive in view of the Hahn's lack of expert support. They produced only one expert witness, a Russian painter, who knew a certain amount about Italian art practices, but very little art history, and had never been employed by a museum or dealer to authenticate pictures. In such circumstances Hahn's lawyers had decided to challenge the right of Duveen's connoisseurs to be treated as "experts" in a court of law, putting pressure on the presiding judge, Judge Black, to disqualify their testimony and on the jury to disregard their opinions.

The use of expert witnesses had, by 1929, become a common feature of American law: to cite but one example, in 1909, 60% of cases brought before the Massachusetts Superior Court used expert testimony (Golan 2007: 301). But the use of expert witnesses was not without its problems in American law. In an adversarial legal system, each party was pressed to bolster its case with expert testimony, and there had been considerable criticism in the first quarter of the twentieth century of how such testimony from different sides seemed contradictory and in conflict, apparently undermining its status as somehow special and yet impartial, i.e. 'expert', as well as raising the question of who or what an expert actually was. A further problem arose with the rules of evidence, which made matters of hearsay and opinion, as opposed to fact, inadmissible, despite the fact that most expert testimony was indeed expert *opinion*. The overall effect of these problems, especially after the decision of the Court of Appeals in the District of Columbia Frye case in 1923, was to place more and more discretionary power in the hands of the judge, who was increasingly in a position to determine whether a witness was truly "expert" and whether his or her testimony was admissible as "expertise" (Golan 2007: 258).

One way that lawyers in American courts sought to establish the bona fides of their experts was to train them as court witnesses, and establish lines by which their credentials "as experts" could be clearly demonstrated. But Duveen's connoisseurs were deeply resistant to this. They refused to be tutored or trained, to work with Levy, Duveen's lawyer, and his legal team to produce a coherent body of evidence. Levy had an elaborate questionnaire prepared for all the expert witnesses. It asked them to provide information about their professional qualifications, and evidence, in the form of publications and consulting by museums, dealers and collectors, of their expertise not just in Italian art but the works of Leonardo in particular. This anodyne request, analogous to asking an expert doctor about his career at medical school, his technical papers and specialty, was greeted with horror by most of the art connoisseurs. Levy was told in a terse cable that "the questions sent would be considered insult to people of such intellectual standing and sensitive temperaments." Only one of Duveen's experts, the Professor of Chemistry, A.P. Laurie, who had volunteered his services to Duveen, provided a *curriculum vitae* for the lawyers. As a result, and despite the fact that many of Duveen's witnesses were eminently competent to speak to the authenticity of a Leonardo, the court was not properly aware of their qualifications (Brewer 2009: 142).

A further problem arose because of the nature of the expert opinion that connoisseurs offered which was inevitably highly subjective. Neither the rule-governed, arid procedures of science, nor the dry-as-dust scholarly investigation of documents could explain the dismissal of an inauthentic work or capture the genius that was the connoisseur's elusive quarry, but only a certain sort of intuition. Thus Sir Martin Conway, justifying his rejection of the Hahn picture, said, "I simply look at the Hahn picture and the impression produced on my mind is that it is not by Leonardo." Maurice Brockwell said, "[I]t is a question of psychology, not of the magnifying glass; it is the mind of the great master that we see, the spiritual content, the psychological correlations." Bernard Berenson spoke of a "sixth sense... difficult to find the vocabulary to express oneself," emphasizing the importance of "accumulated experience upon which your spirit acts almost unconsciously." Very little of the expert testimony was "technical." Thus, even when individual parts of the Louvre or Hahn paintings were discussed—the jewel on the band around the head, the eyes, the tapes at the sleeves—they were subordinated to a discussion of quality—were they, or were they not, worthy of the skills of the master (*New York Times* 1923, 1929).

The court was not impressed with sixth senses and psychological correlations. The presiding Judge Black, who had a lawyerly sense of hard evidence and a strong commitment to proof on the basis of facts, was withering, both in the court and in his written opinion, about Duveen's experts. He also explicitly challenged Duveen's view of the nature of expertise. He told the jury, "there are two ways that experts in this case can help you with their opinions. One is in their study of the authentic history of a painting. The other by their study of the methods used or materials employed by the painters or schools of painting of the period in which it is claimed the pictures were painted." Black, in other words, was pointing to the importance of the historic, documentary record about pictures and the nature of their physical properties as vital to establishing an attribution or determination "founded on facts." He was instructing the jury to take note of the two grounds of attribution that had been explicitly rejected by all of Duveen's experts apart from the chemist A.P. Laurie (*Hahn versus Duveen* 1929).

Black conceded that court experts were not just concerned with "the facts" but were also useful in making inferences from them that would not be immediately apparent to jurors. He went out of his way to express his general support for expert witnesses: "I have profound respect for critics whose conclusions rest upon facts. What they say should be carefully considered by a jury." But then he went on to consider those experts whose claim to expertise he viewed as suppositious. "Beware of experts," Black warned the jury, "because a man claims to be an expert does not make him one." As he wrote in his opinion of the case, "the opinions of any other kinds of experts are as 'sounding brass and tinkling cymbals.' Some of them expound their theories largely by vocal expression and gesture; others wander into a zone of speculation founded upon nothing more tangible than 'psychological correlation.' I do not say that this is as absurd as it sounds to a layman, but it is too introspective and subjective to be the basis of any opinion a jury can pin its faith upon." Alluding to Berenson's remarks, Black added, "There are also experts who admit that they have no formulas, rules or ability to produce any artistic thing, but who claim to have a sixth sense that enables some of them after they have seen a picture even for five minutes to definitely determine whether it is genuine or not." Once again, Black played the legal skeptic: "I do not say that this faculty may not be possessed by some men, but it is not based upon enough objectiveness to convey definite meaning to a jury." Where, Black was tartly asking, was the "science" in "scientific connoisseurship?"

Black was also unimpressed by the way in which so many of Duveen's experts had changed their mind over the Louvre version of *La Belle*, rejecting it as a Leonardo before the 1920s, but accepting it as a Leonardo when questioned in Paris in 1923. This shift in ground was an interesting one, not least because scholarly opinion later in the century shifted back to characterizing *La Belle* as a possible studio picture or as executed by a pupil (Boltraffio was the favorite candidate). It seems that the eagerness of Duveen's experts to call the Louvre picture a work of Leonardo was a response to the pressure to find "an original" in order to label the Hahn picture "a copy." "It required," Judge Black remarked in the opinion which rejected Duveen's attempt to prevent a re-trial, "some mental agility to follow some of the experts from their positive testimony on the stand to the diametrically opposite views they had expressed in their books long before." Here he was tacitly acknowledging the skills of the Hahn's attorney, Lawrence Miller, who had adroitly diverted attention away from the experts' unanimity about the Hahn picture in order to emphasize their disagreement over the Louvre *La Belle*.

There were two bodies of evidence at the trial that were different in kind from that which Duveen's chosen experts offered the court. The first was x-ray evidence admitted towards the end of the trial. The Hahns and their lawyers had placed great store in their case on "facts" rather than opinion. They had tried (not altogether successfully) to trace the provenance of the Hahn painting, and their expert witness, the Russian painter, had dwelt on the nature and techniques of Italian painting. They also introduced an x-ray of the Hahn painting that they claimed supported their case of its authenticity, though they were only able to produce a medical radiographer rather than an art expert to make their point. But the jury, eager for facts rather than opinions, was clearly impressed, prompting Duveen to turn to Alan Burroughs, the Harvard expert, who had x-rayed the Louvre *Belle Ferronniere* a few years before. Though he was on record as being against x-rays, he flew Burroughs and his x-ray from Boston to New York. Burroughs' testimony was revealing. He pointed out that the jewel on *La Belle's* forehead was invisible on the x-ray of the Louvre: "This indicated that the painter of the Louvre *Belle* had first laid down metallic flesh tints (impermeable to x-rays) then painted the jewel over them." The painter of the Hahn *Belle*, however, had painted round the jewel, leaving a hole for it. This suggested to Burroughs that "the Hahn *Belle* had first been carefully sketched then colored in separate sections—flesh, fabric, jewelry. This is a practice favored by copyists." Remarkably, the jury either did not accept or seem to understand Burroughs' remarks, probably because they were so unfamiliar with an x-ray of a painting, and did not know how to interpret Burroughs' explanations. Certainly, there is no evidence that his testimony affected their deliberations (Brewer 2009: 133–4).

The evidence that did affect the jury's view of the case was, however, also scientific, namely the report that Laurie offered of his microscopic examination of the two versions of *La Belle*, made six years earlier in Paris. When asked how they had arrived at their verdicts, the nine jurors who supported the Hahns all pointed to the scientific evidence that, ironically, had been offered by one of Duveen's witnesses, the Scottish chemist. In Paris, after examining both pictures with a microscope and having failed to find what he called "dating pigments" that would prove either of the pictures to be later works or copies, Laurie refused to testify on the authorship of either picture as he did not "pretend to be an expert on technique and did not want to be drawn into artistic questions." But what he did say, which was seized upon by the Hahn's lawyers, was that the Louvre picture contained "neither lapis lazuli, vermillion, Naples yellow or a non-fading green, which were the finest and most prized paints," and that "the red ochre used in the bodice is termed barn painters' paint." He added that the greens in the Louvre picture "are verdigris crystal which have faded." Compared with the pigments used in the Hahn picture, the lawyers claimed, "The paints... are of the most ordinary and inferior kind, and not such as would be used by a master in da Vinci's time." During its final deliberations, the jury requested a copy of Laurie's book, *The Pigments and Mediums of the Old Masters*. While three jurors (not surprisingly they were the most prosperous members of the jury) felt convinced that the cumulative testimony of recognized art connoisseurs had to be heeded, the remainder of the jury relied on the sort of hard scientific testimony that they and the court had always wanted. (Brewer 2009: 184–6.)

As *Time* magazine, a publication that prided itself in taking the pulse of the American people, commented, "There is probably only one way of convincing a twentieth-century jury that a given painting is by a particular 15th Century master. That is by absolute evidence, such as a fingerprint, document or signature known to be valid. Such evidence the Duveen trial did not produce. Arguments on technique, expression, nuances of genius, only served to exhibit the latitude and variance of personal opinion" (*Time* 1929).

The hung jury (9 to 3) in favor of the Hahns and their picture was always an ambiguous verdict. The jurors had asked Black if they could compensate the Hahns but not make a determination about their picture, a compromise the judge would not allow. The issue then became that of whether or not there should be a retrial. In an interesting and important move, Duveen's lawyers, arguing against a retrial, claimed that the law was not competent to rule on what was essentially a matter of opinion rather than fact. Duveen, having first tried to use legal means to confirm connoisseurial expertise, now sought to protect connoisseurship from the courts. He chose exit over voice. This was, of course, a complete reversal of Duveen's original intent which had been to use the law to validate his view of connoisseurship.

The Hahn trial was not the end of the saga of the Hahn's *Belle Ferroniere*. It was used by Harry Hahn to attack the entire art establishment, by a number of European connoisseurs to promote their ideas about connoisseurship, and by some Midwest businessmen as a means of securing loans and raising capital. In January 2011, the picture was sold at Sotheby's in New York for $1.5 million dollars; a veritable bargain for an authentic Leonardo, but a high price for a seventeenth-century copy.

A Site of Dissonance

The case and trial of *Hahn versus Duveen* revealed both contradictory and conflicting values, differing and different mechanisms of truth. At the heart of the case lay the contradictory nature of connoisseurship, its commitment both to aesthetic valuation and to empirical historical inquiry. It is remarkable how little the aesthetes, those committed to judgment through the educated eye, were prepared to concede to their critics. They made no attempt to adapt to the venue of a court of law, a circumstance partly explained by the fact that many of the experts were not present at the trial, but were deposed (but also cross-examined) in private. They also repeatedly disparaged any procedure, like those of Laurie and Burroughs, that used the physical attributes of the painting to make an attribution. They would concede no ground to the empirics. The sense among connoisseurs that treating art attribution as a science meant downgrading or re-evaluating personal acquaintance with a large number of works of art (a form of knowledge that was also only available to a select few) sustained their hostility to any attempt to "scientize" attribution. The roots of this hostility lay partly in connoisseurs' desire to protect their (lucrative) turf from the incursions of self-confessed scientists. But the experts' defensiveness was also a consequence of their profound unease about art works as physical objects. An Old Master may have been made up of panel or canvas and paint, an object that existed in time and space, but what the connoisseur looked for (and knew about)—and, equally important, what the rich collector was collecting—was something altogether more intangible: those timeless signs of the human spirit that emanated from the hand of a great master. Paradoxically it was the ineffable quality of a picture that gave it real economic value. The ability and the right to discern this quality, to convert dross into gold, was therefore hugely valuable. No wonder connoisseurs were sometimes compared to alchemists and magicians.

The traditional connoisseurs were right. Stretchers and pigment were of no use in identifying a quality like "genius." So, when Berenson was asked whether a particular work was painted on canvas or wood, he dismissed the question by saying that it didn't matter whether Shakespeare wrote on paper or parchment. But, of course, the investigation of stretchers and pigment, parchment and canvas—like the historical work of reconstructing a provenance—did have value in establishing a work in time and space. As a number of art trials of the 1930s were to show, scientific testing, though it could not make a positive attribution certain, certainly could positively demonstrate that a painting could not have been produced by the artist to whom it was attributed. It was this capacity to identify fakes—to deprive fraudulent art works of their putative value—that was to enhance the status of the art scientist not just in the courtroom, but also in the art world more generally (Brewer 2009: 186–203).

Hahn versus Duveen also made clear that intuitive connoisseurship lacked value as expertise in a court of law. The law, as Judge Black made clear, placed no value on supposition, inference and intuition. For opinion to have value—and avoid being ruled out of court—it needed to use some sort of identifiable expertise, whether embodied in a technology or instrument, or expressed in a series of professional protocols and procedures. Duveen and his colleagues had none of these.

At one obvious level, the New York courtroom in 1929 was unquestionably a site of dissonance where valuation via experienced human intuition confronted scientific evaluation and was, at least in a legal sense, devalued through the process of the trial. Yet the sources of this dissonance lie in Duveen's (radical and innovative) decision to test "the experienced eye" in a court of law, a stratagem that insiders in the art world, like his fellow dealer, Rene Gimpel, found both baffling and foolish and that, as their conduct demonstrated, connoisseurs and experts found exceptionally uncomfortable and disconcerting (Gimpel 1966: 238–41). Duveen's colleagues found his conduct baffling because they thought they knew how value was (or should be) created: by connoisseurs exercising their accumulated skills within a realm of appreciation defined through transactions between dealers, collectors and experts. A rhetoric of beauty, skill and genius, as elaborated by the expert, affirmed attribution, which in turn then conferred economic value in the form of price. To disrupt or break this virtuous circle was to encourage interlopers—lawyers, scientists, and even ordinary New York citizens—to become part of the process of valuation. Was it not madness to ask a jury with no expertise to determine the authenticity of a work of art? Was it not even madder to link this decision to an evaluation of connoisseurship itself?

But, whatever Duveen's motives and intentions—and we can be sure that he did not predict or expect the outcome of the trial—he did see, far more clearly than his colleagues, that the legitimation of art expertise was changing, not least because of the escalating growth of the art market and the public visibility of the art world that he himself had done so much to promote. Art disputes were going to end up in court; scientific testing was going to play an ever-more important role in evidence and evaluation. His attempted solution to this problem (a very Duveen-like attempt to blow the opposition out of the water) failed; it led some connoisseurs like Berenson to fear for their futures. But what the trial really demonstrated was that a credible art expertise needed to combine the use of the trained eye with what were, in 1929, still only emerging scientific tests and techniques. The expert with the trained eye remains central to the configuration of art history, the creation of canons, and the valorization of particular artists. No scientific technique can substitute this. But the material evidence revealed through scientific investigation is a vital resource both in excluding false claims to authenticity and in detecting error. This modern cooperation is, like all systems of valuation, still fraught with dissonance, but its necessity was made publicly apparent by the spectacular trial of *Hahn versus Duveen*.

References

Bewer, F. G. (2010). *A Laboratory for Art: Harvard's Fogg Museum and the Emergence of Conservation in America, 1900–1950*. Cambridge, MA: Harvard Art Museums.

Brewer, J. (2009). *The American Leonardo. A Tale of Obsession, Art and Money.* New York, NY: Oxford University Press.

Decker, A. (1985). The Multimillion-Dollar Belle, *Art News* 84(6): 87–97.

Eastaugh, N. (2009). Authenticity and the Scientific Method: Past Approaches, Present Problems and Future Promise, *CoRM Journal* 1(1): 3–12.

Friedländer, M. J. (1942). *On Art and Connoisseurship,* trans. Tancred Borenius. London: Bruno Cassirer.

Gennari Santori, F. (2003). *The Melancholy of Masterpieces: Old Master Paintings in America 1900–1914*. Milan: Five Continents.

Gimpel, R. (1966). *Diary of an Art Dealer,* trans. John Rosenberg. London: Hodder & Stoughton.

Gladwell, M. (2006). *Blink. The Power of Thinking without Thinking.* New York and London: Penguin.
Golan, T. (2007). *Laws of Men and Laws of Nature: the History of Scientific Expert Testimony in England and America.* Cambridge, MA: Harvard University Press.
Hahn, H. (1946). *The Rape of La Belle.* Kansas City, KS: Glenn Publishing.
Hahn versus Duveen (1929). Supreme Court of New York 133 Misc. 871; 234 NYS 185; 1929 N.Y. Misc. (Lexis 756, 13 Spr. 1929). Judge Black's Opinion.
Morelli, G. (1900). *Italian Painters. Critical Studies of their works by Giovanni Morelli (Ivan Lermolieff). The Borghese and Doria-Pamfili Galleries in Rome,* trans. Constance Jocelyn Ffoulkes, intro. by Sir A. H. Layard, 2 vols., 2nd impression. London: Murray.
New York Times, 4, 18 September 1923; 20, 21. 27. 28 February 1929.
Padfield, J., Saunders, D., Cupitt, J., and Atkinson, R. (2002). Improvements in the Acquisition and Processing of X-ray Images of Paintings. *The National Gallery Technical Bulletin* 23(1): 62–75.
Schwartz, G. (1993). Rembrandt Research after the Age of Connoisseurship. *Annals of Scholarship* 10(3/4): 313–335.
Secrest, M. (2004). *Duveen: A Life in Art.* New York, NY: Knopf.
Seligman, G. (1962). *Merchants of Art: 1880–1960: Eighty Years of Professional Collecting.* New York, NY: Appleton-Century-Crofts.
Time Magazine (1929). February, 18.
Velthuis, O. (2005). *Talking Prices: Symbolic Meanings of Prices on the Market for Contemporary Art.* Princeton, NJ: Princeton University Press.

6

An Evaluative Biography of Cynical Realism and Political Pop

Svetlana Kharchenkova and Olav Velthuis

Introduction

On April 2nd, 2012 at Sotheby's in Hong Kong, "1993.4," a large canvas by the Chinese artist Fang Lijun sold for 3.7 million USD (Sotheby's 2012a). At the forefront of this canvas is a couple, another five figures are grouped in the lower right corner in the background. All the men are bald and look both mischievous and bored, a type of protagonist that Fang Lijun adopted since his graduation project at the Chinese Academy of Fine Arts in 1988–1989 and that became part of his signature style (Figure 6.1).

"1993.4" had travelled long distances before its sale at Sotheby's. It was created by the artist in Yuanmingyuan, an artist village on the outskirts of Beijing in 1993, as part of a series of six canvases designated for that year's Venice Biennale. The 1993 edition of the Biennale, arguably the world's most prestigious art event and widely considered as the best launch pad for a contemporary artist's international career, was the first in which Chinese artists participated. The year before, Francesca Dal Lago, an Italian art historian who at the time was a cultural attaché of the Italian Embassy in Beijing, had convinced Biennale-director Achille Bonito Oliva to come to China. Subsequently, together with the revered Chinese art critic Li Xianting, Dal Lago had prepared a list of artists for Bonito Oliva to check out. Fang Lijun was one of them. He ended up being selected by the Biennale director.

In the spring of 1993, the painting was transported, together with dozens of other works by Chinese artists, to Venice. Due to the lack of space, however, another work—only one of the six paintings—was exhibited. Later on, a European collector purchased "1993.4" directly from the artist, and then passed it on to another European collector (Sotheby's 2012a). In turn, this

Figure 6.1. Fang Lijun, 1993.4, 1993. Acrylic on canvas, 180x230 cm. Image courtesy of Springs Center of Art

collector, characterized as "important" by Sotheby's, sold the work about 19 years after its creation for the aforementioned amount in Hong Kong, where it was acquired by a Shanghai private museum (Art Radar Journal 2012).

During the travels of the canvas, its worth was assessed and justified by many people, beginning with the artist himself and his companions who frequented his studio, by Li Xianting, Achille Bonito Oliva and the organizers of the Venice Biennale, the two European buyers, one of whom purchased the work at the beginning of Fang Lijun's international career and the other one when it was developing further, by the auction officials of Sotheby's Hong Kong who accepted the work for their auction and subsequently had to come up with a price estimate for it, as well as its current owner when s/he bought it in 2012. Moreover, potential buyers at Fang Lijun's studio in Yuanmingyuan, visitors to the European collectors' premises, and scores of anonymous spectators and the prospective bidders at auction previews, assessed its worth as well. The spatio-temporal unfolding of these valuations, sometimes coherent, often contradictory, some provoked by previous valuations, others relatively independent, some superficial, others in depth, some produced by laypersons, others by experts, some valorizing the work (i.e. attaching value to it), others eroding value, is what we call an evaluative biography.[1]

In this paper we reconstruct the evaluative biography of a style with which Fang Lijun has been associated—Cynical Realism—and its sister-style Political Pop. Artists affiliated with these two styles have been extremely successful, both artistically and commercially (Lü 2010). Moreover, they were probably the first two contemporary Chinese styles to become recognized among audiences in Europe and the United States. By reconstructing this biography, we seek to understand the reasons for their success, and for their wide, global recognition.

Throughout its development in the last 30 years, the Chinese contemporary art scene has been complex and has gone way beyond the several iconic styles such as Cynical Realism and Political Pop. While recognizing the intricacies of Chinese contemporary art, we focus on these movements because of their eventful biography and manifest presence in different discourses about Chinese contemporary art, both in China and abroad.

We study the two styles together because of their similar evaluative biographies; indeed, they have often been evaluated together (e.g., Li 1992/2010; Hou 1994). However, the biography that we reconstruct is not completely unique to Cynical Realism and Political Pop. It also partly pertains to successful artists of roughly the same generation born around 1960, such as Zhang Xiaogang. On the other hand, for reasons that go beyond this paper's scope, some artists who were initially identified as Cynical Realist or Political Pop artists, never became household names and thus do not fully share the evaluative biography that we present.

Regimes of Justification

We draw on the work of Luc Boltanski and Laurent Thévenot in order to make sense of this evaluative biography (Boltanski and Thévenot 2006). Their framework provides the tools to study the particularly rich, complex, contradictory evaluative biographies of art movements (see Eyck and Busch 2012 for a previous application to art criticism). Indeed, Boltanski and Thévenot's understanding of worth as "a site of tensions, risk, and uncertainty, which requires continuing reconstruction and reenactment of agreements" (Lamont 2012: 208) seems particularly apt for the intricate taste-making

[1] Our project bears resemblances to what Appadurai (1986) calls the "social life of things" and reminds us of what Kopytoff (1986) calls a thing's biography. But whereas biographies in Kopytoff's sense are encompassing, describing the social and cultural contexts in which an object circulates, and the way a thing gets commodified and de-commodified, our biography focuses on valuations and their justifications.

machinery of the art world. Moreover, the arguments which various actors, Chinese, foreign and those with mixed identities, have used in order to justify their decisions to give attention to artists working in the styles of Cynical Realism and Political Pop, to buy their works, to select them for a public exhibition, or to do the very opposite (to ignore, reject or to criticize them) can be located in at least five of the worlds which Boltanski and Thévenot have identified: (1) the inspired world, where a work's worth is justified because of the artist's uncompromising creativity; (2) the civic world, where the value of art is seen in political terms, as an attempt to come to terms with China's repressive political regime; (3) the world of the market, where works are auctioned and worth is predominantly talked about in dollar prices and justified by referring to its scarcity or its expected financial return; (4) the world of renown, where artists are concerned about their reputations and buyers boast about their acquisitions of art for the status it may confer upon them; and (5) the industrial world, where artists' studios are compared to efficient art factories.

Before we can apply Boltanski and Thévenot to the biography of Cynical Realism and Political Pop, however, some specifications, reservations and additions need to be made. First of all, while the justifications that figure in Boltanski and Thévenot's work are typically studied synchronically, when looking at the evaluative biography of Cynical Realism and Political Pop it makes sense to consider how justifications emerge diachronically. Related to this, we find that one actor's valuation of Cynical Realism and Political Pop strongly depends not only on the particular context in which a valuation is made, but also on preceding valuations by others. For instance, the negative valuations of Cynical Realism and Political Pop, which have been widespread in recent years, cannot be understood without taking into account the earlier positive valuations of curators and buyers, the changing position of the Chinese government, and the steep rise of the Chinese art market since the first decade of the 2000s.

The pattern we find resembles mediation processes that exist in Western modern and contemporary art, where the worth of a style is recognized in ever widening circles, starting with peers, followed by critics and curators, then by market actors and finally the general public (see e.g., Bowness 1990). As Nathalie Heinich has argued, the work of these intermediaries (critics, curators, dealers, auctioneers etc.) is key to the field of contemporary art: without it "no installation, performance or contemporary painting could move successfully from the artist's studio (or mind) to a public place, let alone to any private home" (Heinich 2012: 701).

In the case of Cynical Realism and Political Pop, the intermediation was not only necessary to move a work from one institutional world to another, but more importantly to move it spatially, from one art world to another. It enabled works of art created in the Chinese art scene, which had for decades been sealed off from most of the rest of the world, to circulate in Europe and the United States.

Our argument is that two factors have been particularly important in enabling the cross-boundary circulation of the two styles. First, all actors embraced a regime of inspiration. Both Euro-American and Chinese audiences saw the two styles in terms of what Heinich has called "personalization," according to which the artist's work and life are deeply entangled and cannot be separated (Heinich 2009). Cynical Realism in particular was seen as an expression of the artist's character.

Second, the cross-border circulation and success of the two styles was enabled by dissonance (cf. Stark 2009): crucial differences exist in the way the worth of these two movements has been assessed. We find, for instance, that European and American actors tend to justify the worth of Cynical Realism and Political Pop in terms of political dissidence, whereas Chinese actors are wary of such justifications. Although geography is not necessarily a factor here, it is a proxy for intimate knowledge of and familiarity with the socio-economic and cultural context in which Chinese contemporary art is produced. This knowledge and familiarity, which apart from native Chinese actors, foreigners who have resided in China or Hong Kong are likely to have at their disposal, matters for the way the worth (or the worthlessness!) of Cynical Realism and Political Pop are justified.

This dissonance among valuations, which resulted from the use of different, frequently conflicting orders of worth, did not harm the artistic and commercial trajectory of both styles. While some Chinese art critics complained that the foreign audiences "misunderstood" Cynical Realism and Political Pop, we argue that this very misunderstanding enabled the two styles to move from one geographical world to another. In other words, the eventual commercial and artistic success stems from a combination of shared understandings and misunderstandings among the styles' various audiences.

Method

In order to reconstruct the evaluative biography, we primarily study texts from the early 1990s up until the early 2010s by a wide range of people who have been (semi)professionally engaged in evaluating Cynical Realism and Political Pop. We supplement the analysis of texts with interviews and participant observations conducted in China and the Netherlands in 2012–2013 in the framework of a larger research project on the globalization of art markets. We use texts and interviews because alternative historical data which are necessary for the reconstruction of an evaluative biography are simply not available. The advantage of historical texts is that they provide more reliable accounts of the way actors justified the worth of art in the past, because such accounts are not constructed a posteriori.

The method we chose obviously comes with strong biases. Some valuations (such as those of art critics) are easier to locate than others; other valuations, such as the way an art collector evaluates his acquisitions or a member of the public evaluates a painting at an exhibition, may be encountered "in the wild," but may not be documented. Some valuations are impossible to publish, even if they exist. For instance, positive valuations, praising "unofficial" or "avant-garde art"[2] for its political critique, could not easily be published in mainland China in the 1990s.

Another objection against our method may be that, given the visual qualities of the object whose value is being assessed, discursive data do not suffice. Instead, more detailed, micro-level observations of the in situ interactions between actor and object could be seen as *sine qua non* for an evaluative biography. Our analysis, however, shows that actors justify the worth of these artworks by means of a wide variety of valuation techniques, of which in situ viewing of the work is only one. Actors frequently do not refer to their interactions with the works of art, but, for instance, to interactions with the work's makers, their inferences of the artist's state of mind when s/he created the work or their assessment of the artist's lifestyle (Heinich 2009). The selection process for the Venice Biennale of 1993, where the international career of Fang Lijun and fellow Cynical Realist and Political Pop artists would be launched, is a case in point. Bonito Oliva, whose knowledge of contemporary Chinese art was limited in those days, stated that the main criterion he used were "the conversations" he had with the artists during his visit to China in 1992 (Sotheby's 2012b).

"A Passion for the West": Looking for New Artistic Directions in Post-Mao China

Until the late 1970s, apart from traditional Chinese painting and calligraphy, Chinese art was dominated by socialist realism. Under Mao, art was meant to serve the people, and functioned predominantly as a propaganda tool (Andrews 2008; Kraus 2004; Lü 2010). After Mao's death and in the slipstream of Deng Xiaoping's Open Door Policy, the stronghold of the government on Chinese art loosened (Andrews 2008). This allowed a Chinese contemporary art scene to develop. One important contribution to this scene was the new

[2] Terms such as "avant-garde" art are contested in the Chinese context, for discussion see e.g., Wu 2000: 11.

possibilities to learn about, and, to a limited extent, see modern and contemporary art created in Europe and the United States (Sullivan 1999). Although training in Soviet style socialist realism still prevailed in Chinese arts schools, gradually throughout the 1980s, materials on, among others, Impressionism and Expressionism became available in libraries and bookstores. Artists devoured and were inspired by them, and many of those who would later be classified as Cynical Realist or Political Pop artists were no exceptions.

Li Xianting characterizes this period as "a passion for the West." "From the late 1970s through the mid-1990s, Chinese art was open to the West in terms of art history; in each period you could find links with Western art history" (Li 2007).

The dynamic atmosphere which resulted from the cultural relaxation manifested itself among other things in the so-called '85 New Wave Movement (Sullivan 1999). Dozens of artists' groups from different parts of the country, in the absence of an infrastructure of contemporary art galleries and museums, organized exhibitions on their own.

In February 1989, only months before the tragedy of Tiananmen Square, many of the artists exhibited at the *China/Avant-Garde* exhibition (Figure 6.2), the first show of contemporary art held at an official art space—the National Art Gallery in Beijing (Andrews 2008; Lü 2010). The exhibition was shut down temporarily by the government after two artists enacted a performance in

Figure 6.2. Wang Youshen. Sample 8–China Avant-Garde, 1989. Image courtesy of Asia Art Archive

which gunshots were fired at a work of art on the opening day. Only slightly more than three months later, the Tiananmen massacre took place. In its aftermath, the sphere of optimism, idealism and engagement that had characterized the 1980s waned. Young artists like Fang Lijun, who graduated from art academies at the time, were disillusioned.

Some of these young artists born around 1960 went to live together in Yuanmingyuan (Lü 2010). Those who made oil paintings used a visual language that was vaguely reminiscent of European and American contemporary art. Some of them made more or less evident references to the Mao era. The work of some of these artists was soon referred to as being part of new movements, which Li Xianting called Cynical Realism and Political Pop. Introducing an exhibition of Fang Lijun and Liu Wei (1992), Li Xianting wrote: "I call the neo-realistic trend that emerged after 1988 or 1989, and that concentrated mainly in Beijing as *Cynical Realism*; 'cynical' is an English word, and we take up its connotations of ridicule, sarcasm and cold views on reality and life" (Artspeak China). In a different essay Li presented Political Pop as a "twin brother" of Cynical Realism in terms of their "focus on apathy" (2010: 164). The artists reportedly lived in abject poverty, without any means to exhibit, let alone sell their work (Figure 6.3). But that would soon change.

Figure 6.3. Xu Zhiwei. Fang Lijun's studio in Yuanmingyuan, 1993. Image courtesy of Xu Zhiwei

Making Art in a Meaningless World

Two figures were fundamental for establishing Cynical Realism and Political Pop: first of all, the Hong Kong art dealer Johnson Chang (Chang Tsong-zung), owner of one of the pioneering galleries for contemporary Chinese art who had been following the local art scene since the 1980s. The location of Chang's gallery was significant: contemporary art that was not welcome on the mainland could easily be exhibited and sold in Hong Kong. At the same time, Hong Kong was close enough to mainland China for Chang to develop and maintain dense ties with artists and intellectuals which are prerequisite for the promotion of new contemporary art (Velthuis 2013). Moreover, Chang, who was educated in the United States, had a strong international network, and managed to arrange exhibitions in various countries.

Early on he teamed up with another person who would turn out to play a crucial role in establishing Cynical Realism and Political Pop: Li Xianting, also known as the "godfather" of Chinese contemporary art. Li had known Fang Lijun since the 1980s (they grew up in the same village and had studied art at the same Children's Palace)[3] and was friends with some other Cynical Realist and Political Pop artists (Smith 2008). This close friendship with the critic would launch Fang Lijun's dazzling artistic career and would last throughout.

In 1989, when Fang Lijun was still studying at the country's most prestigious art academy, the Central Academy of Fine Arts, Li Xianting selected him for the famous *China/Avant-Garde* exhibition in Beijing, which provided early legitimation of his work. In the years to follow, Fang Lijun and other artists would discuss new work with Li and frequently hang out at his house. In 1993, Chang and Li put together the first important show of Chinese art after the 1989 Tiananmen massacre, *China's New Art, Post-1989*. It featured, among others, artworks labelled as Political Pop and Cynical Realism, and included works by Fang Lijun. Both Chang and Li used their networks to increase exposure for the artists, directing, among others, foreign curators and collectors to their studios. As a journalist of the *New York Times* characterized Li Xianting's role: "though he is a fine writer and curator, his main role is to guide artists gently into their own powerful history" (The New York Times 1993). For instance, all fourteen artists, which Bonito Oliva selected for the 1993 Biennale with Li Xianting's assistance, were associated with Political Pop and Cynical Realism. Moreover, when the famous British artist duo Gilbert & George had a solo show in Beijing in 1993, the large group of

[3] Children's Palaces are centers where children engage in extra-curricular activities, for example, learning music, art or sports.

Western art critics and dealers, who had been invited to "follow" them to China, were taken on a tour of the Beijing contemporary art scene. This tour included, among others, talks with Li Xianting and a visit to Fang Lijun's studio. Here, lasting contacts were established.

How did Chang and Li account for the worth of the Cynical Realist and Political Pop artists? Why did they decide to organize a show partly dedicated to them and point foreigners to their work? First of all, and, importantly, given the contrast with future valuations of Cynical Realism and Political Pop in Europe and the United States, Li Xianting praised the artists for turning their backs to politics and society. In doing so, he opposed Cynical Realism and Political Pop to the idealism and optimism of art movements in the 1980s. This art represented a "collective psychological state of loss and apathy" in the aftermath of the Tiananmen massacre (Li 2010: 162). After the crackdown, it became clear that "art no longer possesses a sacred social function," which led artists to turn towards "introversion and self-mockery" (159–161). Li continued: "Their roguish humor and apathy means meaninglessness. [...] Since the real world is meaningless, one need not treat it seriously, which is the essential rationale behind the rogue artists' cynical and mischievous style" (159). Similarly to Li's seminal article, in the catalog of the *China's New Art, Post-1989* exhibition, Chang claimed that these "younger artists" were "free of the idealistic commitments which had charged their seniors" (Chang 1993: 1).

Secondly, Chang and Li positioned them in recent Chinese art history, and related them to the developments which had started in the 1980s and had come to a standstill in 1989. With Cynical Realism and Political Pop, the suggestion seemed to be, Chinese art history was moving again. In the first article introducing the two styles, the first draft of which was published in Hong Kong and the final version in mainland China, Li Xianting (2010: 165) concluded: "The emergence of Political Pop marked a significant turning point in modern Chinese art history."

Thirdly, they specified the worth of these styles in spatial terms. If the 1980s artists had found inspiration in Western art history and intellectual thought, drawing on, among others, Freud, Nietzsche, Sartre or Camus, from the 1990s onwards "people started to question the idealistic belief that Western modern intellectual history could be utilized to rescue and reconstruct Chinese culture" (Li 2010: 158).[4] Li Xianting saw Political Pop in particular as a "milestone" because it indicated "the starting point of Chinese modern art in a direction of its own choosing," away from "the close attention paid to Western modern ideas and art" (Li 2010: 165).

[4] In the catalog of the Hong Kong exhibition Li repeated this argument, but also acknowledged that these artists were still "perfectly willing to learn whatever they can from Western artists such as Lucian Freud and Balthus" (Li 1993: xx).

International Recognition

Li Xianting's and Johnson Chang's promotional activities of the Cynical Realist and Political Pop artists did not go unnoticed abroad. In 1993, Fang Lijun not only showed at the Venice Biennale but also at the prestigious Marlborough gallery in London, whose director, Gilbert Lloyd, was a friend of Chang's. Lloyd's interest had been triggered by an exhibition of Political Pop, organized by Chang at the 1992 Art Asia fair in Hong Kong (Li Xianting wrote an essay for the catalog). In 1994, Chang curated a section of the São Paolo biennale, at the time the world's second most important art biennale, and showed Cynical Realism and Political Pop there. Li and Chang's joint exhibition *China's New Art, Post-1989* travelled to Australia, Canada, the United States and was in 1995 exhibited at the next edition of the Venice Biennale. Adding to his fame among a wide, lay audience, Fang Lijun's painting Series 2, No. 2 (1991–1992), or "Howl," featured on the cover of *New York Times Magazine* in 1993 (Figure 6.4). In 1999, the new director of the Venice Biennale Harald Szeemann once more invited a large group of Chinese

Figure 6.4. Fang Lijun, Series 2–Number 2, 1992. Oil on canvas, 200x200 cm. Image courtesy of Springs Center of Art

contemporary artists, including Fang Lijun. Many other commercial and non-commercial exhibitions in Europe and the United States would follow.

The foreign critics, dealers and curators frequently did not make the selection of these artists themselves, but again relied on Chang and Li. Given their lack of knowledge of the local Chinese art scene, and the novelty of the work, this may hardly be surprising. In terms of the French sociologist Lucien Karpik, these curators reduced their cognitive deficit of the quality of Chinese contemporary art by relying on the judgment device of social networks (Karpik 2010: 183–187). Indeed, in almost all peripheral art scenes that are in the process of integrating into the global system, equivalents of Chang and Li can be identified. It has become a common global art world practice for international curators to depend on conduits who have detailed knowledge of the local art scene and are able to introduce them quickly to upcoming artists (Bydler 2004).

But while foreign-based dealers, critics, curators and journalists frequently adopted their selection of artists from their Chinese counterparts, they mostly interpreted artworks differently. The way they saw the worth of these works diverges in at least two respects. First of all, where Li and Chang emphasized the emotional qualities of the work, and its cynical, apathetic stance towards the post-Tiananmen China, foreign-based actors saw it mostly in ideological terms, as politicized art which was socially engaged, and tried to come to terms, albeit in subtle ways, with the prevailing government censorship and repression (Wang 2009).

Take the title of one of the first foreign feature stories on the Chinese contemporary art scene, published in 1993 in the *New York Times*: "Their Irony, Humor (and Art) Can Save China." The article was written by Andrew Solomon, a foreign journalist who had previously covered the emergence of the post-Soviet contemporary art market (Solomon 1991). Although Solomon rejected the simplistic reading of Chinese contemporary art by Westerners as a quest for democracy, he emphasized the artists' civic engagement and close connection with society. For example, he wrote how artists, including Fang Lijun, signed a petition in favor of a fellow artist who had been beaten up by the police and how their works were "like cries for help" (The New York Times 1993). Other foreign journalists outside China went further than Solomon. They framed Cynical Realist and Political Pop artists as active "dissidents" and mentioned that Li Xianting "is forbidden to publish inside China" (Time 1993; Le Nouvel Observateur 1994). In a 1993 article in *Time* magazine, the image of a painting from the "Great Criticism" series by Political Pop artist Wang Guangyi featuring soldiers in an expressive revolutionary pose next to the article's title "China's New Vanguard" underscored the anti-government reading of these works (Figure 6.5). In the foreign press, the works were characterized as, for instance, "a small but impressive selection of dissident

Figure 6.5. Wang Guangyi, Great Criticism–McDonald's, 1992. Oil on canvas, 120x150 cm. Image courtesy of Wang Guangyi Studio

contemporary art" (The Age 1993). Journalists emphasized that the artists had been "subjected to censorship and exile for their efforts" and constituted a "pro-democracy movement" (Time 1993). They called Fang Lijun's artworks "subversive" (The Age 1993). The fact that Cynical Realism and Political Pop would not and could not be exhibited at home made it additionally interesting for foreign audiences (cf. Barmé 1996; de Kloet 2010; Kraus 2004). In contrast to these journalistic accounts, we do not find such direct references to "dissidence" in Li and Chang's essays.[5]

[5] This cannot be explained by censorship—some of the texts were published in Hong Kong, which was under British rule at the time.

The second difference between "local" and "foreign" justifications concerns the way Cynical Realism and Political Pop were positioned in recent art history. Where Chang and Li juxtaposed the movements to the idealism and optimism of Chinese contemporary art of the 1980s and to official or traditional Chinese art, many foreign commentators were oblivious to or ignorant of other Chinese contemporary art. And where Li Xianting saw Cynical Realism and Political Pop as a turn away from Western art and intellectual history, and as the "starting point of Chinese modern art in a direction of its own choosing" (Li 2010), foreign commentators saw the movements as a reconciliation of two artistic traditions—oriental and occidental. They drew parallels with Western Pop Art, referring to some Chinese artists as "Chinese Warhol" (NRC Handelsblad 2007). At the same time, however, as a critic for the London *Times* phrased it, they emphasized that "the cultural origins of these exiles still show through" (Times 1993). As Cynical Realism and Political Pop were among the first contemporary Chinese artworks shown outside China and Hong Kong that made them especially attractive for the unfamiliar audiences. In short, for "Western" audiences, Cynical Realism and Political Pop struck the right balance between exoticism on the one hand (making it new) and familiarity on the other (rendering it understandable). This was a high art instance of creolism, where, to paraphrase Ulf Hannerz, the mixing of different local traditions within a global ecumene resulted in novelty and diversity which proved to work well for foreign audiences (Hannerz 1992; cf. de Vaan et al. forthcoming 2014). Cynical Realism and Political Pop were also seen by some foreigners as representing or at least offering a glimpse at the "real" unfamiliar China.

Li Xianting facilitated this creole interpretation of the two movements through the labels he had chosen: Cynical Realism and Political Pop prompted foreign audiences to relate these peculiar Chinese movements to their own canon. As a reporter who followed Gilbert & George during their visit to China noted: the labels "might have sounded familiar even to the professional art aficionados that accompanied the two artists to Peking" (Press Association 1993). Grouping the artists and their artworks into two clearly defined movements also made them more recognizable and easier to promote and remember.

Commercial Success in China and Abroad

The international recognition for Cynical Realism and Political Pop was partially accompanied and to a greater extent followed by commercial success. From the beginning, some Western collectors had taken an interest in this work. Usually, they were businessmen or diplomats working in or frequently visiting China, such as the Swiss Uli Sigg and the Belgian couple Guy and

Myriam Ullens. By the mid-1990s the work of some artists originally associated with Political Pop and Cynical Realism was selling for four-digit prices; ten years later this had shot up to five, six digits and in the case of Fang Lijun and Yue Minjun seven-digit prices.

The works of art, once created under shabby circumstances, now circulated in the international auction circuit and in prestigious private gallery spaces of powerful European and American art dealers such as Max Protetch (Fang Lijun) and the Pace Gallery (Wang Guangyi, Liu Wei). In those imposing, white cube spaces, devoid of references to commerce, the world's most influential collectors, such as the former advertising mogul Charles Saatchi, would take an interest in their work. Coinciding with an overall boom in the Chinese contemporary art market which started around 2005, Cynical Realism and Political Pop now appeared in newspaper articles which focused not on the work itself, but on the multimillion dollar prices they achieved at auctions (see Pollack 2010). Investment funds and investment consultancy firms started assessing these styles in terms of their potential return on investment, using the format of analyst reports, indexes and charts which are also common to the valuation of financial products (ArtTactic 2007; see also Velthuis and Coslor 2012).

In these new, highly commercial valuation settings where Cynical Realism and Political Pop started turning up, their worth was to some extent justified in financial or market terms. For instance, the auction catalog entry for Fang Lijun's "1993.4" mentions that "two or three works in this series were lost during their return from the event. 1993 No. 4 is a rare survival," implying that the painting's high worth is justified by scarcity (Sotheby's 2012a). Moreover, one of the first pieces of information that is provided after factual information such as the lot number, the work's title and size and the name of the artist, is the estimate of the sales price: $2.3–3.2 million (see also Herrero 2010). Locating the work in what Veblen called pecuniary canons of taste, the (expected) price itself was now a justification of the work's worth (cf. Velthuis 2005). Some of the collectors we interviewed did the same when they referred to current auction prices in order to account for the worth of artworks hanging in their homes.

Remarkably, however, other regimes were mostly used in market settings to justify the worth of both movements. Collectors would refer to Cynical Realism and Political Pop's "historic importance" and artistic "quality," which has been recognized by true art lovers, unlike other Chinese contemporary artists, who lack such "real base" of collectors and art lovers and are instead favored by speculators (interview, the Netherlands, 2012). Auction catalog as well as gallery exhibition texts make this recognition explicit by compiling detailed curriculum vitae of the artists (e.g., the places where the works have been exhibited or the art critical texts in which they were mentioned). In the 2000s, the works' worth was also increasingly justified in social terms, by referring to owners of other examples of Cynical Realism and Political Pop. They now became status goods, in particular for the nouveau riche in China, who, we were told by the Chinese art world participants, felt like they have to have a Fang Lijun or a Wang Guangyi on their walls if people in their social circle had them as well (informal conversations, Beijing, 2012–2013).

In the slipstream of its international recognition and commercial success, the Chinese government started embracing "avant-garde art." In the days of their emergence, the early 1990s, Cynical Realist and Political Pop artists were unable to exhibit their work in official exhibition spaces because of its potentially subversive meanings (Boucher 2009; Sullivan 1999: 718). But Chinese authorities did allow foreigners to visit the artist's studios, export the works out of the country, and let the artists themselves travel out of the country to attend the openings of museum and gallery shows. The term "dissident" which was used in the foreign press to characterize these artists was, in other words, hardly accurate.

From the late 1990s onwards, after the works had already achieved international recognition, the Chinese government changed its stance and started seeing the "avant-garde" as a soft power tool: through accepting and promoting "avant-garde" art, the Chinese authorities could show an enlightened image of themselves to foreign countries (Nye 2004; cf. Pi 2005). The interest by the Chinese government in contemporary art became manifest when the Shanghai Biennale (curated by Hou Hanru) included international artists and curators for the first time in 2000, and when China organized its own national pavilion at the Venice Biennale in 2003 (Wang 2009). Fang Lijun and others are now courted by government officials (such as those from China's powerful National Development and Reform Commission—participant observation, Beijing, 2013), and have become affiliated with the Chinese official art system organizations, such as research and educational art academies (Kharchenkova, Komarova and Velthuis, forthcoming). To some extent, they have become incorporated into the establishment. But despite the fact that the Chinese government now works with these artists, outside China, some continue to see their art in critical, anti-government terms (e.g., The New York Times 2012). One reason could be that these commentators extrapolate the government's hostile treatment of Ai Weiwei, arguably the country's only truly dissident artist, to the contemporary Chinese art scene in its entirety.

Critical Voices

Alongside the praise, the commercial success and the eventual acceptance by the government, both in China and abroad, there were critical voices, usually of art critics, who disputed both the artworks' authentic, original character and their political engagement. For instance, Meg Maggio, a prominent pioneer dealer of Chinese contemporary art, arts writer and curator, wrote in her review of *China's New Art, Post-1989* exhibition that Political Pop art is not innovative but rather a simplistic combination of communist and capitalist propaganda (Maggio 1993). Maggio, American by origin, had by that time been living in Beijing and Hong Kong for seven years. By comparing the content of this art with T-shirts sold to tourists in Moscow, she implied that this art did not reflect Chinese reality and had no serious ambition to challenge the political regime. Instead, Political Pop was grounded in "Mao craze" (cf. Barmé 1996).

Chinese art critics who, like Meg Maggio, were intimately familiar with the Chinese art scene and Chinese society, also lamented the abundance of attention the artworks were getting. Art critics such as Hou Hanru and Wang Lin expressed their discontent with "Westerners" for appropriating the discourse on Chinese contemporary art; to make things worse, in the Chinese critics' eyes, "Westerners" misunderstood the message or saw Chinese art through the wrong—i.e. political—lens. As Hou (1994), who moved from Beijing to France in 1990, put it, the "Western" art media "concentrate their energies and interests on revealing how 'unofficial' artists suffer from political pressure in the country, as if the significance of both artists and work can only be found in ideological struggles." In his piece "Oliva is not the Saviour of Chinese Art (1993)" Wang Lin likewise reproached the Venice Biennale director for selecting the art on the wrong, ideological grounds: "The limitations of Oliva's selections are not only due to the contingent and opportunistic nature of his engagement with avant-garde Chinese art, but are also located in the covert Eurocentrism concealed in his standards of evaluation" (Wang 2010: 366–367). Both Hou and Wang detested the exotisizing and "othering" of Chinese art and artists by "Western" art critics, curators and press.

Some art critics continued to reject Cynical Realism and Political Pop also after they achieved resounding success. One of the most vocal detractors is Gao Minglu, an influential art critic, art historian and curator, who was one of the organizers of the *China/Avant-Garde* exhibition in 1989. After moving to the United States from China in 1991, in the late 1990s Gao curated *Inside Out: New Chinese Art*, one of the major overviews of contemporary Chinese art touring in the United States. His critique of Cynical Realism and Political Pop followed from their—in his eyes—lack of societal engagement and "sense of social and moral responsibility" (Gao 2012: 218). Moreover, Gao claimed that these artists lack sincerity and innocence and are neither critical, nor self-critical. In response to the successes of Political Pop abroad Wang Hui argued likewise that this movement "has nothing politically incisive to say within the contemporary PRC political context. Rather, it only serves as a basis for Westerners to manufacture an imaginary connection to these Chinese revolutions [Cultural Revolution and Mao Zedong's revolution]" (Wang 1998: 247). Likewise, art critic and dealer Pi Li discards this art (and its makers) as overly "export-oriented" (Pi 2005).

One of the ways in which the worth of these movements was discredited was by referring to the lifestyles of their makers. Chinese contemporary art, according to Gao, "became increasingly devoid of content, stylized and fashion driven" (Gao 2012: 211; see also Pi 2005). In an earlier piece, he responded to Solomon's *The New York Times* article (1993) discussed above in the following way: "a visitor to this revolutionary artist might have been surprised to find him ensconced in a large house in Beijing with a beautiful garden and a big gate between him and the little people... this particular artist has been passionately engaged in creating and participating in a consumer culture ..." (Gao 1998: 37).[6] Texts would start mentioning that Fang Lijun lived in a penthouse, drove an Audi A6 (like Wang Guangyi) and operated several restaurants (Smith 2008: 135).

Cynical Realism and Political Pop also got criticized for being repetitive and mass produced. As some artists became more famous and allegedly started hiring assistants to create their paintings in larger quantities—a practice often seen as illegitimate—other people started taking this into account when assessing their art (e.g., The Guardian 2007).

But for reasons which go beyond the scope of this paper, these critical opinions, whether voiced in the early 1990s or more recently, did not have significant impact on the overall recognition of Cynical Realism and Political Pop: they were hardly listened, let alone responded to.

Conclusion

In the contemporary Chinese art world, Cynical Realist and Political Pop artists are among those who enjoy prestige as they were the pioneers who, for the first time, got international recognition after the Mao period. With them, the international success of Chinese contemporary art and the emergence of a market for contemporary art in China were initiated. These artists, who were described as rogues and outcasts only two decades ago, are now respected and referred to as "teachers" or "masters" (*laoshi*). Most in the Chinese art world recognize that it will be impossible for the new generations of artists to replicate their resounding aesthetic and commercial success.

[6] It is possible that Gao's justification of worthlessness of Cynical Realism and Political Pop is grounded in his own agenda. Gao Minglu coined the term "Apartment Art" to refer to, in his view, the less commercial artworks created by other artists after the Chinese experimental art movement split into two (Cynical Realism/Political Pop and Apartment Art) in the aftermath of the 1989 massacre. Gao constructs the value of "Apartment Art" by juxtaposing it to the Cynical Realism and Political Pop.

In this paper we have tried to understand what enabled the local and global success of some of the artists originally labeled as Cynical Realist and Political Pop. In order to do so, we have reconstructed an evaluative biography: we studied the consecutive acts of valuation and valorization in order to understand how objects created in a context which is distinct in social, economic and art historical respects, started to circulate within and outside China. The approach we are taking therefore pertains to other contexts as well, such as the objects created in the emerging art worlds of Africa, Latin America and the post-Soviet states, which ever more frequently make their appearance in museums, galleries and auction houses in Europe and the United States and have been recognized in these settings as "contemporary art."

This circulation has been possible because of the global diffusion across local art worlds of a regime of inspiration in which the artist is viewed as an authentic individual who, without compromise, and without commercial interests, creates original works of art with which he can be identified. Now even the Chinese authorities, which continue to support a traditional academic system anchored in a regime of craftsmanship (see Kharchenkova, Komarova, and Velthuis, forthcoming), to some degree, accept this regime. Thus the diffusion of the category of "contemporary art" and an emerging shared understanding of what contemporary art is about, has allowed Cynical Realism and Political Pop to be appreciated across borders.

The evaluative biography of Cynical Realism and Political Pop is, however, not just a story of smooth diffusion of cultural objects among like-minded actors (cf. Rogers 1995). It is also a story of dissonance, as different audiences with, in some cases, complex mixed identities located within and outside China were able to apply their own orders of worth and managed to make sense of the work by means of their own understandings of Chinese contemporary art and the political and societal structures it was embedded in. They drew the works in their own worlds of worth (e.g., a civic regime), and by doing so, defined and constituted them as different objects (e.g., an investment vehicle, a status object, a diplomacy instrument). Although these justifications were often conflicting, they nevertheless built upon each other, attracting the attention of an ever widening set of audiences who became enchanted by these art movements, albeit for very different reasons. The success of Cynical Realism and Political Pop was thus intricately related to the fact that none of the actors, neither supporters nor critics, were able to force their justifications upon the others. Moreover, the ambivalent, or better, polyvalent character of the work enabled it to migrate across borders.

These flows were reinforced by labels which stuck because they managed to do two things at the same time: on the one hand, they provided clear categories which distinctly demarcated the artists associated with them from the many other Chinese artists vying for attention and provided the group with a clear, almost branded identity which could be staged in, for instance, internationally travelling exhibitions (Boltanski and Thévenot 2006: 180). At the same time, the labels enabled foreign audiences to make sense of Chinese art and relate it to their own canons. Because these labels were understood in heterogeneous ways, they facilitated the successful travels of the two styles (cf. Stark 2009: 109).

Although we are convinced that dissonances produced by the application of multiple orders of worth to polyvalent works of art are common to global art worlds, and indeed allow such worlds to exist, the recent episode of the evaluative biography of Cynical Realism and Political Pop suggests that the resulting success may not last. Indeed, the interpretation of staunch critics such as Gao Minglu, who is convinced that this work had nothing to do with political critique but is instead an instance of export-oriented art created by smooth operators looking for quick commercial success, seems to become more dominant in circles outside China as well. Although the work continues to be sold for high prices and exhibited regularly, the productive effect of dissonance may turn out to be temporary.

References

Andrews, J. F. (2008). Post-Mao, Postmodernism. In J. M. White (ed.), *Mahjong: Art, Film, and Change in China*. Berkeley, CA: University of California, Berkeley Art Museum and Pacific Film Archive, 29–36.

Appadurai, A. (1986). Introduction: Commodities and the Politics of Value. In A. Appadurai (ed.), *The Social Life of Things. Commodities in Cultural Perspective.* Cambridge: Cambridge University Press, 3–63.

Art Radar Journal (2012). Sotheby's Hong Kong Spring 2012 Auctions: Modern Asian Art Still Outselling Contemporary. *Art Radar Journal*. April 3, 2012 http://artradarjournal.com/2012/04/03/sothebys-hong-kong-spring-2012-auctions-modern-asian-art-still-outselling-contemporary/ (accessed January 28, 2014).

ArtSpeak China. Cynical Realism 玩世现实主义, *ArtSpeak China*, http://www.artspeakchina.org/mediawiki/Cynical_Realism_%E7%8E%A9%E4%B8%96%E7%8E%B0%E5%AE%9E%E4%B8%BB%E4%B9%89 (accessed January 28, 2014).

ArtTactic (2007). *China Art Market Report*, November 2007, London: ArtTactic, http://www.cherryartfoundation.com/pdf/ChineseReportNov07.pdf (accessed January 28, 2014).

Barmé, G. (1996). *Shades of Mao: The Posthumous Cult of the Great Leader.* Armonk, NY: ME Sharpe.

Boltanski, L. and Thévenot, L. (2006). *On Justification: Economies of Worth*. Princeton, NJ: Princeton University Press.

Boucher, M. (2009). Beyond Pop: Imagery and Appropriation in Contemporary Chinese Art. *Columbia East Asia Review* 2: 37–55.

Bowness, A. (1990). *The Conditions of Success: How the Modern Artist Rises to Fame.* London: Thames and Hudson.

Bydler, C. (2004). *The Global Artworld Inc. On the Globalization of Contemporary Art.* Uppsala: Acta Universitatis Upsaliensis.

Chang, T. (1993). Into the Nineties. In: T. Chang and X. Li (eds.), *China's New Art, Post-1989.* Hong Kong: Hanart TZ Gallery; Seattle: University of Washington Press, i–ix.

de Kloet, J. (2010). *China with a Cut: Globalisation, Urban Youth and Popular Music.* Amsterdam: Amsterdam University Press.

de Vaan, M., Stark, D., and Vedres, B. (forthcoming). Game Changer: The Topology of Creativity. *American Journal of Sociology.*

Eyck, T. A. T. and Busch, L. (2012). Justifying the Art Critique: Clement Greenberg, Michael Kimmelman, and Orders of Worth in Art Criticism. *Cultural Sociology* 6(2): 217–231.

Gao, M. (1998). Extensionality and Intentionality in a Transnational Cultural System. *Art Journal.* Winter 1998, 36–39.

Gao, M. (2012). Changing Motivations of Chinese Contemporary Art Since the Mid 1990s. *Journal of Visual Art Practice* 11(2–3): 209–219.

Hannerz, U. (1992). *Cultural Complexity. Studies in the Social Organization of Meaning.* New York: Columbia University Press.

Heinich, N. (2009). The Sociology of Vocational Prizes: Recognition as Esteem. *Theory, Culture & Society* 26(5): 85–107.

Heinich, N. (2012). Mapping Intermediaries in Contemporary Art According to Pragmatic Sociology. *European Journal of Cultural Studies* 15(6): 695–702.

Herrero, M. (2010). Performing Calculation in the Art Market. *Journal of Cultural Economy* 3(1): 19–34.

Hou, H. (1994). Entropy, Chinese Artists, Western Art Institutions: A New Internationalism. In J. Fisher (ed.), *Global Visions. New Internationalism in the Visual Arts.* London: Kala Press, 79–88.

Karpik, L. (2010). *Valuing the Unique. The Economics of Singularities.* Princeton, NJ: Princeton University Press.

Kharchenkova, S., Komarova, N., and Velthuis, O. (forthcoming). Official Art Organizations in the Emerging Art Markets of China and Russia. In O. Velthuis and S. Baia Curioni (eds.), *Cosmopolitan Canvases: The Globalization of Markets for Contemporary Art,* Oxford: Oxford University Press.

Kopytoff, I. (1986). The Cultural Biography of Things: Commoditization as Process. In A. Appadurai (ed.), *The Social Life of Things. Commodities in Cultural Perspective.* Cambridge: Cambridge University Press, 64–91.

Kraus, R. (2004). *The Party and the Arty in China: The New Politics of Culture.* New York: Rowan and Littlefield.

Lamont, M. (2012). Toward a Comparative Sociology of Valuation and Evaluation. *Annual Review of Sociology* 38(21): 201–221.

Le Nouvel Observateur (1994). Peretie, O. Y a-t-il une Culture après Tiananmen? [Is there a Culture after Tiananmen?] Le Nouvel Observateur, Paris. April 1994, 102–103.

Li, X. (1992/2010). Apathy and Deconstruction in Post-'89 Art: Analyzing the Trends of "Cynical Realism" and "Political Pop" (1992). In H. Wu and P. Wang (eds.), *Contemporary Chinese Art: Primary Documents*. New York: Museum of Modern Art, 157–166.

Li, X. (1993). Major Trends in the Development of Contemporary Chinese Art. In: T. Chang and X. Li (eds.), *China's New Art, Post*-1989. Hong Kong: Hanart TZ Gallery; Seattle, WA: University of Washington Press, x–xxvi.

Li, X. (2007). Transcript of an interview of Li Xianting with Asia Art Archive. November 6, 2007 http://www.china1980s.org/files/interview/lxtftfinal_201105091221138398.pdf (accessed January 28, 2014).

Lü, P. (2010). *A History of Art in 20th Century China*. Milan: Charta.

Maggio, M. (1993). Exhibition Review: China's New Art, Post-1989. Hanart TZ Gallery, Hong Kong, *Orientations*, May, 110–111.

NRC Handelsblad (2007). Berkhout, K. Verliefd op een Chinese Warhol. [In Love with a Chinese Warhol]. *NRC Handelsblad,* March 2, 2007 http://vorige.nrc.nl/next/article1774882.ece (accessed January 28, 2014).

Nye, J. (2004). *Soft Power: The Means to Success in World Politics*. New York: Public Affairs.

Pi, L. (2005). Between Scylla and Charybdis: The New Context of Chinese Contemporary Art and Its Creation since 2000. In B. Fibicher and M. Frehner (eds.), *Mahjong: Contemporary Chinese Art from the Sigg Collection*. Ostfildern: Hatje Catz, 35–36.

Pollack, B. (2010). *The Wild, Wild East: An American Art Critic's Adventures in China*. Hong Kong: Timezone 8.

Press Association (1993). Quinn, A. Pop Art Duo Puzzle Peking's Masses, *Press Association,* August 30, 1993.

Rogers, E. M. (1995). *Diffusion of Innovations*. New York: Free Press.

Smith, K. (2008). *Nine Lives: The Birth of Avant-Garde Art in New China*. Zurich: Scalo Verlag.

Solomon, A. (1991). *The Irony Tower: Soviet Artists in a Time of Glasnost*. New York: Knopf.

Sotheby's (2012a). Contemporary Asian Art, Lot 816. *Sotheby's*. http://www.sothebys.com/en/auctions/ecatalogue/2012/contemporary-asian-art-hk0382/lot.816.html (accessed January 28, 2014).

Sotheby's (2012b). Contemporary Asian Art, Lot 818. *Sotheby's*. http://www.sothebys.com/en/auctions/ecatalogue/lot.pdf.HK0386.html/f/818/HK0386-818.pdf (accessed January 28, 2014).

Stark, D. (2009). *The Sense of Dissonance: Accounts of Worth in Economic Life*. Princeton, NJ: Princeton University Press.

Sullivan, M. (1999). Art in China since 1949. *China Quarterly* 159: 712–722.

The Age (1993). Heathcote, C. Impressive Show of Dissident Work from China. *The Age*. April 7, 1993. Arts, 16.

The Guardian (2007). Honigman, A. Don't Believe the Hype about Chinese Art. *The Guardian*. February 16, 2007 http://www.guardian.co.uk/artanddesign/artblog/2007/feb/16/dontbelievethehypeaboutch (accessed January 28, 2014).

The New York Times (1993). Solomon, A. Their Irony, Humor (and Art) Can Save China. *The New York Times*, December 19, 1993 http://www.nytimes.com/1993/12/19/magazine/their-irony-humor-and-art-can-save-china.html (accessed January 28, 2014).

The New York Times (2012). Lankarani, N. The Many Faces of Yue Minjun. *The New York Times*. December 5, 2012 http://www.nytimes.com/2012/12/06/arts/06iht-rartminjun06.html?_r=0 (accessed January 28, 2014).

Time (1993). Burton, S. China's New Vanguard. *Time*. February 8, 1993, 46–47.

Times (1993). Cork, R. Eastern Eyes Look to the Future. *Times*. July 9, 1993.

Velthuis, O. (2005). *Talking Prices: Symbolic Meanings of Prices on the Market for Contemporary Art*. Princeton, NJ: Princeton University Press.

Velthuis, O. (2013). Globalization of Markets for Contemporary Art: Why Local Ties Remain Dominant in Amsterdam and Berlin. *European Societies* 15(2): 290–308.

Velthuis, O. and Coslor, E. (2012). The Financialization of Art. In K. Knorr Cetina and A. Preda (eds.), *The Oxford Handbook of the Sociology of Finance*. Oxford: Oxford University Press, 471–488.

Wang, H. (1998). PRC Cultural Studies and Cultural Criticism in the 1990s. *Positions* 6(1): 239–251.

Wang, L. (1993/2010). Oliva is Not the Saviour of Chinese Art (1993). In H. Wu and P. Wang (eds.), *Contemporary Chinese Art: Primary Documents*. New York: Museum of Modern Art, 366–368.

Wang, M. (2009). Officializing the Unofficial: Presenting New Chinese Art to the World. *Modern Chinese Literature and Culture* 21(1): 102–140.

Wu, H. (2000). *Exhibiting Experimental Art in China*. Chicago, IL: David and Alfred Smart Museum of Art, The University Of Chicago.

Part II

Devices Valorizing Uncertain Aesthetic Experiences

7

Playing Nice, Being Mean, and the Space In Between: Book Critics and the Difficulties of Writing Bad Reviews

Phillipa K. Chong

Introduction

Violence. Revenge. And a long-standing rivalry. This sounds like the stuff of Medieval or heroic literature but actually describes a dinner party at which Norman Mailer, the famed American novelist, journalist, and essayist, punched writer and critic Gore Vidal in the mouth. The reason? Retaliation over a negative book review.

Critics are an example of market mediators: third parties responsible for offering professional recommendations about new products that buyers consult to inform their consumer choices (Hirsch 1972; Zuckerman 1999). Like analysts in the financial market, literary critics assess and evaluate which products (i.e., books rather than stocks) are worth consumers' time and money. Readers consult book reviews to inform their reading selections. But critics are not simply messengers. They do not just report on the underlying value of products; otherwise they would not be of interest to sociologists of valuation (Sauder, Lynn, and Podolny 2012). Critics actively participate in constructing the worth of the objects they review.

And reviews matter. In the literary world, if a book attracts the attention of critics, amidst the hundreds of new fiction titles released each week, then this is in itself a small triumph (Heilbron 1999; Janssen 2009). Simply being reviewed in the *New York Times Book Review (NYTBR)* is a good predictor of whether an author will go on to publish another novel (Ekelund and Borjesson 2002), thus lending empirical weight to the trade wisdom: "[I]t is better to get a negative review in the NYTBR than to get none at all" (Ekelund and Borjesson 2002: 354). Getting the attention of newspaper and magazine critics is also a first and necessary step on the road to being consecrated as a high-culture novelist (Bourdieu 1993; Janssen 1998; van Rees 1983).

Yet, as the opening vignette suggests, critics' evaluative practices carry consequences not only for the books they review, but also the evaluators themselves. While the violence of the Mailer-Vidal feud may be an extreme example, it clearly makes the point that reviews can have real social, professional, and on rare occasion even physical consequences for the critics who write them. This chapter explores how critics perceive the consequences of their judgments and how these perceptions shape the final evaluations they produce.

Work elsewhere has pointed to the social character of evaluative practice. In his theory of symbolic fields, Bourdieu (1993; 1996) focuses on the strategic self-serving consequences of evaluation. He emphasizes how critics can use reviewing as a vehicle for reproducing their tastes and cultural authority as gatekeepers and agents of consecration in the cultural field. Critics achieve this, for instance, by representing their personal taste as "good" taste or using reviews as a venue for displaying their cultural capital; though they may be more or less conscious about these processes. Reviews, then, reflect not only critics' evaluations of aesthetic quality, but also the larger project of competing with people occupying similar positions in the field, namely, other critics.

Lamont (2009), in contrast, argues that we need to move beyond considerations of self-interest to examine the neglected aspects of evaluation, including how evaluators understand their role and the emotional consequences of their work. Using the world of scientific peer review as her case study, Lamont finds that peer review represents more than just an opportunity for panelists to advance their research agendas or reproduce their positions in the academic field. Panelists are driven by the desire to contribute to collective problem solving and through the process, they derive feelings of pleasure and validation of their self-concepts as fair judges and experts whose opinions matter. Lamont's analysis demonstrates the value of considering actors' subjective experiences for a fuller portrait of evaluative practice.

I follow cues from both scholars and take for granted that power, pleasure, self-concepts and competition all come to bear on evaluation. The task of this chapter is to examine how critics' subjective experience of evaluation can help us understand why they review the way they do. I address these questions using contemporary American fiction critics as my case study. This is a particularly rich case because of the context in which reviewing occurs. The contemporary fiction market has a switch-role structure (Aspers 2008) meaning that market actors are not fixed to a single position (e.g., a novelist) but can switch into different roles (i.e., from producers of fiction to reviewers of fiction and back again). Indeed, most critics whose by-lines appear in major American newspapers are themselves working writers (like Mailer and Vidal).

A widespread assumption in the literary field is that authors who switch roles make for excellent critics because they possess a unique insight into the creative works of fellow writers. The case of book reviewers is similar to Lamont's (2009) analysis of peer review wherein many of the panelists populating her study are esteemed academics asked to take time (or switch) from their regular professorship duties to serve as evaluators on peer review panels. But there is no indication that these participants experienced difficulty when making the switch. Not all role-switching is so seamless. In the case of book reviewers, there are many moral, emotional, and professional frictions which arise from switching back and forth between the reviewer and author role. I argue that examining such frictions is crucial for making sense of how reviewers understand their work as critics and what they put in their reviews.

The analysis is based on in-depth interviews with 30 fiction critics who reviewed for prominent American newspapers including *The New York Times, Los Angeles Times*, and *The Washington Post*. In what follows, I explain how the perceived consequences of critics' reviews directly shape the contents of their criticism. While none of the reviewers I interviewed admit to being blatantly dishonest about a book's quality, many did report experiencing a mix of emotional and professional tensions, amplified by their role switching, that alternatively pushed and pulled them towards being more ruthless or forgiving in their reviews.

Data and Methods

I selected book critics for my interviews by first compiling the names of people who published a fiction review in one of three newspapers in 2007. Respondents were then randomly selected from the list of names and invited to participate in the study. I do not specify which papers I used to generate the original list of names to preserve the anonymity of my respondents, but the publications were selected based on a combination of criteria including their having: i) among the largest national circulation numbers, ii) comparable targeted audiences, iii) a reputation for paying attention to fiction. Although I used only three newspapers to generate my list of reviewers, the majority of my respondents have published reviews in three or more different outlets.

Although in everyday parlance we speak of critics writing reviews, there are actually multiple forms of literary criticism. Van Rees (1983) offers a three-branch model of criticism consisting of: i) academic critics who offer close analysis of predominantly high-culture forms of literature and whose discourse is geared towards fellow academics; ii) literary essayists who publish longer-form reviews in specialized magazines or quarterly literary reviews; and iii) journalistic reviewers who report on the range of newly-published fiction for the general audiences of newspapers. As one moves across these branches of criticism—from academic critics, to literary essayists, and finally newspaper reviewers—the pool of critics, the range of books discussed, and the intended audience become more general (van Rees 1983). I focus on the world of literary journalism because it is at this level that the role switching occurs most frequently, and I use the terms "critics" and "reviewers" interchangeably to refer to those people who write reviews for a general newspaper audience.

There were 14 women and 16 men in the sample. The majority were "occasional" reviewers, meaning that in addition to writing reviews they make their living through freelance journalism, teaching English or creative writing at the college level, writing books—or more likely a combination of all the above. Only four of my respondents had full-time staff positions with a newspaper as book critics or book section editors at the time of the interview. Discussions with industry informants (publicists and book section editors) suggest that this is a typical sample. Newspapers have greatly diminished the space allotted for book reviewing, with many stand-alone book sections being folded into general "Entertainment" sections due to the changing economics of print publishing. With less space for book reviews, there are fewer review assignments and a lower demand for full-time book critics on staff. The editor for a major east coast newspaper book section estimates there are probably only twelve full-time critics left in America; other informants estimate it cannot be more than a handful.

Interviews were conducted over the telephone and typically lasted 60 to 90 minutes, during which time I asked critics to describe how they went about the task of book reviewing. Despite widespread interest in valuation, little work has investigated valuation as a phenomenological practice—that is, the concrete steps and considerations actors make when assessing quality or worth (Boltanski and Thévenot 2006; Lamont 2009; Shapin 2012; Hutter and Stark, in this volume). The advantage of interviews is that it provided critics with the opportunity to reflect upon processes and practices that are not observable in the final published reviews the general public receives.

Recipe for a Good Review(er): One Part Writer, One Part Critic

What is the job of the book critic? Instructional books on reviewing emphasize that the typical newspaper review aims to answer two fundamental questions: (1) What is a book about? And (2) How well is the book written? (Drewry

1966; Hooper 2010). A book review can be likened to an essay wherein a critic argues for the success or failure drawing upon evidence, which comes in the form of direct passages from the book or other insights from literary culture (Chong 2013; Pool 2007).

For the editors of book review sections, always on the lookout for contributors, it is common practice to seek out novelists to review other novelists. This is an efficient recruitment strategy since editors can be reasonably sure that this pool of talent has adequate knowledge of fiction and facility with the written word to produce usable reviews (Chong 2014).

Beyond considerations of efficiency, editors are particularly interested in the perspectives of other novelists because of the common belief that the skills and insights people gain through creating original pieces of fiction enhance their abilities as reviewers. A long-time book section editor, and one of the few remaining full-time staff book critics in America, explains his preference for using novelists for these reasons: "I think there is more a sense of process... more of a sense of what a writer might have been going for, whether they succeeded or not, so I think there's a little bit of a more interior vision of how a book works."

It is reasonable to presume that the act of writing a book sensitizes a reviewer to the intricacies surrounding character development, plotting, structure, and a myriad of other elements that the general reader and people who have never tried their hand at writing may not be attuned to. But the editor, himself a published author, goes further, implying that writers possess an artistic empathy, by virtue of the writing experience, which allows them to take the perspective of the writer and see inside their creative process (i.e. knowing "what a writer might have been going for"). This is a quality that makes them especially desirable reviewers.

Writers, too, believe that their experience as novelists is beneficial to their work as reviewers. One critic explains she takes a "more writer-ly perspective" to reviewing meaning that she is able to perceive "mechanics of creating a story, character and all of these things." Alongside the artistic empathies described above, this critic explains that her work as a novelist enriches her work as a reviewer because she can perceive and convey the mechanical successes or failings of a book: "I think that I can see what's happening when it's not working well. Definitely, you know, obviously my own work as a writer has made that possible."

Here the reviewer explains that beyond simply deciding that a book is not working, she can pinpoint what ingredients are missing by virtue of her own work as a writer. This is advantageous for reviewers since critics are expected to go beyond simply conveying like or dislike of a book; they should be able to identify specific features of the book as evidence to support their judgment.

The benefits of the writer-reviewer flow not only from the writer to the review, but also in the reverse direction. Gaining and sustaining the attention of a reviewing community is very important for the longevity of a writer's career. Studies have demonstrated that participating in sideline activities, such as publishing creative work in forms other than novels, is an effective strategy for maintaining the attention of relevant gatekeepers in the literary field (Janssen 1998).

Today, book reviewing is itself considered a sideline activity novelists engage in to draw attention to their own creative work. Reviewing offers authors the opportunity to promote their own creative work as writers in a variety of ways. For instance, at the bottom of book reviews are by-lines, which typically include the following information: [*Reviewer's Name*] is the author of "[*book title*]," which won [*award name, if any*]. Her latest book, "[*title*]," will be published in June.

Reviewing thereby offers writers a way to keep their name in the public eye while they are between books, which can span several years. It also serves as a way to attract readers to written works they have published elsewhere or to publicize upcoming projects. In theory, then, the writer-reviewer is a reciprocally enhancing combination of talents and responsibilities.

Writer-Reviewers in Action

During my interviews with critics, I asked them to reflect upon their review process, including specific instances when they had written positive and negative reviews. They told me about what they liked or disliked about the books and how they felt about writing those reviews.

Writing a positive review is a relatively straightforward task for reviewers since it allows them to serve multiple goals simultaneously: (i) critics get to read a good book and apply their literary talents to discuss the merits of this worthy material; (ii) critics fulfil their function as mediators and get to tell people about a new book to read; and (iii) critics benefit from the publicity attached to publishing another piece of writing (perhaps even currying favor with the author they praise). The perceived interests of all parties involved are also served when a review is positive: the book under review gets positive publicity; the reviewer gains a venue to demonstrate his/her literary acumen; and readers learn about a new novel.

When faced with writing a negative review, however, critics report experiencing a range of professional and emotional dissonance. As one critic states, "writing a negative review for me is full of anxiety." Another critic describes the experience as "a whole hornets' nest of conflicting emotions" amplified by the fact so many reviewers are also working writers. Below I illuminate some of the sources and contents of these perceived tensions.

The Upside of Disappointment: Bad Reviews as an Opportunity for Distinction

"There are nice ways to put derogatory comments," one reviewer tells me, but critics are also acutely aware of the benefits of being a little bit nasty. In so much as reviewers try to write articles that engage and entertain their readers, negative reviews are rich with possibility. For instance, recommending that a writer should "give up writing and take up pottery" captures readers' attention in a way that a straightforward laundry list of a book's technical failings does not.

In critics' vernacular, these types of intensely negative reviews are called "hatchet jobs" or "take downs." When I ask critics how they draw the boundary between a negative or "mean" review, they explain that the defining feature of a mean review is that the writer is trying to "score points off a failure by making himself *[sic]* look more clever or witty, or sharp, or talented in their own writing of the review."

The temptation to score points off a failed novel is a common occupational hazard. One reviewer comments on this tension in his own practice: "I feel like, you know, there's constantly this tension between wanting to sort of show off your own erudition and write some snappy kind of funny lines at the expense of the work, on one hand, and then awareness that that's a little morally debase."

What is "morally debase" here is an implied violation of disinterestedness (Bourdieu 1993). Briefly, disinterestedness refers to the symbolic value of disavowing economic and material profits. Bourdieu argues that it is by demonstrating disinterest in profit that individuals accrue prestige and authority in the cultural field. For artists, the principle of disinterestedness requires that they make "art for art's sake." For reviewers, this means their criticism should be guided primarily by aesthetic considerations. Sacrificing literary substance for the sake of adding a few snappy lines to one's review clearly violates this core value in the field.

Yet, even reviewers who shy away from being overtly negative in their own practice confess to enjoying mean reviews written by others. One reviewer reveals, "I don't like to do a hatchet job [but] I like to read them." Another critic, who describes herself as "too timid" when it comes to negative reviewing, is nonetheless drawn to the work of one critic because, "He's not afraid to be a bitch... He's not afraid to 'go there.'"

Some reviewers are famous for "going there." *New York Times* reviewer Michiko Kakutani is known for her polemical style: being equally passionate in her praise and criticism of novelists. She is perhaps most famous for her take downs of writers, having once been described by the equally controversial Norman Mailer as a "one-woman kamikaze." Although the critics I spoke with do not always (or even regularly) agree with Kakutani's literary opinions, many still check her reviews to see what she has to say because her willingness to be so damning in her reviews does "keep the conversation lively."

Another potential benefit of writing a negative review comes from when the reviewer is contradicting the critical consensus (de Nooy 1999). Being the singular voice of dissent is a way of distinguishing oneself from the crowd. The same effect can be achieved by being the one reviewer who positively reviews a book that other critics deem a failure; however, critics often describe such instances as embarrassing rather than beneficial.

For instance, one critic sheepishly recalls a time he wrote a glowing review about a book that most other people hated. He "felt a little embarrassed" by it and tells me that other critics still tease him about that review. How does he make sense of this experience? He reasons that in this instance, "more serious critics, perhaps more judicious critics, saw the book more clearly" than he did. He casts his enthusiasm as the opposite of more careful reviewing (i.e., reviewing that is "judicious" and "serious").

In contrast, a critic with a reputation for writing very negative reviews recounts a time when he wrote such a devastating review of a book that he and his editor decided not to publish it. The book went on to win the prestigious *Man Booker Prize for Fiction*. Yet this critic does not experience relief that his opposing judgment was not revealed, but instead expresses regret at a missed opportunity: "I really wish it had come out because I ended up being one of the few—or maybe the only—dissenting voice."

What is the benefit or significance of being the "only dissenting voice"? In addition to the publicizing effects, this critic explains that it is part of his responsibility to readers: "This is self-aggrandizing, but there's like an 'Emperor's New Clothes' kind of feeling, like you're going to be the little boy that points out that this is actually not so good."

The "Emperor's New Clothes" reference elicits the image of a person courageous enough to point out what other people know, but are too frightened to say for fear of appearing foolish; hence, the critic equates his criticism with the bravery and public good provided by the young boy in the fairy tale. He suggests that his negative reviews act as a corrective or counterbalance to the "nimbus of excessive praise" he sees created by newspaper reviewers who may be working too quickly or are inclined to "go with the flow" rather than point out when a book is simply not very good. Thus, critics recognized that negative reviews have their professional benefits and functional purposes in the reviewing landscape.

The Pitfalls of Negative Reviews: Risking Reputation and Relationships

Given the potential benefits, one might assume that critics would be enthusiastic when they came across a disappointing novel that offers the opportunity to write a negative review. But pursuing such gains does not come without risk, invoking considerable uncertainty amongst reviewers about how best to proceed.

Alongside the potential benefits of writing a very negative review, critics risk being perceived as *writers in critics' clothing*: novelists who disingenuously adopt the role of the critic so they have a venue for advancing their own interests as novelists. Few critics openly admit to taking shots at writers or turning phrases just for the sake of it. Or they explain that these are things they did in the past when they were "younger" or less "mature." Whether these reported tendencies are accurate accounts of their own practices or efforts at positive self-presentation, reviewers are quick to identify and condemn other writers whom they perceive as giving into the temptation of self-promotion at the cost of fair criticism.

Sometimes this offense is seen as a matter of house style. *The London Review of Books* was singled out as a repeat offender on this count: "[I]t's presumably about books but it always turns out to be about the writer of the review." One writer reports that "it generally tends to be sort of mid-rate writers" who "use reviews as an opportunity to make a name for themselves, to make a big splash, to do something bold." Thus, one consequence or risk of writing mean reviews as a mode of distinction is being recognized as doing so, which can hurt one's legitimacy as a critic and one's status as a writer if one is grouped in such an unflattering and disparaging category of writers (i.e., "mid-rate" or "immature").

Another consequence of writing negatively about a book is the risk of *retributive reviewing*. Critics who are too cavalier with their criticism run the risk that the favor will eventually be returned since the switch-role structure of the field assures that the review*er* today will be the review*ee* tomorrow.[1]

Other forms of retribution include *burning bridges* within the literary community. For instance, writing a devastating review in the *New York Times* of a writer who is represented by a literary agent or publishing house that one might want to work with in future can be seen as a short-sighted manoeuver. Writer-reviewers are, in effect, "critics for a day." But what they do with that day—how they convey their judgments in their reviews—can have long-term consequences.

In an ironic turn, recall that writers and review editors agree that having gone through the process of writing a novel equips writer-reviewers with an artistic empathy that enriches their criticism. Reviewers' artistic empathies,

[1] There is an expectation amongst critics that they should recuse themselves if there is some conflict of interest preventing them from giving a fair and unprejudiced review. Many respondents agree that reviewing a novel by someone who has previously reviewed their own work is a sufficiently corrupting factor. Yet, while review editors may do their best to weed out biasing relations, not all such conflicts are reported or easily identifiable.

however, do not stop at the writing process. It also extends to the feelings of devastation that can come from receiving a negative review.

Most respondents report reflecting upon their own painful memories of getting negative reviews when faced with the prospect of writing one themselves. This type of identification with the author could render critics' experience as writers more of a liability than an advantage. One critic describes the conflict this way:

> [T]here's no way to separate my experience with books as a writer...from the process of talking about them in a review...I think *I try to be more generous* just because I absolutely know what it is to get a bad review...I know that there's a human being that probably spent three to seven years working on this thing.

The reviewer explains she is unable to bracket her experiences as a writer from her review practice. She explains that her identification with the author, the "human being" at the other end of the review, often leads her to skew her comments in a more positive direction ("to be more generous") rather than being completely forthcoming in her criticisms. She was not alone in this tendency.

Many critics report very unpleasant stories about when facing the hurt and anger caused by a negative review. Incidents such as the punch-out between Norman Mailer and Gore Vidal, which opens the chapter, circulate amongst writer-reviewers as a kind of cautionary tale. For some, this and all of the other risks described above are enough to make them think twice about how direct they should be when writing a negative review. Some critics opt to "play nice."

Playing it Safe by Playing Nice in Reviews

Just as there are potential benefits to being ruthless in one's criticism (i.e., "making a splash") there are also benefits to following the opposite impulse and playing nice in a review. "Playing nice" as a discursive practice is accomplished in multiple ways but its overall intended effect is to downplay critics' own negative feelings towards a book and skewing the overall valence of a review in a more positive direction.

Playing nice can mean softening negative language. One critic, for example, expresses his strong distaste for a book he recently reviewed negatively. He concludes that the book was "a mess and all over the place!" The intensity of his criticism is noticeably muted in the print version of the review, where he suggests rather benignly that the reader may feel a little "lost" at some points of the book. Another means of playing nice is to fill the review with descriptive details of the book's plot to crowd out any explicit evaluative statements about the book's overall quality.

Several critics admit that when they are faced with writing a negative review, they hedge by emphasizing how the book might appeal to other readers. One critic explains: "If I ever have to say something bad about a book... I kind of try to be—to *play nice about it* and say, you know, I just didn't like it, but maybe someone else might like it?"

Acknowledging the variety of readers' tastes is an effective strategy critics can employ to sidestep being overtly negative and avoid potential backlash from an angry author. Critics emphasize, however, that they are not being dishonest about the absolute quality of a book, they are not writing a positive review of a very negative book, but eliding the true intensity of their criticism.

For some reviewers playing nice is not only a defensive measure, but also a matter of professional service and responsibility. One critic challenges, "Why are you telling people about nine books they shouldn't be reading?" He sees playing nice as serving the needs of his readers, which means guiding them towards books they actually might want to read (Drewry 1966).

Playing nice is also seen by some critics as a way of taking care of the literary field at a time when reading culture is thought to be on the decline. Critics cite struggling publishing houses, the retrenchment of dedicated arts coverage, and the allure of alternative entertainment media like film and television towards this conclusion.

One critic reflects, "[P]eople are not reading novels nearly as much as they were 20 or 30 years ago" which "puts you in a very different position [as a reviewer]... than it would if [fiction] had this more prominent place." Similar to the comments of the previous reviewer, this critic recognizes that telling the reading public about books they should not read seems like a wasted opportunity and an unnecessary blow to an already ailing industry.

Against this backdrop, the same critic reflects upon a time she wrote a very negative review for the *Los Angeles Times*. She later learned: "[The author's] publisher lost confidence, not only in that particular book but maybe in future books. In other words, to write that negative of a review had real repercussions. I never felt terribly that I was wrong in my critical assessment of the book... but it did raise for me some kind of deeper issues of responsibility as a critic."

There are two points worth making about this critic's comments. First, the reviewer is confident in her critical assessment about the book: she did not like it. But she treats the issue of aesthetic judgment ("I never felt terribly... wrong") separately from concerns about the consequences of her review ("deeper issues of responsibility"). Second, the consequences she refers to regard the author's career, and taken with her previous comments about the decline of the book industry in general, she reasons that the right or responsible thing to do is to be gentler with her criticism. It is questionable whether a single review can generally make or break a writer's career (let alone the entire literary enterprise), but this critic's comments are representative of the emotional conflicts and sense of stewardship other reviewers expressed towards books and their fellow writers.

The Good, the Bad, and the Space In Between

A central motif in this chapter is that a lot of interpretive space exists between highly critical or positive reviews on one end and merely lukewarm reviews on the other. Critics confront different types of consequences as they move across this range, and these perceived risks or benefits inform how they choose to position their criticism in terms of its intensity.

By way of summary, book review editors seek out individuals who are knowledgeable about fiction and good with words to review for their papers. Fiction writers fit the bill nicely; however, writers bring not only their literary expertise, but also their literary ambitions, social ties, and ethical commitments to the task. These findings support previous research on the social nature of reviewing (Bourdieu 1996; Lamont 2009; Lamont and Huutoniemi 2012; van Rees 1983). But a contribution of the analysis is that it clarifies how critics' emotions, interests, and reviewing practices are related.

First, this chapter shows how critics' reported experiences of pleasure or anxiety, competition and stewardship, empathy alongside self-preservation are related to the *valence* of their valuations: the degree to which their assessments are positive or negative. When critics favor a novel, writing the review proceeds with minimal personal or professional anxiety. A few critics mention that being too "gushy" in one's tone comes with the risk of "looking like a shill" or being perceived as "uncritical" by their peers. Recall, for instance, the reviewer who is still being teased for a "glowing" review he had written years earlier. In contrast, when writing a negative review the overwhelming majority of critics report conflicts and anxieties regarding their reputation and relations within the literary field. These include feelings of sympathy that come from identifying with the author at the other end of the review, the temptation to "score points" off of a novel's failing, and the fear of retribution. Most interesting is the case when critics feel that they must write a negative review, despite the anticipated repercussions. Here is the moment of dissonance, the situation that demands their most careful valuation (Hutter and Stark in this volume).

Second, this chapter reveals how critics' social position in the field shapes what they put in their reviews. My analysis enters the critics' review process at the point when they have already decided upon a book's qualities and are contemplating how to convey those judgments in written form. Critics actively reflect upon the consequences of their judgments when deciding what to put in their reviews; for instance, the perceived costs and benefits of penning a negative review entices some critics to be very damning in their criticism whilst others choose to "play nice." Hence, critics' concerns about the social consequences of their reviews are not just (at least, consciously) about reproducing or legitimating their tastes (Bourdieu 1984), but also how to frame and compose their reviews, which in some cases (i.e., "playing nice") meant obscuring how much they liked or disliked a book.

References

Aspers, P. (2008). Analyzing Order: Social Structure and Value in the Economic Sphere. *International Review of Sociology* 18(2): 301–316.

Boltanski, L. and Thévenot, L. (2006/1991). *On Justification: Economies of Worth*. Princeton, NJ: Princeton University Press.

Bourdieu, P. (1984). *Distinction: A Social Critique of the Judgment of Taste*. Cambridge, MA: Harvard University Press.

Bourdieu, P. (1993). *The Field of Cultural Production: Essays on Art and Literature*. New York, NY: Columbia University Press.

Bourdieu, P. (1996). *The Rules of Art: Genesis and Structure of the Literary Field*. Stanford, CA: Stanford University Press.

Chong, P. (2013). Legitimate Judgment in Art, the Scientific World Reversed? Maintaining Critical Distance in Evaluation. *Social Studies of Science* 43(2): 265–281.

Chong, P. (2014). Getting it 'Just Right': Goldilocks Positioning as Professional Boundary Work. Paper presented at the Eastern Sociological Society Annual Meeting. Baltimore Hilton. Baltimore, MD Feb 21, 2014.

De Nooy, W. (1999). A Literary Playground: Literary Criticism and Balance Theory. *Poetics* 26(5): 385–404.

Drewry, J. E. (1966). *Writing Book Reviews*. Boston: The Writer.

Ekelund, B. and Borjesson, M. (2002). The Shape of the Literary Career: An Analysis of Publishing Trajectories. *Poetics* 30(5–6): 341–364.

Heilbron, J. (1999). Towards a Sociology of Translation. *European Journal of Social Theory* 2(4): 429–444.

Hirsch, P. M. (1972). Processing Fads and Fashions: An Organizational-Set Analysis of Culture Industry Systems. *American Journal of Sociology* 77(4): 639–659.

Hooper, B. (2010). *Writing Reviews for Readers' Advisory*. Chicago, IL: American Library Association.

Janssen, S. (1998). Side Roads to Success: The Effect of Sideline Activities on the Status of Writers. *Poetics* 25(5): 265–280.

Janssen, S. (2009). Foreign Literatures in National Media. *Arcadia: International Journal for Literary Studies* 44(2): 352–374.

Lamont, M. (2009). *How Professors Think: Inside the Curious World of Academic Judgment*. Cambridge, MA: Harvard University Press.

Lamont, M. and Huutoniemi, K. (2012). Comparing Customary Rules of Fairness: Evaluative Practices in Various Types of Peer Review Panels. In C. Camic, N. Gross, and M. Lamont (eds.), *Social Knowledge in the Making*. Chicago, IL: University of Chicago Press, 209–232.

Pool, G. (2007). *Faint Praise: The Plight of Book Reviewing in America*. Columbia, MO: University of Missouri Press.

Sauder, M., Lynn, F., and Podolny, J. M. (2012). Status: Insights from Organizational Sociology. *Annual Review of Sociology* 38: 267–283.

Shapin, S. (2012). The Sciences of Subjectivity. *Social Studies of Science* 42(2):170–184.

Van Rees, C. J. (1983). How a Literary Work Becomes a Masterpiece: On the Threefold Selection Practised by Literary Criticism. *Poetics* 12(4–5): 397–417.

Zuckerman, E. (1999). The Categorical Imperative: Securities Analysis and the Illegitimacy Discount. *American Journal of Sociology* 104(5): 1398–1438.

8

Structures of the Tasted: Restaurant Reviews in Berlin Between 1995 and 2012

Sophie Mützel

> Chef Gary Hoopengardner uses his connections with local producers to create an up-to-date German cuisine that places both vegetarian and offal-based dishes on the table. Respect for the ingredients partners with a sense of purpose in a way that should delight the diner: the calf sweetbreads with carrot marmalade and sage aïoli (10 Euros) was fantastic, although the accompanying frisée salad, a bit overambitious with its house-made sour-cherry compôte, distracts from it. Very good, and a nice appetizer to recommend for the organ-meat-challenged diner, was the calf tongue with marrow semolina and turnip marmalade (9 Euros). The mâche salad with red beets, pears and tonka bean vinaigrette (9 Euros) will appeal to more than just vegetarians as a light appetizer. (translated from Roeger 2012)

As the introductory quote suggests, restaurant reviews are a particular literary genre. For us as customers, reviews offer evaluations and may help us make decisions: Would we want to join in eating vegetarian and offal-based regional and seasonal dishes? For us as analysts interested in valuation, restaurant reviews offer further insights. As written notices, they call attention to the qualities of what was tasted and experienced at a particular time. When looking at them over time and removed from the actual practices of tasting and critiquing, such written reviews exhibit a map of a changing field of attention to culinary practices and to that which was tasted. What experiences are worthy of paying attention to in reviews? What are the tastes being tasted in a city's restaurants? What are the structures and the dynamics of the tasted? This chapter inquires into the dynamics of what was tested, tasted, and experienced by restaurant reviewers in Berlin between 1995 and 2012.[1]

Restaurant reviews speak to particular situations of valuation. Based on a test, namely eating at a particular restaurant, a review is an assessment, i.e., an evaluation of different aspects in relation to expectations and prior experiences (e.g., a meal, a chef, an atmosphere, a style). At the same time, such a review also produces value, for example as the evaluation may influence future customers' decision to go there. Moreover, a review is also a valuation that ascribes potentially multiple qualities to a restaurant at a particular moment. A restaurant may disappoint a reviewer's expectations, but may offer decent food for little money. Restaurant reviews thus present overlapping and at times intermingled "registers of valuing" (Heuts and Mol 2013) of evaluation and valuation. They speak to that which was being experienced and tasted in the test of a restaurant.

This chapter inquires into that which was tasted, noted as that which caught the reviewers' attention and as that which was worthy of reporting. The focus is not on individual reviews, individual reviewers, or reviews of particular restaurants, but rather on an entire field of restaurant reviews. Of interest here are systematic patterns of what caught reviewers' attention. What was tasted and experienced together? Pizza and antipasti? Can I reveal larger patterns of what was being tasted and experienced together? Different cuisines, different tastes? In order to detect that which was tasted as patterns over time and thus to gain insights into, what I call, the *structures of the tasted*, I use a unique data set of all restaurant reviews published in the two Berlin city magazines from 1995 through 2012. I "follow the reviews" over time. That is, I do not enter the field with an ex post facto knowledge of analysts of the culinary world, but rather follow developments as they take place.

Rather than focusing on the rankings of restaurants and chefs at the very top of the gastronomic ladder, this study looks at the written reviews of an entire site over time, ranging from the next door eatery to Michelin-starred restaurants. As a site, the city of Berlin was chosen for two reasons. First, after political unification the city of Berlin offers a laboratory for processes of social unification. What does the gastronomic field of unified Berlin look like? Second, Berlin became Germany's political capital in 1999/2000. Did turning Berlin into the political capital of Germany affect what was tasted?

[1] I would like to thank Reini Joosten, Julian Jürgenmeyer, and Jessica Haas, who helped me with data gathering and data cleaning; Annette König of the Zentral- und Landesbibliothek Berlin, who generously opened the archives for digitization; my colleagues at the WZB and the research unit "Cultural Sources of Newness" for their support to make this project possible. I especially thank the three editors of this book for their insightful comments and suggestions.

To study the structures of the tasted from the past to the near present, this chapter uses topic modeling, a text mining method, to gain insights into large textual corpora from a macroscopic perspective.[2] What is of interest here are larger patterns of similarity as evidenced in clusters of terms used in reviews when describing that which was tasted. Over the span of eighteen years, I am able to trace how different registers of the tasted develop and evolve across all reviews based on a systematic analysis of actual data. This work not only contributes to the formal study of culture and meaning making and to the use of quantification for the study of valuation processes, but also substantively to the study of reviews and the study of culinary fields.

The study of culinary fields, culinary cultures, and the emergence of culinary styles is well established in the sociology of culture and in organizational analyses. Scope and focus of these studies are diverse. For instance, ethnographic analyses have explored the organization of creativity of kitchen life (Fine 1996). Others have analyzed the role of institutional entrepreneurs, such as influential Spanish chef Ferran Adrià (Svejenova et al. 2007), for changes in culinary styles. Survey-based cluster analyses show regional culinary cultures in Europe based on food types, eating habits, and diet (Askegaard and Madsen 1995). Network analyses of ingredients listed in recipes and their chemical flavor compounds yield a worldwide flavor network, which shows overlaps and differences (Ahn et al. 2011; Ahnert 2013). Analyses of U.S. gourmet food magazines have indicated distinct frames of culinary discourse underlying larger trends of omnivorousness (Johnston and Baumann 2007). Studies of culinary valuation regularly use entries in gastronomic guides such as Gault Millau and Guide Michelin to extract data on chefs, restaurants, rankings, and price (Karpik 2000). Studies using gastronomic guides in this way have detected shifts in culinary styles (Rao et al. 2003) and the spanning of culinary categories (Durand et al. 2007; Rao et al. 2005). Others understand gastronomic guides as "taste makers" since they attribute aesthetic and economic value to restaurants and chefs (Lane 2013).

Indeed, taste has been studied from different perspectives. Bourdieu has focused on taste in his work as the result of social and cultural determinants (Bourdieu 1984). Others view taste as socio-cognitive patterns from a historical perspective (Leschziner 2006). Yet again others have shown taste not to be the result of socio-structural determinants. Instead, they have, from a pragmatist approach, shown that taste is a reflexive activity of a heterogeneous set of actors (Hennion 2007 and in this volume). Building on these works, this chapter seeks to make the structures of the tasted and the experienced

[2] To be explained later on in more detail. The method clusters together terms that co-occur in documents, and are thus used in similar contexts of meaning.

explicit by analyzing written reviews. They allow for insights into some of these activities after the tasting has happened.

This chapter inquires into that which was tasted, noted as that which caught the reviewers' attention and as that which was worthy of reporting. As I will show, what reviews note changes systematically over time. One big change in what was tasted started in the early 2000s. It is a movement away from cafés serving breakfast and brunch to the prominent role of red beets in the new German cuisine.

The Cultural Genre of Restaurant Reviews: Evaluating and Valuing

As a genre, restaurant reviews are almost as old as restaurants. The first restaurants opened just after the French Revolution in the 1780s (Ferguson 1998, 2004), and the first restaurant review was published in 1803 as the *l'Almanach des gourmands* (Blank 2007: 45). As a genre, restaurant reviews typically entail a description of atmosphere, service, and food. The basic idea of a review is that it is a public summary and evaluation that assists readers to be more knowledgeable in their choices, understanding, or appreciation of products or performances. A review answers two questions: What is it about? Is it any good? Thus, restaurant reviews are involved in the assessment of value as well as in the production of value. Moreover, they also present insights into that which is worthy of being paid attention to (Hennion in this volume).

Restaurant reviews in gastronomic guides such as Gault Millau, Guide Michelin, or as published in newspapers or magazines, have been referred to as "connoisseurial reviews" (Blank 2007): ideally, the reviewer brings with her unique skills, sensitivities, perhaps even training, particular experience for the reviewing.[3] Surveys into the profession of restaurant critics reveal, however, that most of them do not consider themselves food experts. Instead, "many review restaurants as a sideline to regular news reporting duties" (Jolson and Bushman 1978: 74). While reviews thus are a matter of publishing a reviewer's particular perspective, they may lack consistency, and are shaped by social and institutional influences as to why particular restaurants are chosen and others are not, they are nevertheless evaluating restaurants.

As in other economic fields, reviewers as critics are key knowledge brokers: they analyze what is happening and bring it together into organized

[3] In his work on the sociology of reviews, Blank (2007) distinguishes between "connoisseur," and "procedural reviews," e.g., consumer reports, which are anonymous and speak with institutional authority. There are also generalized, public opinion reviews, e.g., Zagat publications or online reviews in which a system of amateur connoisseurship exists.

theories (White and White 1965). They categorize and assign identities necessary for the structuring and making of a market, as studies on film (e.g., Hsu 2006; Zuckerman and Kim 2003), financial securities (e.g., Zuckerman 1999), and wine (e.g., Hsu et al. 2012) have shown. As in other economic fields, gastronomic consumers want "assurance of value for money" and thus "critics are investment counselors" (Shrum 1996: 57). At the same time, reviewers also shape what counts as aesthetically pleasing and what is legitimate.

Many studies understand reviewers as themselves influenced by the social context in which they are embedded, such as institutions, practices, and other actors (e.g., Blank 2007; Ferguson 1998). Taking in particular Bourdieu's perspective on the field of cultural production (Bourdieu 1993), with a focus on the struggle for positions and position-taking, such studies examine individual restaurant critics, often famous, their written reviews, as well as chefs and their restaurants to map the field of gastronomy (e.g., Davis 2009: 239). They show that "reviews are a mechanism through which social status is made publicly visible" (Blank 2007: 1) or that reviews in gastronomic guides assert a significant amount of influence over chefs and their considerations of what constitutes "good taste" (Lane 2013). According to those studies, reviews shape the culinary field as well as are being shaped by it. Furthermore, the evaluative practices of gastronomic guides themselves have been analyzed to show their institutionalized practices and logics necessary to maintain their legitimacy in the field (Christensen and Petersen 2013). Reviews need to resonate between writers and readers, between testers and future consumers in order to be a credible source of evaluation—be it that they entice and stir interest or that critics' evaluations are adopted in own judgments (Davis 2009: 248).

The view taken here is much more agnostic than Bourdieu's notion of a field and its constituting power struggles. The written reviews are mediators (Hennion 1997) between the experiences, i.e., the food, the tastes and atmosphere in a restaurant, the creators of the experienced good, i.e., the chefs and owners of restaurants, and future consumers. They serve as insights on what competitors do. Moreover, and following Hennion's pragmatist approach (in this volume), taste is a reflexive activity, involving many actors and things. Reviews then offer some traces of the reflexive activity of tasting. They are the written reports of what is worthy of paying attention to in a particular restaurant: what were the reviewers experiencing, what were they tasting?

Using restaurant reviews as the reports of what was tested, tasted and experienced in particular situations and at particular moments allows for insights and comparisons across time. What was tasted in the 1990s and what in the 2000s? Did the structures of the tasted change over time? To capture these shifts and changes I use computational, large scale textual analysis. It offers a macroscopic perspective into the structures and the dynamics of the tasted.

Restaurant Reviews: Analyzing Texts as Data

The data set consists of 1,623 reviews of the two local Berlin magazines *Tip* and *Zitty* between January 1995 and December 2012 (n = 790 *Zitty*, n = 833 *Tip*).[4] Data were gathered in public libraries, because the publishing houses themselves do not hold paper copies available for research and have only recent issues as digital copies. Archival work involved using existing microfilm scans, which often consisted of photographs of already bound volumes, so that many patches of texts were darkened and useless for further analysis. These texts were then gathered manually in an additional step by taking new photographs of bound volumes or by directly typing the information into the data gathering spread sheet. Most restaurant reviews were scanned, sent through optical character recognition (OCR), read into a text program, and then corrected for common spelling/OCR mistakes.

With the help of student researchers, I was able to gather the entire corpus of all Berlin restaurant reviews published in *Tip* and *Zitty* that included some sort of evaluation. Evaluations started in 1995. Before that time, listings of restaurants were published without further evaluative comments. All reviews were entered into tabular form, with addresses, ratings, and price range in separate cells from the review text.

Altogether the data set of 1,623 articles consists of 401,016 words, with reviews ranging between 256 and 657 words in length. When reading through all the texts, I noticed two things: Only few reviewers are regulars and have been contributing over longer periods of time. Indeed, what seems to be common, and thus in tune with empirical evidence gathered elsewhere, is that most of the reviewers are not full-time restaurant critics. Also, contrary to more famous examples of restaurant reviewing, such as in *The New York Times* or in *The New Yorker,* the reviews' prose is very colloquial.

Given the large amount of data and my interest in the dynamics of what was tasted, I did not start developing coding guides for the documents a priori on the basis of a prior qualitative understanding of what had happened in the culinary field in Berlin over those 18 years, and then proceed with content analytic procedures. Instead, I used the method of topic modeling.

Developed in the fields of computer science, machine learning, and natural language processing (Blei 2012; Blei and Lafferty 2009; Blei et al. 2003), topic modeling has received heightened attention in the humanities (e.g., Jockers 2013; Meeks and Weingart 2013; Moretti 2013) and the social sciences (e.g., Grimmer

[4] Each magazine is published every other week, so that each magazine has 26 issues per year. In some issues, there were no restaurant reviews at all, for instance because focus in the food section was on local farmers markets, bars, or bakeries. *Tip* and *Zitty* are glossy magazines, comparable to *Time Out* city guides.

2010; Grimmer and King 2011; Grimmer and Stewart 2013; Ramage et al. 2009), as the recent special issues of the *Journal of Digital Humanities* (Vol. 2, No.1) and of *Poetics* (Vol. 41, No. 6) also attest. Topic models provide an automated procedure for sorting the contents of a text corpus into a set of meaningful "topics." Topic modeling algorithms involve minimal human intervention: researchers need to specify a number of topics for the algorithm to find within the given corpus; the algorithm then yields that number of topics and the probabilities of words being used in a topic as well as the distribution of those topics across the text corpus.

Substantively, the idea is that each textual corpus is made up of several topics, in which particular words sort together as a constellation. A restaurant review may be about quite obvious and expected topics, such as the food served, the quality of the service, the style of cuisine, the types of customers, and less obvious topics, such as the level of *gemütlichkeit* or a sense of coziness, all in the same article. While automatic coding for frequencies finds the spread of terms across an entire textual corpus, e.g., the term "*neu*" (i.e., new), frequency counts cannot provide automatically for relational meaning of terms co-occurring in similar constellations in documents. Topic modeling works with this idea of relational meaning of words within a document, an idea similar to those developed in network analytic studies using texts in delineating deeper thematic structures (e.g., Bearman and Stovel 2000; Moody and Light 2006; Smith 2007). Moreover, topic modeling is able to capture multiple meanings of a word, since a word may belong to different topics, embedded in different meaning structures in each. Thus, topic modeling looks for "deep structure" and patterns of meaning in a corpus of documents as it clusters together semantically related words.

More technically, topic models are fully automated clustering techniques, allowing for mixed membership of terms, analyzing the document level. The goal of topic modeling is to discover topics from a collection of documents in an unsupervised way. A topic is a group of words that often co-occur with each other across all documents. The documents and their words are observed while the topic structure is a hidden structure. Computationally, topic modeling finds the hidden structure that likely generated the observed collection of documents, in a reverse generative process. To do so, it uses Bayesian statistical techniques. The underlying idea of topic modeling and its Latent Dirichelet Allocation (LDA) algorithms is that each document consists of different words and multiple topics—this is the generative model of each document. The idea is then to reverse the process and infer, given all documents and topics already generated, what the topics are that each document draws on.

The inferred hidden structure resembles the thematic structure of the collection. The LDA algorithm yields distributions of topics over the corpus of documents as well as lists of top terms making up topics. Thus, it generates topics from the documents rather than imposing categories or keywords a priori on the texts and provides for a way of tracking the evolution of topics over time—over large bodies of text. It turns out that the resulting topics are interpretable, and indeed may be viewed as "frames (semantic contexts that prime particular associations or interpretations of a phenomenon in a reader) and employed accordingly" (DiMaggio et al. 2013: 578).

One of the methodological issues raised in discussions on topic modeling is how to validate the results. In a systematic discussion of using text as data, Grimmer and Stewart (2013) underline the need for topics to be labeled. They suggest two methods for validating topic labels. One evaluation is based on semantic validity: do the topics identify coherent groups that are "internally homogeneous yet distinctive from other topics" (Grimmer and Stewart 2013: 21)? Another evaluation is based on predictive validity. Quinn et al. (2010) argue that if topics are valid, the external events should explain sudden increases in attention to a topic.

To be sure, topic model analysis has its limitations given its "bag of words" assumption, which disregards the order of words within a text. Grammar and rhetoric cannot be taken into account, the analyses are run on very simplified versions of the documents. Following the practices developed in natural language processing, the words are stemmed, stop words are removed, as are the most frequent words and those most rarely used.

In the ensuing analysis, topic modeling is used as a tool for the tracing of dynamics in restaurant reviews. It provides a lens for viewing the evolution of a field from a macroscopic view, suggesting different structures of the tasted. After the algorithm has sorted the corpus,[5] interpretive work is necessary to read the topics and to think about the phenomena under scrutiny.

Mapping the Field of Restaurant Reviews

I used the Stanford Topic Modeling Toolbox (TMT),[6] a publicly available software (Ramage et al. 2009), to train and infer my models. I ran several models. Checking back and forth between model results and the actual reviews as well as using insights from the tuning procedure in the TMT, I settled for 30

[5] For the analysis, I developed a stop word list on the basis of common German stop words and frequency counts of the actual texts to be analyzed. I cleaned out variations in declination and conjugation. I used n-gram analysis to pull common n-grams together. I retained some unique terms in the texts, such as "schmecken" (to taste) or "atmosphaere" (atmosphere). After cleaning, the dataset consists of 145,065 words, with 43,187 different words in 1,623 reviews, ranging in length from 24 to 266 words, with the word "neu" (new) being the most frequent in the entire corpus (used 678 times).

[6] Accessible here: http://nlp.stanford.edu/downloads/tmt/tmt-0.4/

topics that I found to be semantically meaningful. Table 8.1 presents an overview of those 30 topics and their 20 top terms. The 20 top terms are listed in the order of importance for the topic. To be sure, the clusters of terms result from the algorithmic procedure; the titles of those topics are based on my interpretation of those clustered terms. Often, the assignment was fairly simple, as in topic 4 "Italian." Other topics needed more thought, for instance, topic 17, which I called "cigars, drinks and the world."

Of the 30 topics, 12 sort together terms referencing particular *types of cuisines*: three German cuisines (Nordic, Bavarian, Swabian), four other European cuisines (Italian, Austrian, Spanish, French), two Asian cuisines (Vietnamese/Thai, Japanese/Korean), and the Middle Eastern cuisine (with some Indian references). In addition to those very regional types of cuisines, one topic also grouped together vegan/vegetarian cuisine. The results of the automatic clustering yield no surprises. For example, the Italian cuisine consists of "delicious, pasta, pizza, parmesan, antipasti" (topic 4), the Bavarian cuisine is about "sauerkraut, schnitzel, beer on tap, hearty, sausages" (topic 9), while the French cuisine is noted to be about "crème, oysters, French fries, entrecôte, brulée" (topic 29).

Of the 30 topics, five topics group together what the *things being tasted* were. Reviewers tasted "filled things" ("pleasant, fruity, spicy, ginger," topic 2), "sandwiches and burgers" ("bread, French fries," topic 10), "cocktails and drinks" ("vodka, bar, drinks, music," topic 11), "baked and filled" ("desserts, cream, cake," topic 13), and "small bites" ("tapas, bratwurst," topic 22).

One topic refers to a *particular time of the day*: "lunch break" (topic 15). Here, terms of sites such as "street, snack bar, cafeteria, hotel, window" are grouped together with terms describing the state of being of those needing food ("hungry") and those describing the food or servings ("liquid, generous").

Three topics sort together terms describing *particular sites and situations* reviewers found worthy of paying attention to: "new, local restaurant," "outside terrace, summer evening" and "outside in garden." The topic of "new, local restaurant" (topic 1) groups together a theme in restaurant reviews on neighborhood restaurants, in the districts of Kreuzberg and Neukölln, which are "comfy" and offer hearty food from a weekly changing menu. The topic "outside terrace, summer evening" brings together the theme of "young, beautiful" people, enjoying "grilled things, burgers" and "cocktails" outside during the summertime and along the "river Spree" (topic 24). The third topic denoting a site and a situation is that of "outside in garden." It suggests a theme of eating food in Berlin outside: in "hidden, difficult-to-find places," on "islands," or in "beer gardens" (topic 28).

The topic modeling procedure yields four clusters of terms, which can be labeled to be about flavors. These are about the *flavors that the reviewers tasted*: "surprising flavors" (topic 14), "perfect flavors" (topic 18), "spicy"

Table 8.1. 30 Topics

Topic	1	2	3	4	5	6	7
titled topics	**new local restaurant:** neighborhood, homemade, comfy	**filled things:** pleasant, fruity, spicy, ginger	**moments of tastings:** quality, orderliness, feeling, nice, scrump-tious	**Italian:** pasta, pizza, antipasti, homemade, delicious	**local quality & ambiance:** lunch, breakfast	**pizza parlor:** pizza, German, Italian, chefs	**Vietnamese & Thai:** tofu, duck, hot, authentic
clusters of terms (top 20)	neu	gefuellt	qualitaet	italienisch	qualitaet	pizza	vietnamesisch
	kreuzberger	teigtaschen	glas	pasta	ambiente	koch	tofu
	kiez	angenehm	hauptgang	pizza	halb	deutsche	asiatisch
	neukoelln	kulinarisch	schwarz	nudeln	tageskarte	italienisch	thailaendisch
	besitzer	gewuerzt	ordentlich	rucola	dezent	koeche	curry
	betreiber	schmeckt	australisch	parmesan	menschen	kuenstler	ente
	hausgemacht	ingwer	ueppig	italien	schwarz	fest	thai
	kreuzberg	jung	gefuehl	antipasti	mittags	gott	zitronengras
	gebraten	verschieden	holz	hausgemacht	kiez	neu	koriander
	gemuetlich	kartoffeln	moment	italiener	wechselnd	duenn	scharf
	qualitaet	note	rinderfilet	spaghetti	aufmerksam	knusprig	gebraten
	praesentiert	magen	zander	gefuellt	fruehstueck	problem	garnelen
	mischung	angerichtet	flasche	trattoria	betreiber	haengt	vietnam
	wert	delikat	koestlich	tagliatelle	ergaenzt	augen	schmeckt
	wechselnd	aroma	welt	mozzarella	lachs	verstehen	nudelsuppe
	klassisch	eindruck	einzig	ravioli	guenstig	thema	nudeln
	deftig	hausgemacht	oben	tomaten	abendkarte	dick	bruehe
	wochenk arte	mild	form	koestlich	terrasse	besten	authentisch
	nachbarschaft	bohnen	erscheint	spinat	gehalt	schweizer	ingwer
	atmosphaere	kartoffel	medium	tiramisu	ordentlich	anfang	koestlich

8	9	10	11	12	13	14	15
German & Nordic: salmon, ambiance, potatoes, surprising, satisfied	**Bavarian:** sauerkraut, schnitzel, beer on tap, sausages, hearty, happiness	**Sandwich & burgers:** bread, French fries, English, crunchy, vegetarian	**Cocktails & drinks:** customers, bar, vodka, sofas, young, beautiful	**new German cuisine:** perfect, broiled, classic, marinated, excellent, skilled, seasonal, red beets, riesling	**baked & filled:** desserts, cream, cake delicious, around the corner, gemuetlich	**surprising flavors:** strong, beef, chicken, tea, Chinese, Mexican	**lunch break:** snack, cafeteria, hungry best, generous
koch	bayerisch	sandwich	cocktails	neu	schmeckt	chinesisch	neu
lachs	sauerkraut	burger	neu	perfekt	griechisch	gefuellt	schoen
deutsche	wirt	fruehstueck	trinken	gebraten	gefuellt	china	strasse
ambiente	woche	brot	cocktail	klassisch	gruen	chinesen	mittags-pause
kulinarisch	schnitzel	pommesfrites	publikum	kuechenchef	nachtisch	mexika-nisch	espresso
versucht	fass	verschieden	jung	gruss	kuchen	neu	sorten
neu	deftig	bio	fenster	rotebeete	nett	mexika-nische	imbiss
schmecken	kraut	hausgemacht	drinks	viergaenge-menue	gemuetlich	ver-schieden	lunch
klasse	treffen	schmeckt	musik	deutsche	kreuzberg	mexika-nischen	fluessig
gebraten	wuerstchen	kuchen	schoen	sehrgut	koestlich	salsa	qualitaet
sommer	deutsche	amerikanisch	mischung	hauptgang	brot	burritos	menschen
friedrichshain	wurst	englisch	kiez	gang	freundlich	huhn	hotel
herrlich	glueck	bunt	helmholtzplatz	gaengen	sahne	spezialitaet	mittags
kartoffeln	aehnlich	food	freundlich	riesling	griechischen	gaumen	fenster
solide	moabit	broetchen	rot	regional	tuer	kulinarisch	hunger
angenehm	senf	snacks	cocktailbar	kombination	gebacken	typisch	kantine
wirt	hauptstadt	vegetarisch	dunkel	harmoniert	glas	deutlich	mama
zufrieden	sympathisch	knusprig	drink	kraeftig	naehe	gewuerzt	aussehen
leisten	bayern	paprika	wodka	kulinarisch	orange	gebraten	schmalen
mischung	schweins-braten	belegte	amerikanisch	schoen	russisch	sucht	besten

(continued)

Table 8.1. Continued

Topic	16	17	18	19	20	21	22
titled topics	**Austrian:** wiener schnitzel, tafelspitz, quality, crispy, wonderful	**cigars, drinks, and the world:** beautiful, classic, Cristiano Rienzner, cigars, Russian, drinks	**perfect flavors:** taste, tongue, mint, roast, pickled, asparagus, mixed, wonderful	**Indian, Turkish & Arabic:** lamb, vegetarian, yoghurt, hummus, spiced, tasty	**Spanish:** roasted, atmosphere, tourists, drinks	**vegan & vegetarian:** tofu, soy, marinated, love, organic, important	**small bites:** tapas, bratwurst, filling, happy, ham, mother
clusters of terms (top 20)	oesterreichisch	neu	perfekt	indisch	spanisch	vegan	tapas
	wienerschnitzel	koch	zwiebeln	lamm	gegrillt	vegetarisch	spanisch
	klassiker	schoen	eis	tuerkisch	atmosphaere	gefuellt	satt
	klassisch	klassisch	zunge	orientalisch	tapas	koch	neu
	gruen	cristianorienzner	vegetarisch	joghurt	grill	produkte	hauptgerichte
	schnitzel	russisch	gebraten	arabisch	lecker	verde	mutter
	tafelspitz	gewuerzt	tischdecken	hummus	glas	tofu	gefuellt
	kartoffelsalat	kulinarisch	rot	falafel	geschmack	kunst	gluecklich
	neu	trinken	hauptgang	vegetarisch	kulinarisch	fleischlos	halb
	ungarisch	welt	schoen	spezialitaeten	trinken	spezialitaeten	kleinigkeiten
	oesterreich	remake	tomaten	zwiebeln	gemischt	ravioli	vegetarier
	wiener	kuba	empfiehlt	tomaten	perfekt	schmecken	flasche
	veltliner	koestlich	aromatisch	fladenbrot	kartoffeln	wichtig	schinken
	paniert	gang	pro	auberginen	schoen	biologisch	wunderbar
	schmeckt	verschieden	spargel	gegrillt	freundlich	hauchduenn	buergerlich
	qualitaet	kochkunst	gefuellt	gefuellt	dorade	europa	hervorragend
	knusprig	asiatisch	blau	huhn	einzig	mano	geschaefts-fuehrer
	kaiserschmarrn	stil	flasche	gewuerzt	rot	veganer	stuben
	alpenlaendisch	italienisch	minze	schmeckt	jung	verschieden	durchschnitt-lich
	deutsche	zigarren	wunderbar	israelisch	neu	soja	haeppchen

23	24	25	26	27	28	29	30
new best cooking: superb, wonderful, beef tenderloin, lobster, sweetness	**outside terrace & summer evenings:** artists, cocktails, beautiful, young, river Spree, drinks, burger	**Swabian:** spaetzle, homemade, maultaschen, venison, spinach, vegetarian, comfy	**spicy:** Asian, hot, chilli, curry, ginger, exotic, tasty	**Japanese & Korean:** sushi, salmon, tuna, tea, beef, sesame, soy sauce, marinated, tasty	**outside in garden:** green, island, weekend, beer garden, beach, hidden	**French:** oysters, creme, French fries, entrecote, au gratin, perfect, delicious, juicy	**new flavors:** palate, strong, tasty, elegant, unobtrusive, fresh, enjoyable, duck, goat cheese, balsamico, fruity
neu	terrasse	spaetzle	asiatisch	sushi	garten	franzoesisch	gebraten
deutlich	jung	maultaschen	gruen	japanisch	potsdamerplatz	creme	geschmack
gegart	schwarz	hausgemacht	neu	roh	britischen	austern	neu
kuechenchef	hoch	vegetarisch	curry	sesam	schoen	gebraten	rot
adresse	schoen	schwaebisch	rot	lachs	gruen	perfekt	perfekt
beste	leer	gemuetlich	chili	asiatisch	getraenke	pommes-frites	geschmacklich
stil	gebraten	gebraten	scharf	staebchen	schwarz	entrecote	knackig
geschmack	koch	schwaben	ente	thunfisch	insel	brulee	schoen
hummer	cocktails	rot	gebraten	eingelegt	eingang	klassisch	kraeftig
kulinarisch	bunt	kamin	schaerfe	nigiri	versteht	schmeckt	blau
rinderfilet	neu	koch	schmeckt	japaner	rot	frankreich	ente
filet	spree	schmeckt	garnelen	koreanische	atmosphaere	brasserie	schmeckt
hervorragend	schmecken	baden	koch	sojasauce	bedeutet	fischsuppe	ziegenkaese
qualitaet	rot	mittags	kokosmilch	maki	lage	saftig	angenehm
suesse	burger	kartoffelsalat	suess	sake	versteckt	deftig	spinat
funktioniert	newyorker	schraeg	mango	gruen	sommer	flasche	balsamico
trueffel	lecker	kreuzberger	thai	japan	neu	hervorragend	modern
wunderbar	brunch	wild	huhn	frittiert	chef	rosa	dezent
gelungen	sorgen	geniessen	europaeisch	schmeckt	biergarten	bratkartoffeln	geraten
geschmacklich	klassisch	atmosphaere	knoblauch	sashimi	historisch	tarte	gaumen

(topic 26), and "new flavors" (topic 30). "Surprising flavors" brings together elements of the Chinese and Mexican cuisine with a focus on "spices" and the "palate." "Perfect flavors" are tasted on the "tongue," food is "roasted, beautiful, flavorful," and with "mint." In "spicy" a theme of that which was tasted as spicy comes together. This ranges from Thai curry to chilies and garlic. The topic "new flavors" brings together an almost exclusive list of adjectives on how the tested was tasted: there were "new tastes" for the "palate," things were "strong," "beautiful," "delicious," "tasty" and "perfect."

Three topics group together terms, which describe how things tasted and experienced are done at the tested restaurants. These topics speak to *elements of style*: styles of the sites as well as styles of the tasted. The topic "local quality and ambiance" brings together a theme on local cafés, "breakfast, lunch," where "quality" was tasted, where things were found to be "orderly" and "low priced" (topic 5). Another topic related to styles of what I called "new German cuisine" (topic 12). It groups together food tasted "red beets, Riesling, four courses" with adjectives of how food was prepared, e.g., "broiled, combined," and where it comes from ("regional"), and evaluates what was tasted as "new, perfect, excellent, beautiful." Another topic of style is that of "new best cooking" (topic 23). It brings together terms of evaluation and quality "superb, wonderful, successful, it works" with terms of what was tasted, e.g., "sweetness, tasty." "Beef tenderloin, lobster, truffles" are types of food tasted.

One topic is about *first impressions* and the "moments of tastings" (topic 3). Terms in this theme group together descriptions of what was experienced when first entering the restaurant. Upon first impressions what was tasted and experienced was "orderly, abundant, unique" and about "quality." Indicators of stationary elements are "wood, glasses, bottles," but this topic also combines elements of "feeling," "moments," and "appearance."

Lastly, two topics are about *sites, food, things, people, and interactions*. One is the theme "pizza parlor" (topic 6) that combines "pizza, German, Italian, chefs" with "artists, god." It captures themes of conversations while eating. The other topic "cigars, drinks and the world" (topic 17) also combines food, things, people, and interactions. It lists Cristiano Rienzner, a chef using elements of molecular cuisine, who was chef at several Berlin restaurants during the period studied. It also lists adjectives that reviewers were paying attention to: "beautiful, classic, delicious, different, spicy." Things such as cigars and drinks are also part of the interactions described.

The Big Shift: From Brunch to Red Beets

When mapping out the dynamics over time of all topics, one change is strikingly evident: During the mid to late 1990s, the largest topic was topic 5 "local quality and the ambiance of restaurants, in particular at breakfast and lunch" (13% of all reviews spoke of this topic in 1997). This topic waned over time. Slowly rising and taking a sharp increase beginning with 2006, "new German cuisine" (topic 12) went to the top: perfect, broiled, marinated, red beets, and *Riesling* wine. What was tasted had apparently changed. Figure 8.1 shows these two strong opposite trends.

What are possible explanations for these opposite trends? One explanation could be that the restaurants in Berlin changed. Another is that the *selection* of restaurants tested in Berlin changed. When looking at the data, both explanations apply. While the cafés offering breakfast and brunch did not disappear from Berlin's streets, the attention paid to "new German cuisine" by the reviewers as well as by restaurants had increased by the late 2000s. Indeed, until the late 1980s, Berlin was not known for its restaurants. This started to change a bit in the mid-1990s. At that time, the notion "new German cuisine" was used as an empty label (Matthies 2001). With the move of the political capital of Germany from Bonn to Berlin in 1999/2000 "new German cuisine" started to appear more regularly as a concept in gastronomic writing (e.g., Studnitz 1999). By the 2010s, "new German cuisine" as a concept, encompassing regional and seasonal food, is claimed and marketed by German chefs (Lane 2014). The analysis of reviews reveals the rising trend in German and regional food.

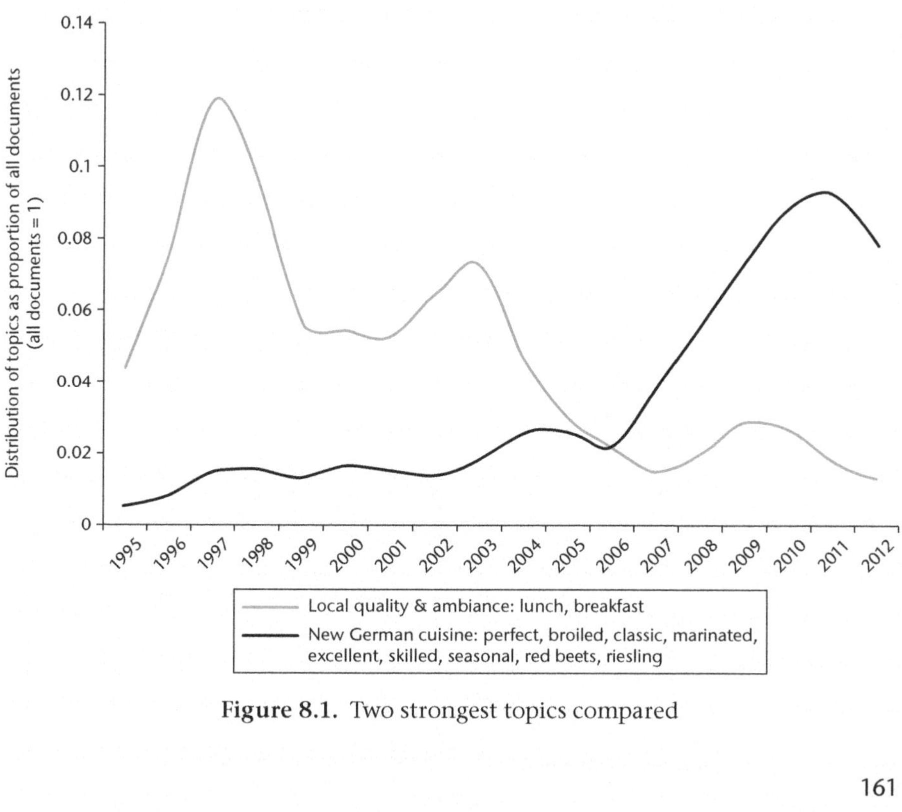

Figure 8.1. Two strongest topics compared

Dynamics of the Tasted

The analysis of topics over time reveals further dynamics, three of which I will discuss in the following. One is that the two opposing topics of that which was tasted (local quality and ambiance vs. new German cuisine) are accompanied by other supporting topics. In the 1990s and until the mid-2000s Vietnamese and Thai cuisine was tasted a lot, beginning in 2000 with the opening of a much reviewed and acclaimed restaurant in the central district of Berlin. Many Chinese restaurants, founded in the 1990s and run by ethnic Vietnamese, began to copy this success story and expanded their menus to offer Vietnamese and Thai food. The topic features the term "authentic" and speaks to those issues of identity as evaluated and tasted by the reviewers (topic 7). In addition to Asian cuisines, Mediterranean cuisines were predominantly tasted during the 1990s until the mid-2000s. Only since then have topics of Swabian, Austrian, and French cuisine been rising, although less so than the topic of "new German cuisine." The latter topic is thus accompanied by other cuisines tasted, which are somewhat related.

A second dynamic is evident in the group of five topics on the *things being tasted* (Figure 8.2). Reviewers tasted "filled things" ("pleasant, fruity, spicy, ginger", topic 2), "sandwich and burgers" ("bread, French fries," topic 10), "cocktails and drinks" ("vodka, bar, drinks, music, sofas," topic 11), "baked and filled" ("desserts, cream, cake," topic 13), and "small bites"

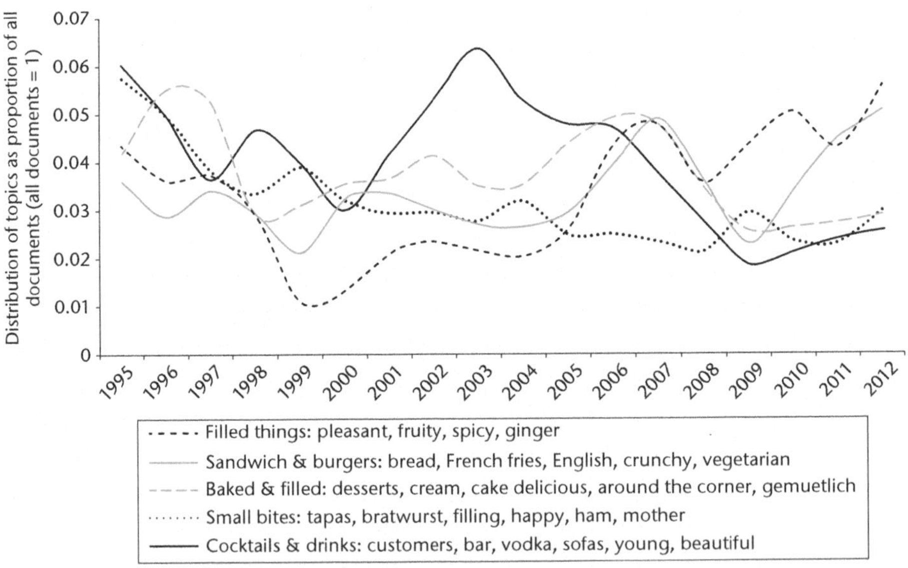

Figure 8.2. Things tasted

("tapas, bratwurst," topic 22). Two trends are particularly noteworthy. The topic model analysis reveals that the topic of "cocktails and drinks" peaks in the early 2000s; 6% of all documents showed this topic in 2003. The topic "filled things" (the topic lists different types of stuffed, savory pastry) was very seldom found in the late 1990s, though in comparison to the other topics on things tasted, "filled things" were more often tasted in the late 2000s.

The group of topics on *flavors that the reviewers tasted* (Figure 8.3) "surprising flavors" (topic 14), "perfect flavors" (topic 18), "spicy" (topic 26), and "new flavors" (topic 30) reveal almost rhythmic patterns of ups and downs. Over the 18 years studied, the flavors tasted show changing trends of the palate. At the beginning of the time frame of analysis, the late 1990s, the topic "surprising tastes" was dominantly tasted. It then was replaced by "new flavors" in 2000. "Spicy" then was dominantly tasted in 2005, followed by a drop and a replacement of attention to "perfect flavors" in 2007, which again was replaced by "spicy" in 2008, and then by "surprising tastes" in 2010. At the end of the period under analysis, "spicy" and "surprising tastes" are decreasingly tasted, while in particular "perfect flavors" is on the rise. Thus, the analysis reveals that the flavors tasted alternate, almost rhythmically.

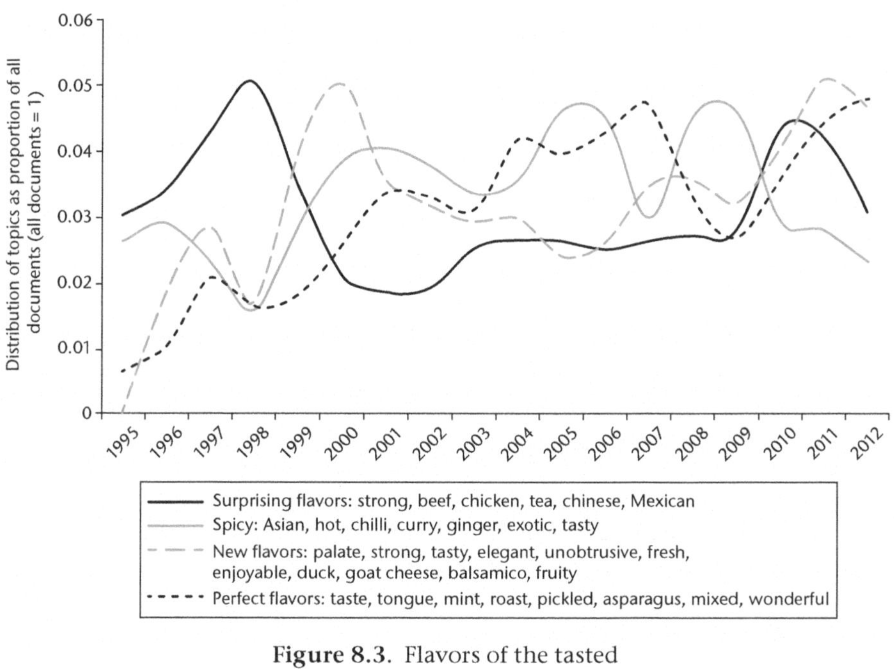

Figure 8.3. Flavors of the tasted

Conclusion: The Rise of Red Beets as the New Brunch

Empirical studies of valuation and evaluation processes have shown the insightful use of qualitative methods, using for instance ethnography or case studies. My work suggests using topic modeling as a more formal analysis to study culture through texts. The method used fits the theoretical idea that meaning is relational and depends on context.

Substantively, this chapter explored the dynamics of what was being tasted and experienced as noted down in restaurant reviews from a macroscopic perspective. First, it showed expressions of the tasted as different topical structures. The topic model analysis helped to form constructs about that which was being paid attention to in reviews. It then explored the dynamics of these structures of that which was tasted over time. The chapter thus provides an account of changes in that which was tasted in Berlin's restaurant reviews.

My analysis reveals a dominant dynamic: a shift from a breakfast and brunch topic to the steady rise of a new German cuisine, featuring red beets and *Riesling* wine. The introductory quote captures a flavor of how the new German cuisine tastes. This shift may not only reflect upon that which was tasted, but indeed indicate changes in the larger gastronomic and cultural field in Berlin. The topic model analysis of the written reviews gives some further support to larger, detectable changes, for example when particular cuisines become more prominent in reviews. Further inspection is needed to support the idea that this shift in that which was tasted is making visible larger cultural change. Moreover, the analysis also revealed smaller, more rhythmic patterns of change, when flavors of the tasted alternate over the time frame studied. These patterns of change suggest particular trends alternating over time. When the new turns old, a new "new" starts being tasted.

References

Ahn, Y.-Y., Ahnert, S. E., Bagrow, J. P., and Barabási, A.-L. (2011). Flavor Network and the Principles of Food Pairing. *Scientific Reports* http://dx.doi.org/10.1038/srep00196 (accessed December 13, 2013).

Ahnert, S. (2013). Network Analysis and Data Mining in Food Science: The Emergence of Computational Gastronomy. *Flavour* 2(1): 4.

Askegaard, S. and Madsen, T. K. (1995). *European Food Cultures: An Exploratory Analysis of Food Related Preferences and Behaviour in European Regions*. Århus: MAPP.

Bearman, P. and Stovel, K. (2000). Becoming a Nazi: A Model for Narrative Networks. *Poetics* 27(2): 69–90.

Blank, G. (2007). *Critics, Ratings, and Society: The Sociology of Reviews*. Lanham: Rowman & Littlefield.

Blei, D. M. (2012). Probabilistic Topic Models. *Communications of the ACM* 55(4): 77–84.

Blei, D. M. and Lafferty, J. D. (2009). Topic Models. In A. Srivastava and M. Sahami (eds.), *Text Mining: Classification, Clustering, and Applications*. Boca Raton, FL: Chapman & Hall, 71–93.

Blei, D. M., Ng, A. Y., and Jordan, M. (2003). Latent Dirichlet Allocation. *Journal of Machine Learning Research* 3: 993–1022.

Bourdieu, P. (1984). *Distinction: A Social Critique of the Judgment of Taste*. Cambridge, MA: Havard University Press.

Bourdieu, P. (1993). *The Field of Cultural Production*. New York, NY: Columbia University Press.

Christensen, B. T. and Petersen, J. S. (2013). Restaurant Rankings in the Culinary Field. In B. Moeran and B. T. Christensen (eds.), *Exploring Creativity: Evaluative Practices in Innovation, Design, and the Arts*. Cambridge: Cambridge University Press, 235–259.

Davis, M. (2009). *A Taste for New York: Restaurant Reviews, Food Discourse, and the Field of Gastronomy in America. Program in Food Studies*. New York, NY: New York University.

DiMaggio, P., Nag, M., and Blei, D. (2013). Exploiting Affinities between Topic Modeling and the Sociological Perspective on Culture: Application to Newspaper Coverage of U.S. Government Arts Funding. *Poetics* 41(6): 570–606.

Durand, R., Rao, H., and Monin, P. (2007). Code and Conduct in French Cuisine: Impact of Code Changes on External Evaluations. *Strategic Management Journal* 28(5): 455–472.

Ferguson, P. P. (1998). A Cultural Field in the Making: Gastronomy in 19th-Century France. *American Journal of Sociology* 104(3): 597–641.

Ferguson, P. P. (2004). *Accounting for Taste: The Triumph of French Cuisine*. Chicago, IL: The University of Chicago Press.

Fine, G. A. (1996). *Kitchens: The Culture of Restaurant Work*. Berkeley, CA: University of California Press.

Grimmer, J. (2010). A Bayesian Hierarchical Topic Model for Political Texts: Measuring Expressed Agendas in Senate Press Releases. *Political Analysis* 18(1): 1–35.

Grimmer, J. and King, G. (2011). General Purpose Computer-Assisted Clustering and Conceptualization. *Proceedings of the National Academy of Sciences of the United States of America* 108(7): 2643–2650.

Grimmer, J. and Stewart, B. M. (2013). Text as Data: The Promise and Pitfalls of Automatic Content Analysis Methods for Political Texts. *Political Analysis* 21(3): 267–297.

Hennion, A. (1997). Baroque and Rock: Music, Mediators and Musical Taste. *Poetics* 24(6): 415–435.

Hennion, A. (2007). Those Things That Hold Us Together: Taste and Sociology. *Cultural Sociology* 1(1): 97–114.

Heuts, F. and Mol, A. (2013). What Is a Good Tomato? A Case of Valuing in Practice. *Valuation Studies* 1(2): 125–146.

Hsu, G. (2006). Evaluative Schemas and the Attention of Critics in the US Film Industry. *Industrial and Corporate Change* 15(3): 467–496.
Hsu, G., Roberts, P. W., and Swaminathan, A. (2012). Evaluative Schemas and the Mediating Role of Critics. *Organization Science* 23(1): 83–97.
Jockers, M. L. (2013). *Macroanalysis: Digital Methods and Literary History*. Champaign, IL: University of Illinois Press.
Johnston, J. and Baumann, S. (2007). Democracy Versus Distinction: A Study of Omnivorousness in Gourmet Food Writing. *American Journal of Sociology* 113(1): 165–204.
Jolson, M. A. and Bushman, F. A. (1978). Third-Party Consumer Information Systems: The Case of the Food Critic. *Journal of Retailing* 54(4): 63–79.
Karpik, L. (2000). The Red Michelin Guide. *Sociologie du Travail* 42(3): 369–389.
Lane, C. (2013). Taste Makers in the "Fine-Dining" Restaurant Industry: The Attribution of Aesthetic and Economic Value by Gastronomic Guides. *Poetics* 41(4): 342–365.
Lane, C. (2014). *The Cultivation of Taste: Chefs and the Organization of Fine Dining*. Oxford: Oxford University Press.
Leschziner, V. (2006). Epistemic Foundations of Cuisine: A Socio-Cognitive Study of the Configuration of Cuisine in Historical Perspective. *Theory and Society* 35(4): 421–443.
Matthies, B. (2001). Funkturm-Restaurant. Flacher Lulatsch. [TV Tower Restaurant. No Giant]. *Der Tagesspiegel*, December 9, 2001, S06.
Meeks, E. and Weingart, S. B. (2013). The Digital Humanities Contribution to Topic Modeling. *Journal of Digital Humanities* 1, http://journalofdigitalhumanities.org/2-1/dh-contribution-to-topic-modeling/ (accessed October 10, 2013).
Moody, J. and Light, R. (2006). A View from Above: The Evolving Sociological Landscape. *The American Sociologist* 37(2): 67–86.
Moretti, F. (2013). *Distant Reading*. London: Verso.
Quinn, K. M., Monroe, B. L., Colaresi, M., Crespin, M. H., and Radev, D. R. (2010). How to Analyze Political Attention with Minimal Assumptions and Costs. *American Journal of Political Science* 54(1): 209–228.
Ramage, D., Rosen, E., Chuang, J., Manning, C. D., and McFarland, D. A. (2009). Topic Modeling for the Social Sciences. *Workshop on Applications for Topic Models, NIPS* http://vis.stanford.edu/papers/topic-modeling-social-sciences (accessed December 13, 2013).
Rao, H., Monin, P., and Durand, R. (2003). Institutional Change in Toque Ville: Nouvelle Cuisine as an Identity Movement in French Gastronomy. *American Journal of Sociology* 108(4): 795–843.
Rao, H., Monin, P., and Durand, R. (2005). Border Crossing: Bricolage and the Erosion of Categorical Boundaries in French Gastronomy. *American Sociological Review* 70(6): 968–991.
Roeger, K. (2012). Modern Deutsch: Lokal. [Modern German: Lokal]. *Zitty*, January 26–February 8, 2012, 33.
Shrum, W. (1996). *Fringe and Fortune: The Role of Critics in High and Popular Art*. Princeton, NJ: Princeton University Press.

Smith, T. (2007). Narrative Boundaries and the Dynamics of Ethnic Conflict and Conciliation. *Poetics* 35(1): 22–46.

Studnitz, P. v. (1999). In Mitte lockt die Neue Deutsche Küche. [In the Center of Berlin, the New German Cuisine Is Tempting Guests]. *Welt am Sonntag*, October 31, 1999, 106.

Svejenova, S., Mazza, C., and Planellas, M. (2007). Cooking up Change in Haute Cuisine: Ferran Adria as an Institutional Entrepreneur. *Journal of Organizational Behavior* 28(5): 539–561.

White, H. C. and White, C. A. (1965). *Canvases and Careers: Institutional Change in the French Painting World*. New York, NY: Wiley.

Zuckerman, E. W. (1999). The Categorical Imperative: Securities Analysts and the Illegitimacy Discount. *American Journal of Sociology* 104(5): 1398–1438.

Zuckerman, E. W. and Kim, T. Y. (2003). The Critical Trade-Off: Identity Assignment and Box-Office Success in the Feature Film Industry. *Industrial and Corporate Change* 12(1): 27–67.

9

Making Things Precious: A Pragmatist Inquiry into the Valuation of Luxury Perfumes

Anne-Sophie Trébuchet-Breitwiller[1]

We are at 21 rue du Mont Thabor in the 1st district of Paris, in a Parisian luxury shopping hub between the Place de la Madeleine and the Place Vendôme. On the very short portion of road linking the Rue Cambon to the tiny Rue Mondovi, which has few passers-by, the Editions de Parfums Frédéric Malle is a discreet boutique. It is not possible to see inside through the two shop windows to the right of the front door. Displayed in the one is a series of identical bottles of the brand's latest perfume, Portrait of a Lady, and in the other, also strongly silhouetted against a dark background, a series of identical red glasses containing the brand's candles. Nothing else. One looks at these shop windows, but not for long. The single door itself is transparent. Inside the shop, the door leads into a front room containing a desk with chairs for visitors. On the wall to the right of this desk, above a marble mantelpiece, is a series of photo portraits; on the left, three glass columns; and on the desk itself a series of perfume bottles which differ only in the writing on their labels. Against the wall at the back of the room are large display cabinets containing numerous perfume bottles, all very similar. There is no particular acoustic atmosphere but the colors are warm and the furniture

[1] An earlier version of this chapter was presented at the 2nd Interdisciplinary Market Studies Workshop, Dublin, June 7–8, 2012. I wish to thank Luis Araujo, Franck Cochoy, Liliana Doganova, Susi Geiger, Hans Kjellberg, Alexandre Mallard, and Kevin Mellet for their comments. I acknowledge helpful reviews by David Stark and his PhD students, Maxwell Foxman, Kathrin Nistrup Madsen, and Edvin Sandström. Finally, I am deeply indebted to the participants at the Berlin authors' conference, and to the editors of this volume whose comments and advice have been of great help.

comfortable. The space is neither cluttered nor seemingly empty; the effect is that of an inhabited space. Today is Saturday; a salesman and saleswoman are present in the shop. The saleswoman seated at the desk in the front room is on the phone when I enter. She interrupts her call very courteously to greet me, and the salesman who welcomes me leads me into the second room. In this room, adjacent to the first, one finds the same cabinets against the back wall. Additionally, immediately to the left as one enters, there is a low shelf with the same collection of perfume bottles as on the desk in the first room.

The entire collection of fragrances is there. The brand's offer consists of 100ml and 50ml bottles, as well as body lotions, perfumed oils and pocket sprays. All these products are available to be presented to customers by the salespersons, yet their presence is discreet, and the boutique uncluttered. The prices are not displayed in either the shop window or the shop itself. There is actually little to see or look at. The tall columns which take up a substantial amount of space in the first room show nothing, only a potential fragrance: they are "smelling columns." In this second room I notice imposing pierced cylinders high upon the wall: a ventilation system, explains the salesman. I realize that although I am at a perfumer, the space is not saturated with fragrances; in fact the room has no smell. This absence of odors aside from the product itself enables one to smell what there is to smell, that is, the perfume—or rather, each perfume separately.

Before giving me anything to smell, the salesman asked which perfume I was wearing or had worn, that I liked. He thought for a moment and chose a perfume that he allowed me to smell by spraying it onto a "blotter," a fairly wide white rectangular card embossed with the brand name "Editions de Parfums Frédéric Malle—Paris." At first he said nothing, apart from the name of the perfume; he waited for my reaction. The fragrance brought words to mind and, like the vast majority of customers, I commented on what I had smelled. The salesman repeated what I had said, corrected me, explained. He also gave me indications on how this perfume would evolve. Then he got me to smell a second perfume, the name of which he wrote in pencil on the bottom of the card. The same exchange of words took place. These two perfumes corresponded to very different olfactory sensations, yet both of them were familiar to me and I liked them. The salesman then gave me a third perfume to smell, Portrait of a Lady, the brand's latest release. He explained that this perfume contained a "huge" dose of rose, in the dual form of Turkish and Bulgarian Rose essential oil and absolute. This rose component constituted 20% of the formula. In addition there was a large dose of patchouli, prepared with a particular distillation technique. The perfume itself, he said, was concentrated "like a perfume extract" at 17%. He turned around the bottle and showed me the label: "17% PV".[2]

[2] This is the concentration of perfume, that is, the ratio of olfactory concentrate contained in the liquid, to alcohol. Traditionally, this concentration is expressed by a weight/volume ratio (*poids/volume* or *PV*, i.e., weight of the concentrate over the total volume of liquid), which is the

In the meantime, and as the conversation proceeded, the salesman suggested testing the two perfumes that I preferred and sprayed each of the two products on the inside of my wrists. After walking through the fragrances of the candles, I noticed that the fragrance of the second perfume on my wrist, called Lys Méditerranée, became very comfortable and warm. I mentioned that it "smelled like the beach, like my skin after I've left the beach, when it also smells of suntan lotion." This made him smile and almost laugh, although his reaction was in no way unpleasant, and he agreed with me, saying that it was linked to white flowers, lilies as well as orange blossoms. He explained that as white flowers develop they take on a creamy character. Finally, after asking for the price of these products, I left the shop, telling the salesman that I would take the time to see how these perfumes evolved and to get used to them, before possibly returning to choose one the following week. The next day, I could still smell the two perfumes applied to my wrists twenty-four hours earlier (and, interestingly, their fragrances were richer than on the blotters that I had also kept). I was able to let my friends and colleagues smell them as well, and express their opinions.[3]

Introduction: A Pragmatist Turn in Luxury Studies

Luxury is a classical theme in sociology, especially with regard to the topic under consideration in this collective volume: valuation. From the very beginning of the sociological tradition, luxury was recognized as a prime candidate to demonstrate the social reasons of economic values (up to the point that these social reasons themselves were held as the only serious explanation for luxury prices for example). Thorstein Veblen's seminal work has been of great importance here: with his theory of "conspicuous consumption" this economist, breaking away from the classical dogma of "utility," contended that rivalry and honor were the primary and most powerful reasons for spending, and thus for the valuation of goods. In this approach the material richness

strictest ratio used in the perfume industry. The well-known perfumer Jean-Claude Ellena gives the following indications: "*Eau de Cologne* contains between 2% and 4% of perfume concentrate [...]. *Eau de toilette* contains between 5% and 20% of perfume concentrate. *Eau de parfum* contains between 10% and 20% of perfume concentrate. An extract contains between 15% and 35% of perfume concentrate." (Ellena 2007: 100). With a concentration of 17%, *Portrait of a Lady* can effectively be described as an *eau de toilette*, an *eau de parfum*, or an extract.

[3] Note here that I had taken care not to wear any perfume the day I went to the boutique. I have become accustomed to this type of exercise through all my work on perfumes. However, I have no specific olfactory skills. On occasion I have needed them, but the lack thereof has also enabled me to take the position simply as an experimenter, a "normal" customer of perfume shops. During my field work on perfumes, I made over 30 visits to perfumeries (all brands and channels included), usually presenting myself as a customer. The account given here, from 2010, draws on observations made between 2004 and 2010.

of the objects produced and bought matters a great deal (Veblen, 1899). By contrast, in the last years of the 20th century, sociological research focused mainly on demonstrating that the value of luxury goods is symbolic. For example, in their famous article on French haute-couture, Pierre Bourdieu and Yvette Delsaut encourage analysts to overlook and even actively ignore the materiality of the objects (and of perfumes in particular), so that they concentrate on the "symbolic work," meaning the way given objects are made socially valuable or "consecrated" (Bourdieu and Delsaut, 1975). In so doing, sociologists certainly reaffirmed the social nature of luxury markets: they convinced sociological analysts and, to some extent, by highlighting specific activities, they also persuaded marketers (perfume marketers in particular) of the primary importance of symbolic communication and sociology in their own work. In the end, the idea that luxury corresponds to the sign or the "image" much more than to the thing became a sort of common knowledge.

I, however, argue that by splitting the symbolic from the material as they did, and by overlooking the materiality of the objects, these sociologists missed something essential in the building of luxury markets. And the point they missed is perhaps the one that has the most to teach us about valuation. So, without disregarding a sociological inheritance that has its importance, this chapter proposes to take the objects seriously and to implement a pragmatist turn in luxury studies.

Empirically the chapter is an inquiry into the qualification and valuation of the luxury products developed by an alternative perfume company, the Editions de Parfums Fréderic Malle. We follow the economic sociology approach which considers that markets are performed and that they are practical socio-technical achievements (Callon et al. 2007; McKenzie et al. 2007; Çalişkan and Callon 2010; Araujo et al. 2010; Zwick and Cayla 2011). In this approach, markets are first considered in their elementary form, that is, where the empirical work of qualification-valuation is able to attach an individualized buyer to an individualized product, up to the actual purchase.[4] Here, qualifying-valuing a product, building a market and attaching consumers to products are a single process. Hence, the empirical work of qualification-valuation of the product or of producing attachments (which amounts to the same thing) becomes the place from which to observe and understand the formation of any market, be it a mass market or luxury.[5]

[4] This essential individualization of markets was first highlighted by economist E.-H. Chamberlin (1933), quoted by Callon (2002; Callon et al. 2002).

[5] Regarding the use of the notion of "valuation," in this chapter I follow the pragmatist approach that is promoted by the editors of this volume in their introduction and that builds on John Dewey's *Theory of Valuation* (1939). Specifically, in line with recent works in economic sociology that are part of this tradition, I will generally consider valuation as an action and

Until now empirical studies have highlighted the role and importance of market intermediaries in the construction of contemporary mass markets in particular, together with their continuous and extremely laborious work to produce this attachment (Barrey et al. 2000; Callon et al. 2002; Dubuisson-Quellier 2010; Grandclément 2008; Grandclément and Gaglio 2011). Similar observations have been made on the mainstream perfume market (Muniesa and Trébuchet-Breitwiller 2010; Trébuchet-Breitwiller 2011).[6] By contrast, very few empirical studies have been dedicated to luxury. I however argue that luxury is an ideal field for valuation inquiries, not only because it has been concerned with attachment and valuation for centuries (Berry 1994), but also practically because it somehow puts attachment and valuation under the microscope. Luxury ethnographic studies reveal things about attachment and the making of attachments that can be seen nowhere else in markets[7]—provided that we use a pragmatist approach that takes objects seriously and does not split the spiritual from the material, as I have noted above.

The perfume example investigated in this chapter will make this clear. To begin with, we will scrutinize the shopping scene that opened this chapter.

A Retail Device that Takes Fragrance and Consumers Seriously

The Editions de Parfums currently appears to be an important actor and one which has contributed to the development of a new alternative perfumery.[8] This perfumer was born in 2000 with the opening of a small boutique in rue

consequently propose to overcome the distinction between "valuing" and "evaluation" (Callon 2009; Bidet et al. 2011; Stark 2011; Muniesa 2012; Hennion in this volume). To clarify the theoretical context, it is important to note that "valuation" thus considered is very close (if not equivalent) to the notion of "qualification" that was first coined by French economist François Eymard-Duvernay (1986) and has been worked on by Michel Callon in particular. This notion of "qualification" shifted the focus from the noun "quality" to the action of "qualification" precisely in order to depart from the naïve view that quality is technically made on the one hand (by the producer), and socially judged or evaluated on the other (by the consumer) (Callon 2002; Callon et al. 2002). That is why I use both notions, and sometimes the compound noun "qualification-valuation" proposed by Callon (2009).

[6] In the globalized mainstream perfume market the designers of perfume bottles, advertising agencies or brand naming agencies, qualitative and quantitative survey institutes, and professional marketers, are ubiquitous. They are as indispensable to the process of qualification-valuation of the product as are the perfumers.

[7] On the notion of "attachment" and its arrival in the anthropology of markets, see Hennion (2010b).

[8] Since the late 20th century an entirely new and remarkable phenomenon in the contemporary luxury perfume market (in the broad sense) has been the rapid development of what is commonly known as "niche perfumery" (because of the small or medium size of the markets concerned). The actors who, within ten years or so, generated the emergence of this new type of perfumery, have departed from many of the codes and practices of the globalized modern perfume industry that seemed to prevail after the mid-1970s. At the same time their development

de Grenelle, in the 6th district of Paris. The shop immediately attracted specialized journalists as well as leading distributors, including the American department store Barney's that was to be instrumental in the rapid and regular growth of the enterprise.[9] Pragmatically "following the actors themselves" our inquiry into the valuation of the perfumes developed by Frédéric Malle begins by paying attention to every detail of the shop and of the selling and buying process that have been so differentiating.

Luxury, Beyond the Rhetoric...

The store's address, situated in a Parisian luxury shopping hub, is in itself significant. Given the elegance and opacity of the storefront, opening the door is certainly an act that "ritually" introduces one into a place that is more than a shop. The interior purposely arranged to resemble a bourgeois apartment reflects the kind of social class that any sociologist (or non-sociologist) would associate with luxury (Bourdieu and Delsaut 1975; Bourdieu 2010). The tall glass and metal columns, the series of black and white photo portraits (almost exclusively males), and finally the perfume bottles themselves, exhibit an austere and masculine design, conveying ideas of quality and status also commonly identified with luxury goods (Karpik 2010; Aspers 2005). Thus, even though the atmosphere is quite pleasant inside the shop (neither cold nor arrogant), an omnipresent rhetoric of luxury makes it clear that the perfumes sold here are expensive.

"The products you find here are precious," says the shop setup as a whole, the salesman insisting on the perfumes' material richness, and the cabinets standing against the back wall (they prove to be not only showcases but also cooling cabinets designed for the safe storage of the fragrances contained in the bottles). At the same time, entering such a confidential boutique, being courteously welcomed by a smart and qualified saleslady or salesman, and finally enjoying the place make one feel valuable, not only as a rich and cultured (socially speaking) consumer (Veblen 1899), but with something more personal. In the middle of the room the antique chair facing the desk, even when unused, materializes this valuing of the customer.

appears to be closely linked to that, equally recent and rapid, of a new public of perfume amateurs, evidenced in the upsurge and proliferation of dedicated blogs. Based primarily on the latter point, which anchors the phenomenon in a very contemporary form of consumer culture, I define this perfumery as a "new alternative perfumery" (Trébuchet-Breitwiller 2010).

[9] In 2011, the Editions de Parfums Frédéric Malle had opened two other shops in Paris and one in New York. It was distributed in almost all major cities of the United States through specific "shops in the shops" set up in Barney's department stores. Its products were sold in almost all major cities of the developed world through new multi-brand boutiques specialized in niche perfumery, or through particularly selective department stores.

This goes beyond rhetoric: the retail socio-technical setup seems here to give intense value to both the products and the customers, inseparably. Attachment takes place in this valuing process: it is made by it and it is part of it. What we see here, I argue, is more than external flattery, it is a process that builds strong attachments (possibly measured by high prices) and thus makes things "precious" meaning both highly "valuable" and "valued."

Among other potentially attractive examples the Editions de Parfums Frédéric Malle is interesting because the novelty of the market it is building refreshes this valuing process while enhancing our ability to grasp it. Building on an analysis of the making of things precious that is evidenced in this particular example, we will observe how attachments are made, and concretely see valuing in action in these luxury markets. We can thus make an original contribution to the general knowledge and theory of valuation in economic contexts, which is the aim of this chapter. We will do so by examining how actors and objects make the whole thing work. To this end, we will have to go through every moment of valuation that we can scrutinize, beginning with the product valuing and the correlated client's situated experience (examined in the next two paragraphs), before turning our attention to the work of the perfumer and the producer (considered in the last section of the paper).

Displaying Smell

In 2000, and still today, a consumer could but be surprised when discovering the Editions de Parfums Frédéric Malle shop and products. Whereas in the 20th century perfume industry the shape of the bottle, together with the advertising, became as crucial as the fragrance itself in the constitution of a perfume's identity, here, in a completely dissonant way, the bottle is plain and identical for all perfumes. Rhetorically, the bottle becomes neutral, purely functional, whereas the content, meaning the scent or fragrance itself, becomes the thing worthy of attention.

Placed next to one another on the desk or shelf, perfumes are arranged for an olfactory tasting. Yet the consumer is not left alone with this new object. In an unprecedented way, the selling devices promote and support the olfactory exercise that is to take place. First, as noted above, everything seems arranged to avoid hindering the sensorial exercise: very few goods exposed, no music, no ambient smell, intimate interior spaces and, finally, no displayed prices (which can be uncomfortable but can also release one's attention). By contrast, the perfume scent is foregrounded. Since only the writing on the labels varies from one bottle to another, the indications stand out: the name of the perfume, the name of the perfumer who created it (a complete novelty), and some figures related to the physical concentration of the fragrant liquid (that the salesman showed and commented on)—all information that is enriched with the descriptions of fragrances and the portraits of perfumers displayed on the website.

Next, the most prominent feature is the "smelling columns" that occupy almost half of the space in the front room at rue du Mont Thabor (as in any of the brand's shops). They proved to be so new and emblematic that they became a sort of logo for the Editions de Parfums. Technically they are shower cabinets that have been adapted and equipped with a mechanism for diffusing a fragrance throughout the air inside them, from the bottom up. A window has been installed in the upper part of the cabinet, and when opening it and putting one's head at the aperture, one can smell the "aura" of the perfume diffused inside (its "wake" in a sense). This type of small volume is used by professional perfumers to smell and enable their professional clients to smell a perfume other than on a blotter or a particular person's skin. Frédéric Malle, who worked for years as an olfactory evaluator in the perfume industry before creating his own business, had this idea to offer the same opportunity to final consumers.[10] The process was in itself extraordinary. It endowed consumers with an olfactory skill (experienced as they put their head at the aperture). Concretely, such columns not only make scent physically present and visible in the shop, they also constitute a unique device that allows this olfactory object to spread and develop so that one can take a complete picture of it, so to speak. The fragrance seems to be amplified here; it is more legible and clear: dominant scents are evident, while details gradually appear all around, for example. When used, the smelling columns allow and even provoke an olfactory valuation that is equipped and becomes more elaborate.

While the product is ascetically and aesthetically redefined as an olfactory object, the generosity of the olfactory equipment turns the exercise into a rich, interesting and pleasant personal experience—surreptitiously changing the consumer into an amateur, actively involved in tasting (Hennion 2010a).

Individualizing Consumer

Except when selecting the three or four perfumes he will propose to the customer, the salesman is not exactly "introducing" perfumes to her; he is rather watching over an encounter. As mentioned above, my conversation

[10] An "evaluator" is a professional with the required olfactory skills (usually trained in the same schools as perfumers), whose job is to be an intermediary between "noses" (or perfumers) and brands (or marketers).

with the salesman was granted all the time required. It was not disrupted by the presence of other customers in the boutique (although there were some, occupied by the saleslady in the other room, but I noticed them only when I left), and no consideration given to anything aside from the fragrance itself interfered with it. In these circumstances the conversation develops fruitfully. Concretely, encouraged by the courtesy and friendliness of her interlocutor, the client becomes more confident; she can rely on her own olfactory skills to express words and feelings that she associates with the scents. The salesman's olfactory expertise seems to serve as a guardian allowing the client to proceed with reflexivity. Whenever she ventures personal but eccentric or somehow dissonant comments, he endeavors to identify and describe how the object is possibly responsible for them, and even recognizable, so that in the end the client is closer to this object; that is, she has a more personal and better knowledge of the fragrance, which she values with more precision and confidence.

The whole experience takes a particularly individual turn, one that is even more interesting and valuing for the client, as the salesman proposes to test fragrances "on skin." In the scene described above, it is at this point that a personal taste for Lys Méditerranée distinctively emerges, as new words come out which are both the means and the expression of a forthcoming attachment (Hennion, 2007). For precisely on skin the particular comfort of the white flowers reveals itself after a few minutes, along with an attachment to this comfort. Sustained by discussion, this sensation becomes evident and grows in importance in the valuation of the perfume. After leaving the store, an additional twenty-four hours wearing it would, in this case, confirm the taste and appreciation.

Mid-Term Discussion

So, the luxury retail setup gives prominence to a tasting device. When, as in the story presented here, everything is going well, it actually promotes a highly individualized bodily attachment process.

The originality of the retail socio-technical setup implemented by the Editions de Parfums has to be emphasized, first. It focuses neither exactly on the promotion of the object's "aura"[11] nor on the promotion of the designer's or creator's "aura" (the valuing name)—all things usual, or usually analyzed, in fashion and luxury studies (Bourdieu and Delsaut 1975; Karpik 2010; Dion and Arnould 2011; Assouly 2011). Instead the retail setup implemented by the Editions de Parfums focuses on making one individualized thing precious to one individualized consumer; and this is achieved by giving prominence to

[11] The word "aura" refers here to Walter Benjamin (1936), frequently quoted when it comes to luxury.

an olfactory tasting device. Even though it is certainly not unique in contemporary retailing, in current perfumery and the luxury context it is unusual.[12]

Second, we note that the tasting device in question proves to be a valuing device from beginning to end. The olfactory tasting exercise is taken seriously enough for a moment of sensorial valuation to take place where both objects (fragrances) and consumers (bodies) are actively involved, along with the salesman who is constantly evaluating both perfumes and customer's tastes so that he can skillfully promote the encounter. While developing over a short space of time, this valuation process allows individual taste to emerge, evolve, and possibly change: this is where value or attachment is concretely being made. Attachment and value stem from the valuation exercise (rather than preceding it).[13]

Yet, the other way round, the tasting-valuation exercise itself involves valuing processes. As we have seen: inquiring about the perfumes that the customer wears and appreciates in order to offer her fragrances that will please her, for example, is valuing her as a perfume consumer; allowing a fragrance to spread and develop in a dedicated space is valuing this fragrance as a fragrance; at the same time, implementing equipment which is usually reserved for professional perfumers, in a retail shop, is valuing consumers' olfactory skills (and possibly transforming them into perfume amateurs); caring with painstaking attention about every detail of the tasting device that is implemented in the shop (salespeople, columns, blotters, ventilation system, photo portraits, etc.) is valuing the olfactory exercise, which itself values both fragrances and consumers, and so on. In any of these circumstances, valuing is obviously more than flattering or even "prizing": it is "valorizing" things in the sense of making them valuable or "precious" (a word I prefer as it is more precise and fits better with the context of luxury).

Finally—and this will lead us to the second and last part of this chapter—while olfactory tasting is promoted as the cornerstone of the luxury retail setup, we see a marketer (the éditeur) who ultimately innovates by delegating to consumers, and most of all to objects themselves, the work of valuing and attaching. Fragrance (that is, the olfactory object) is first foregrounded. And fragrance, put on skin, will have the last word. So, having considered how the retail socio-technical setup is wound around an olfactory tasting device, and

[12] In the contemporary period, "wine cellars" dedicated to both the tasting and the selling of wines, have developed in wineries from California to the old European vineyards; or, to take a famous urban example, the Nespresso shops provide an important means of tasting their coffee collection.

[13] See Hennion in this volume. This stands in sharp contrast with the quite quick fragrance tasting that is promoted in traditional selective distribution (duty free shop, for example) or, even more conspicuously, in some marketing consumer sniff-tests where, on the contrary, everything seems to be arranged to provoke an immediate valuation or "liking," that is, the mere repetition of an existing taste (Muniesa and Trébuchet-Breitwiller 2010).

having seen the perfume consumer transformed into a fragrance amateur, we will now turn to the third and crucial term of this market construction, the perfumes themselves. We will examine how these olfactory objects are made.

Performing the Product's Olfactory Quality: The Example of the Creation of Carnal Flower

When recounting his initial motivation for creating his own business, Frédéric Malle says it was because of the "noses" he admired and the perfumes he valued.[14] At that time (in the 1990s), he actually felt some dissonance between the perfumes that he as an olfactory professional highly valued, and the perfumes that were valued and exclusively sold on the markets. But, instead of just declaring that his own taste is "old-fashioned" (as he put it, talk 2008), he sought to take advantage of this dissonance (Stark 2009). He reasoned that perhaps the type of perfumes he valued could simply not make their way to potential consumers through the existing perfume market devices, especially through the existing retail distribution networks. Malle claims he therefore became a perfume éditeur[15] (allowing perfumers to sign their creation almost for the first time), and re-established the traditional perfumery shop[16] (with the completely new olfactory means examined above) so that he could produce and put on the market some specific perfumes that he considered to be particularly interesting and of a high quality.

To understand what he means, let us take an example. Carnal Flower was developed and put on the market by Editions de Parfums in 2005, and rapidly became one of the brand's best-sellers. A milestone in the history of the young company, this perfume has been amply commented on by both the perfumer and the éditeur. We can therefore draw on these comments (specifically their technical aspects, which are significant) to have a quick look at the creative process or the way in which the quality of the product is constituted.

The éditeur's Concern

When relating the circumstances of the creation and development of *Carnal Flower*, Malle explains how initially there was a wish to do "a tuberose."

[14] As regards Frédéric Malle himself, my main sources are a public talk that I attended (Talk by Frédéric Malle, Ecole de Paris du Management, ENSMP, 13 May 2008), as well as an in-depth interview which enabled me to explore a number of technical points (Interview with Frédéric Malle, 2008).

[15] The French word *"éditeur"* refers here to both the functions of "editor" and "publisher."

[16] In the 2008 interview, Malle readily referred to the "original" Guerlain model (the family business that dominated French perfumery up to the post-WWII period).

Tackling tuberose perfume was interesting and somehow challenging, he argues, because of famous examples in classical perfumery: he mentions *Fracas* by Piguet, composed by perfumer Germaine Cellier after WWII, and which enjoyed some "mythical" success during those years in Europe and the U.S. Another reason that made tuberose interesting in the *editeur*'s eyes was, paradoxically, the high cost or preciousness of the raw material involved. Tuberose absolute has been one of the most costly components of luxury perfume for a long time. It is seen by "connoisseurs," Malle argues, as a type of "caviar" of perfumery—the extent that the tuberose perfumes currently developed and sold on markets are largely devoid of the tuberose natural raw material, which is much too expensive (talk, 2008). By contrast, the Editions de Parfums "tuberose" was to incorporate a "huge dose" of the flower absolute (interview, 2008).[17] Put on the market at 200 Euros for a 100ml bottle (50% more than the other perfumes of the collection at the time), *Carnal Flower* was to represent a test in terms of quality and price positioning.[18] And it was actually an opportunity to assert or re-assert the Editions de Parfums' positioning in luxury, and to establish a price policy.[19]

From an olfactory point of view, *Carnal Flower* was the result of a long and complex process of "reconstructing a tuberose flower." The impetus of this work, Malle recounts, stemmed from a new raw material produced by the laboratory in which Dominique Ropion, the perfumer-author of *Carnal Flower*, was employed.[20] This new material was a "molecular distillation tuberose." Here the term denotes a tuberose absolute[21] redistilled to get rid of any flaws. Molecules carrying odors unwanted by the perfumer are separated from the olfactory core by means of a technology used to fractionate the odorous material with extreme precision.[22] The absolute thus

[17] The saleslady of a Parisian boutique told me that there is no less than 10% of absolute tuberose in the formula of *Carnal Flower*. A professional I interviewed confirmed that this was "huge" (2010).

[18] In 2005 *Lys Méditerranée* was sold at 135 Euros for a 100ml bottle. By comparison, at the time, a 100ml bottle of Chanel *eau de parfum* for women was sold at a little under 100 Euros (Marionnaud, Paris).

[19] The selling prices are supposed to vary according to the raw material constituting the perfumes in terms of quality and quantity of the ingredients (talk by F. Malle, 2008).

[20] The Editions de Parfums Frédéric Malle does not employ creative perfumers. Like most of the perfume brands, it works with professional perfumers most of whom are employed in the creative centers of the big laboratories that furnish the industry with raw materials or completed perfume concentrates (namely IFF, Givaudan, Firmenich, etc.). However, the way Malle works with his perfumers is quite different from the way the perfume industry usually proceeds.

[21] In the perfume industry, extraction is an operation consisting in extracting the perfume of a natural raw material. In the case of the tuberose flower, the extraction uses a volatile solvent which produces a form of perfumed wax called *concrete*, which is itself reprocessed and transformed into a fragrant liquid called *absolute*.

[22] These unwanted odors are often linked to the extraction process itself: the odor of the distilled product systematically differs from the odor of the fresh product, either because delicate nuances have disappeared, or because other, generally darker nuances have been over-developed. They may also have been added unintentionally as artefacts due to the device used for the processing.

corrected appeared to be more precise from an olfactory point of view and, to some extent, closer to the perfume of the live flower. The tuberose perfume expressed itself more legibly and "purely." All this aroused both the perfumer's and the *editeur*'s curiosity and interest. Moreover, because the natural raw material made available in this way is sufficiently "clean," one can, technically, envisage putting larger amounts of it into the formula. A richer product can thus be made.

Yet Malle explains that synthetic molecules also played an important part in constituting that tuberose perfume. The main advantage of these molecules, and notably of the very sophisticated and delicate ones available in the 21st century, is their precision. They made it possible to express a "fresh" tuberose, reproducing "the green side," "the transparent side," "the aqueous side" of the live flower's perfume—all facets that the product of extraction, even the most refined one, never completely reveals. He likewise emphasizes the importance of the periods and duration of "maturation" (mixing the oils) and "maceration" (mixing the concentrate thus obtained in alcohol) that immobilize the product for one or two months, to obtain a well-integrated, harmonious, united and stable product, one "that's very natural," as Malle put it (interview 2008). Finally—and we understand this is a critical point—he explains that synthetic materials are required as well as natural raw materials, for their advantages differ, for the technical finishing of the perfume, i.e., the fact that it lingers, is comfortable, or merges with the skin.

Thus, in this example of *Carnal Flower*, "natural" becomes virtually synonymous with quality, achievement, perfection of the scent. It refers to the quality of the expression of the tuberose flower of course, emphasizing the accuracy of the processing together with the freshness of the scent that can be obtained through contemporary technological means.[23] But "natural" refers also to the quality of the perfume as a product, specifically the fact that it is comfortable and that it merges with the skin. It refers to the perfume's ability to naturally bodily attach its forthcoming users.

The Perfumer's Explorations

Dominique Ropion explains that, in his creative work, an important line of research consists in analyzing the composition of the flowers themselves.[24]

[23] In the contemporary period, the development of *"vins de terroir"* or *"vins naturels"* (natural wines) (Teil et al. 2011), or in haute cuisine, the "New Nordic Cuisine" (Christensen and Strandgaard 2013; Byrkjeflot et al. 2013) are good examples of such a redefinition of high-quality around natural components, enhancing legibility and transparency—especially in contrast with classical Bordeaux oenology of the post-WWII, or with classical *chef*-oriented French cuisine (Ferguson 2004).

[24] As regards the creative work of Dominique Ropion, I draw on interviews made by Le Guérer in 1999 and 2001 (Le Guérer 2005: 313–316), and on an interview published in a documentary

He asks the researchers "to carry out a study with as much precision as possible, on the rose, jasmine, tuberose, etc., to examine all the details and to smell all the molecules, even the weirdest" (Le Guérer, 2005: 315–316). Thanks to electronic microscopes (i.e., chromatograph and mass spectrometer) allowing for detailed observation of complex odors, the chemists of his laboratory provide him with what resemble formulae of essences, absolutes, and the aerial perfume of the flower.[25] The perfumer explains further that these very detailed analytical translations make it possible to observe "structures that no one would have imagined." In particular, he discovers singular combinations here that will elicit the "emergence" or "appearance" of new "accords,"[26] thus stimulating his creativity.

According to Ropion (and to many other perfumers), the constitution of accord is the essence of the perfumer's creative work. In the example studied here, it is based on preparatory analytical work that is highly equipped. "The fact of studying all these accords elaborated by nature enables you to understand how a flower functions, with its very precise rules," explains the perfumer. "And when you want to create a perfume that's more sophisticated, the fact of having grasped these rules enables you to play around with them. You then look for elements that will illustrate the complexity and richness of a flower." Regarding the composition of *Carnal Flower*, Ropion gives the following example: "In the tuberose you have a molecule called eucalyptol, one of the components of the eucalyptus, and it is very tempting to think that if nature put it there, there's a reason. So why not use eucalyptus absolute, in addition?" The effect goes beyond the illustration: "It opens the tuberose and reveals it," the perfumer says (Soyer and Molloy 2007).

The close continuity between the work of the flower (to produce its perfume) and that of the perfumer is interesting and noteworthy. "There are flowers that are real compositions of perfumers," confides the perfumer who stresses rather than conceals his admiration for roses and tuberoses in particular. As if to demonstrate the point, the detail of the perfumer's formula that is explained in the comments we quoted here is in itself a reflexive gesture. Valuing the tuberose, the perfumer commits to a scientific and technical exploration of the flower. This leads him to imagine a new olfactory accord, the eucalyptus accord, which is elaborated adjacently (from associations that are observed in the flower essence, reproduced, and derived). This accord

book (Soyer and Molloy 2007). These interviews are quite extensive and close enough to the period of creation. The English translations are my own.

[25] *Headspace* is the advanced technology with which chemists capture and transcribe the aerial perfume given off by a live flower or any other object or place.

[26] Here, the term "accord" denotes a balanced blend of olfactory notes that create a new singular scent which is very specific and recognizable.

highlights certain ignored and beautiful olfactory details of the tuberose. The effect, if we are to believe the perfumer, is to "open and reveal" the flower's perfume.

The scent of the tuberose is thus explored, reflected and meditated through that creative exercise. From this meticulous and somehow extreme valuation work stems a unique and original fragrance, one which derives its beauty and seductiveness from its highlighting of the primary perfume of the tuberose flower (rather than from reproducing, or even trying to reproduce it).

Conclusion

To complete this chapter I will try to summarize and theoretically clarify what I have said on luxury and valuation through my case study. I will then conclude with a few additional remarks on today's luxury.

The pragmatist approach considers valuation as an action, preferring the verb to the noun. As the editors of this volume explained in their introduction, this approach refuses both the naturalistic view (things have intrinsic value), and the culturalist view (values are socially attributed to things). Yet, as I pointed out earlier, this also transcends the distinction between "valuing" and "evaluation." To make his point clear, Dewey (1939) insisted on what he called the "the continuum of ends-means." Regarding luxury, Veblen's theory of conspicuous consumption (1899) can be seen, in retrospect, as exemplifying this ends-means continuum, for it precisely highlights how "that which is valorizing is exclusively valued," so to speak; meaning, in Dewey's language, that consumers value first and foremost means of valuing.

Yet this is not enough. The luxury perfumes studied in this chapter offer an illustration of the continuum "evaluation-valuation-valorization," or of the spiral of valuation (Bidet et al. 2011) that goes beyond this first twist. Commenting, at the end of the first part, on the analytical description we made of the Editions de Parfums socio-technical retail setup, while admiring its capacity to intensely attach consumers to products, we remarked how this attachment was primarily produced through an olfactory tasting device itself based on a concrete valuing of both fragrances and consumers. Valuation thus involved valuing, and resulted in valuing; meaning, valuation was end-to-end a valuing process. Drawing on the last part of our chapter, we see now that the fragrance valuing work (creation and production) builds on valuation work: it is based on continuums, just like the one evidenced for the selling. Actually we saw an *éditeur*-entrepreneur who is an olfactory evaluator, literally. Even though, for several reasons, he has not implemented the formal valuation process that is pervasively used in the perfume industry (i.e., fragrance pre-tests), the *éditeur*-evaluator undoubtedly draws on his professional experience and skills to value, wish, select, ask for reworks, require accurate technical finishings, and so on. Yet at the same time we saw a perfumer drawing on technical and sensorial investigation and valuation means to look for the valuable accord, which is of course the valuing one, the one that will "open and reveal" the flower scent. The "accord" itself is made of a valuation spiral.

Making the valuation-valuing continuum reflect in all directions in this way is characteristic of the precious; it is what makes it possible and what makes it valuable. That it does so in a market context, with the exacerbated valuation spiral possibly nurturing and supporting the prices, makes the form of luxury that I call "precious" so unique and interesting.

While this chapter certainly does not claim to offer a comprehensive overview of today's luxury (the pricing issue with many others has to be further investigated), it does profess to give at least a fair insight into it, and a taste of it. So, to conclude, I would like to emphasize the modernity of the Editions de Parfums: how experience and senses are requested for example, and how consumers are turned into amateurs; how that which is finally valued as natural makes its way through the most sophisticated technologies and know-hows; or how the most spiritual quest of the perfumer (as a perfumer) or the customer (as a perfume lover) proves to be the one that is most involved in the material. All of these actually hold on to a contemporary and challenging form of luxury.

The contemporaneous *Louis Vuitton's Murakami* handbags which Michael Hutter examines in this volume constitute another case of the "value spiral" (Hutter 2011): a luxury goods producer builds on a symbolic pattern that has proven its value in contemporary art, to enhance the attachment to his handbags (and to his monogram); in turn, the artist builds on his collaboration with this very well-known luxury brand to develop his own work and career. When we focus on the handbags' value, the valuation spiral points to a specific direction (contemporary art) which is quite different from the essentially nature-oriented valuation observed in the examples of *Carnal Flower*, New Nordic Cuisine or "*vins de terroir*" (that provide another wonderful case study). Yet in all cases, I would argue that it is making the luxury machine all the more efficient and powerful.

References

Araujo, L., Finch, J., and Kjellberg, H. (eds.) (2010). *Reconnecting Marketing to Markets.* Oxford: Oxford University Press.

Aspers, P. (2005). *Status and Standard Markets in the Global Garment Industry.* Köln: Max-Planck Institut für Gesellschaftsforschung. Discussion Paper 05/10.

Assouly, O. (ed.). (2011). Le luxe. *Essais sur la fabrique de l'ostentation.* [Luxury. Essays on the Making of Ostentation]. Paris: IFM/Regard.

Barrey, S., Cochoy, F., and Dubuisson, S. (2000). Designer, packager et merchandiser; trois professionnels pour une même scène marchande. [Designer, Packager and Merchandizer: Three Professionals on the Same Market Scene]. *Sociologie du travail* 42(3): 457–482.

Benjamin, W. (1936/2008). *The Work of Art in the Age of Mechanical Reproduction*. London: Penguin.

Berry, C. J. (1994). *The Idea of Luxury. A Conceptual and Historical Investigation*. Cambridge: Cambridge University Press.

Bidet, A., Quéré, L., and Truc, G. (2011). Ce à quoi nous tenons. Dewey et la formation des valeurs. [The Things we Hold Precious. Dewey and the Shaping of Value]. In J. Dewey, *La formation des valeurs*. Translated and introduced by A. Bidet, L. Quéré and G. Truc. Paris: Les empêcheurs de penser en rond/La Découverte, 5–64.

Bourdieu, P. (1979/2010). *Distinction. A Social Critique of the Judgement of Taste*. London and New York: Routledge.

Bourdieu, P. and Delsaut, Y. (1975). Le couturier et sa griffe: contribution à une théorie de la magie. [The Couturier and his Brand: Contribution to a Theory of Magic]. *Actes de la recherche en sciences sociales* 1: 7–36.

Byrkjeflot, H., Strandgaard Pedersen, J., and Svejenova, S. (2013). From Label to Practice: The Process of Creating New Nordic Cuisine. *Journal of Culinary Science and Technology* 11: 36–55.

Çalişkan, K. and Callon, M. (2010). Economization, Part 2: A Research Programme for the Study of Markets. *Economy and Society* 39(1): 1–32.

Callon M., Méadel, C., and Rabeharisoa, V. (2000/2002). The Economy of Qualities. *Economy and Society* 31(2): 194–217.

Callon, M. (2002). Pour en finir avec les incertitudes? [To Finish with Uncertainties?] In "Dossier-débat: La qualité" [Special File-Debate: Quality]. *Sociologie du travail* 44: 261–267.

Callon, M. (2009). La formulation marchande des biens. [The Market Formulation of Goods]. In F. Vatin (ed.), *Evaluer et valoriser. Une sociologie économique de la mesure*. Toulouse: Presses Universitaires du Mirail, 247–269.

Callon, M., Millo, Y., and Muniesa, F. (eds.) (2007). *Market Devices*. Oxford: Blackwell Publishing.

Chamberlin, E. H. (1933). *The Theory of Monopolistic Competition. A Re-Orientation of the Theory of Value*. Cambridge, MA: Harvard University Press.

Christensen, B. T. and Strandgaard Pedersen, J. (2013). Restaurant Ranking in the Culinary Field. In B. Morean and B. T. Christensen (eds.), *Exploring Creativity. Evaluative Practices in Innovation, Design and the Arts*. New York, NY: Cambridge University Press, 235–259.

Dewey, J. (1939). *Theory of Valuation (Vol. II, No. 4)*. Chicago, IL: The University of Chicago Press.

Dion, D. and Arnould, E. J. (2011). Retail Luxury Strategy: Assembling Charisma through Art and Magic. *Journal of Retailing* 87 (4): 502–520.

Dubuisson-Quellier, S. (2010). Product Tastes, Consumer Tastes. The Plurality of Qualification in Product Development and Marketing Activities. In L. Araujo, J.

Finch and H. Kjellberg (eds.), *Reconnecting Marketing to Markets*. Oxford: Oxford University Press, 74–93.

Ellena, J.-C. (2007). *Le parfum*. [Perfume]. Paris: PUF.

Eymard-Duvernay, F. (1986). La qualification des produits. [The Qualification of Products]. In R. Salais and L. Thévenot (eds.), *Le travail. Marché, règles, conventions*. Paris: INSEE, Economica, 239–247.

Ferguson, P. P. (2004). *Accounting for Taste: The Triumph of French Cuisine*. Chicago, IL: The University of Chicago Press.

Grandclément, C. (2008). *Vendre sans vendeurs: sociologie des dispositifs d'achalandage en supermarché*. [What Sells in a Serve-Yourself Store? A Sociology of the Supermarket's Devices]. PhD thesis. Paris, Ecole des Mines de Paris/CSI.

Grandclément, C. and Gaglio, G. (2011). Convoking the Consumer in Person: The Focus Group Effect. In D. Zwick and J. Cayla (eds.), *Inside Marketing. Practices, Ideologies, Devices*. Oxford: Oxford University Press, 87–114.

Hennion, A. (2007). Those Things That Hold Us Together: Taste and Sociology. *Cultural Sociology* 1(1): 97–114.

Hennion, A. (2010a). Music Lovers: Taste as Performance. In A. Warde (ed.), *Consumption Vol. 3, Appropriation*. London: Sage, 187–208.

Hennion, A. (2010b). Vous avez dit attachements?... [You said attachments?]. In M. Akrich, Y. Barthe, F. Muniesa, and P. Mustar (eds.), *Mélanges en l'honneur de Michel Callon*. Paris: Presses de l'Ecole des Mines, 179–190.

Hutter, M. (2011). Infinite Surprises: On the Stabilization of Value in the Creative Industries. In J. Beckert and P. Aspers (eds.), *The Worth of Goods. Valuation and Pricing in the Economy*. Oxford and New York: Oxford University Press, 201–220.

Karpik, L. (2007/2010). *Valuing the Unique. The Economics of Singularities*. Princeton, NJ: Princeton University Press.

Le Guérer, A. (2005). *Le parfum des origines à nos jours*. [Perfume, from its Origins to the Present]. Paris: Odile Jacob.

McKenzie, D., Muniesa, F., and Siu, L. (eds.) (2007). *Do Economists Make Markets? On the Performativity of Economics*. Princeton, NJ: Princeton University Press.

Muniesa, F. (2012). A Flank Movement in the Understanding of Valuation. In L. Adkins and C. Lury (eds.), *Measure and Value*. Oxford: Wiley-Blackwell, 24–38.

Muniesa, F. and Trébuchet-Breitwiller, A.-S. (2010). Becoming a Measuring Instrument: An Ethnography of Perfume Consumer Testing. *Journal of Cultural Economy* 3(3): 321–337.

Soyer, C. and Molloy, C. (2007). *22 parfumeurs en creation*. [22 Perfumers Creating]. Paris: Editions Pirate.

Stark, D. (2009). *The Sense of Dissonance. Accounts of Worth in Economic Life*. Princeton and Oxford: Princeton University Press.

Stark, D. (2011). What's Valuable? In J. Beckert and P. Aspers (eds.), *The Worth of Goods. Valuation and Pricing in the Economy*. Oxford and New York: Oxford University Press, 319–338.

Teil, G., Barrey, S., Floux, P., and Hennion, A. (2011). *Le vin et l'environnement*. [Wine and the Environment]. Paris: Presses de l'Ecole des Mines.

Trébuchet-Breitwiller, A.-S. (2010). L'émergence d'un marché d'amateur. Le cas du parfum. [The Emergence of an Amateur Market. The Case of Perfume]. In O. Assouly (ed.), *L'amateur. Juger, participer, et consommer.* Paris: IFM/Editions du Regard, 69–97.

Trébuchet-Breitwiller, A.-S. (2011). *Le travail du précieux. Une anthropologie économique des produits de luxe à travers les exemples du parfum et du vin.* [The Making of Things Precious. Economic Anthropology of Luxury Products Drawing on the Examples of Perfume and Wine]. PhD thesis. Paris, Ecole des Mines de Paris/CSI.

Veblen, T. (1899). *The Theory of the Leisure Class.* New York, NY: The Macmillan Company.

Zwick, D. and Cayla, J. (eds.) (2011). *Inside Marketing. Practices, Ideologies, Devices.* Oxford: Oxford University Press.

Part III

Valuation in Fields of Practice

10

When Principles of Evaluation Clash: Assessing the Value of a Demonstration in Artificial Intelligence

Claude Rosental

How are claims and proofs of theorems assessed in artificial intelligence (AI)? In particular, what happens when AI scientists use heterogeneous principles of evaluation to produce antagonistic judgments about the validity and value of a demonstration expressed in symbolic or ordinary language? In order to address these issues, I study how a well-known paper in AI has been publicly assessed in the framework of a recent debate. I analyze how multiple orders of worth have been mobilized in this scientific debate, and I examine its dynamics and outcome. Why study this?

Scientific controversies have been abundantly studied since social studies of science have developed (Bloor 1976; Collins 1985; Latour 1987). However, AI and its debates have not attracted much attention compared to the experimental sciences and their controversies (Collins 1990; Guice 1998). Besides, recent studies of assessment practices suggest that analyzing tensions between principles of evaluation represents a fruitful approach if one wants to grasp the ins and outs of science in development (Lamont 2009; Rosental 2010). This approach may allow us to compare assessment practices in different social spaces, in particular in the artistic and scientific fields (Quemin 2005, Brewer in this volume). It is also helpful to understand how antagonistic references to separate orders of worth (Boltanski and Thevenot 2006), dissonances between different evaluative principles (Stark 2009), or irritations between value systems (Hutter and Throsby 2008) may be a source of innovation. Finally, it represents an opportunity to investigate if conflicts or dissonances between different evaluative principles are always overcome, or if, on the contrary, they may last—even in science.

The debate under study here originated in a paper written in the 1990s by an assistant professor at the University of California San Diego, Charles Elkan (1993). This paper was published originally in the proceedings of the annual conference of the American Association for Artificial Intelligence (AAAI). Elkan denounced the "paradoxical success" of electronic and computer applications of a theory called "fuzzy logic." Fuzzy logic is a subfield of AI that has been developed since the mid-1960s (Zadeh 1965). Its technological applications, such as fuzzy cameras and washing machines, may be found in numerous industrial sectors (Rosental 2004).

Elkan's denunciation was based on the proof of a theorem stating that fuzzy logic, characterized by a system of four specific axioms, is in fact nothing but classical binary logic. The author presented this result as a direct challenge to one of the founding ideas of fuzzy logic, which is supposed to allow for the expression of an infinite number of degrees of truth along a continuum with poles true and false (like half true for example).

The paper was published in a context of strong competition between proponents of various approaches to AI. It attracted a great deal of attention and generated a major controversy in this field. Initially, points of view were exchanged in an electronic bulletin board (or newsgroup) devoted to fuzzy logic, called *comp.ai.fuzzy*.

Authors and readers of messages posted on *comp.ai.fuzzy* were, for the most part, academics, researchers, and engineers working for universities and industrial organizations in the field of AI, and in particular of fuzzy logic. Due to its absence of selective or editorial constraints, the forum helped to open relatively broad debate, allowing the public expression of many points of view. *Comp.ai.fuzzy* became the main locus of an exchange of opinions on Elkan's paper for a period of 6 months, until these first reactions were consolidated in the form of articles published by specialized journals.

Public access to the electronic forum's archives allowed me to reconstruct the discussions. I also interviewed the actors involved in the debate and had access to the content of private mail. Besides, I carried out ethnographical observations in places such as an AI conference and I analyzed works published in AI journals. All this allowed me to relate the authors' written production to various actions conducted outside the electronic forum. I could also grasp the dynamics of the debate and its outcome.

Participants in the debate used different principles of evaluation to publicly assess Elkan's demonstration, as well as fuzzy logic itself. Public evaluations took the form of demonstrations supporting, or counter-demonstrations against, Elkan's views. These demonstrations were themselves subject to public assessment. Principles of evaluations of Elkan's demonstration, of the demonstrations of his interlocutors and of the properties and success of fuzzy logic were based on logical, technological, economic, business, organizational, psychological, linguistic, sociological and cultural considerations. Participants invoked various orders of worth of fuzzy logic and of the arguments produced during the debate. These orders of worth corresponded to different value systems that were shaped among others by cultural or anti-culturalist approaches. I will examine some of the demonstrative registers used by the arguers before I analyze the dynamics of the debate and its outcome.[1]

Technological Assessments of a Paradox

Elkan put forward the contradictory character of fuzzy logic in his article in order to denounce the paradoxical dimension of the success of its putative applications. The display of properties of fuzzy logic in computer systems thus came to be opposed in *comp.ai.fuzzy* to the results of symbolic demonstrations, or else those properties were evoked, on the contrary, in order to underline the correctness of these demonstrations. In other words, the same message could include intersecting considerations on Elkan's proof and the causes of the working of a given electronic device. The way a particular machine worked could for example be attributed to quite general properties of fuzzy logic, and presented by that very token as calling back into question the validity of Elkan's overall demonstration (including the proof of his theorem regarding the contradictions of fuzzy logic).

The debates dealt in fact not only with the reality and the nature of the properties of fuzzy logic, but also with the modalities of the staging of those properties in the texts of the various participants. Readers could then witness a sequential linking of constructionist and deconstructionist undertakings, the appearances of properties attributed to fuzzy logic in certain messages being deconstructed to the profit of others during later exchanges.

It is now useful to study closely, with an eye to pinpointing their specificity, the principles of evaluation of fuzzy logic properties and of demonstrations weighing against each other in the forum. I will begin by analyzing a first excerpt from a message written by J. Wiegand from Temple University that puts the properties of fuzzy logic in electronic and computer systems on display in the face of the critiques developed by Elkan:

> [M]ost of the applications in the control area using FL [Fuzzy Logic]
> have been too simple to date to show the inherent
> weaknesses of the technique.

[1] For further analysis of the debate and its dynamics, as well as for preliminary versions of parts of this paper, see Rosental (2003, 2008).

> Like the fuzzy predictive controller that handles the Sendai trains? Or the Otis elevator scheduler that uses both predictive and interactive inputs? How complicated does it need to be? 100,000,000 rules?
>
> > Once these weaknesses become apparent then FL will
> > turn out to be just as problematic to use as other
> > knowledge-based techniques (and he seems to imply
> > possibly more so).
>
> Well, it IS problematic if you don't understand it. It has taken me well over three years to grasp the philosophical basis of FL. What other AI technique has the properties of universality and proximality? Just neural nets, and they are closely related to FL. The Japanese tried AI in their Fifth Generation project, and it got them nowhere. But now they lead the world in fuzzy patents. Comparison? Just the opposite of what Elkan is saying.... The properties of FL should be obvious to anyone who has investigated it with an open mind.[2]

This excerpt is an example of calling Elkan's demonstration back into question without going through a critique of the details of the symbolic proof of the theorem. The evaluation of fuzzy logic properties in play stems from a practice of *bringing to the fore* (*faire valoir*): It consists, in the face of an enterprise of devalorizing fuzzy logic, of seeking to display, magnify and valorize properties of fuzzy logic in electronic equipment. To this end, Wiegand exhibits commercial products emblematic of brands and of a technology, fuzzy logic in this case. The author presents these devices as literally embodying certain properties of fuzzy logic. The successful operation of controllers used by the Sendai train and an Otis elevator is thus put into the balance against Elkan's demonstration.

The author uses additional demonstrative registers here. Paralleling the reference to the operating principles behind trains and elevators, Wiegand evokes elements as heterogeneous as Japanese leadership in the realm of fuzzy patents and the so-called impasse of a research program in AI mobilizing a logic with two truth values (the "fifth generation project") to exhibit very general properties of fuzzy logic, characterized as "philosophical" (like the property of "universality").

Wiegand goes on to mobilize an additional demonstrative register: a discourse about the self-evidence of the properties of fuzzy logic taken as an object. Wiegand insists on the ideal self-evidence of certain properties of fuzzy logic; then, to account for the fact that this self-evidence may not appear as such in Elkan's eyes, he invokes contingencies and, above all, psychological resistances. For the author, such blindness may stem from a biased attitude that originates in the mind of its victim: a lack of open-mindedness

[2] J. Wiegand, *comp.ai.fuzzy*, message 767. Use of the symbol ">", generated by e-mail software in the formulation of answers to former messages, corresponds to quotes.

would be its cause. Moreover, Wiegand evokes the long process of decantation that may prove necessary to accede to that self-evidence, however immediate it may be from an ideal viewpoint. That affirmation is shored up by testimony, since the author makes much of the three years he needed to apprehend it as such.

Elkan formulated a response to these criticisms and to other participants to the debate in an elaborate message he posted on the electronic forum. The set of demonstrative registers he mobilized for the occasion needs to be analyzed with special care:

> FUZZY LOGIC IS USEFUL IN ENGINEERING The paper is not meant to be "damning of FL for productive engineering purposes." Fuzzy logic has been and will be very successful in heuristic control applications, for example for Sendai subway train braking and for elevator speed control. However these systems are small compared to other knowledge-based systems. They use less than 100 rules, compared to many thousand for many expert systems. The fact that Japan leads the world in fuzzy patents and in fuzzy controllers has several plausible non-mystical explanations: (a) Japan leads the world in manufacturing high-technology consumer products. These are the largest natural application areas for fuzzy controllers. (b) Japan leads the world in patents overall. The number of patents per year that a company chooses to take out is a business decision influenced by many concerns and is not perfectly correlated with the company's overall success in research OR development. (c) Fuzzy logic controllers are engineered in an iterative, heuristic process of incremental improvement. This is congruent with traditional Japanese strengths in incremental quality improvement. (d) The reasons fuzzy controllers work well are that they (1) are rule-based and (2) have many tunable numerical coefficients. More applications outside Japan may use these two features without using the keyword "fuzzy."[3]

Here, Elkan accompanies his original demonstration by translating it, that is, by reformulating it thanks to new demonstrative registers. In particular, he refuses to attribute the success of fuzzy systems to properties of fuzzy logic. He deconstructs that causal chain in order to make a new one. He associates the large volume of fuzzy controllers produced in Japan with a general economic dynamics, Japanese leadership in the realm of production of high-technology goods (the cause of the success evoked is thus displaced a first time: it stems from a more general phenomenon). Then he dissociates the data concerning the volume of patents held by Japanese companies from the success of these companies in the realm of research and development. He then associates the creation of fuzzy controllers in Japan to a Japanese cultural specificity in working methods: the latter would rest on step-by-step improvements in the devices that are being developed. Finally, he associates

[3] C. Elkan, message 794.

the successful operation of fuzzy controllers not with specific properties of fuzzy logic, now, but with the fact that they are developed on the basis of rules and numerical coefficients capable of being optimized.

Elkan thus deploys a set of heterogeneous demonstrative registers to defend his original demonstration: socio-techno-economic considerations on the market in high-technology goods, putative expertise in the organization of labor, production, and management in Japanese businesses, as well as a paternity trial bearing on the causes of the proper functioning of fuzzy controllers—or in other words, a trial bearing on what engenders this proper functioning. Elkan ends up advocating that the emblem "fuzzy logic" no longer be used to account for these mechanisms.

In the face of this work of respecification of the causal chains, several participants in *comp.ai.fuzzy* plunged into analogous paternity trials. Some sought to bring the emblem of fuzzy logic to the fore once again by presenting the properties of fuzzy logic as embodied in technological devices. The electronic forum was thus the theater for a series of reevaluations based on appearances.

Cultural and Anti-Culturalist Evaluations

Some participants in *comp.ai.fuzzy* attempted to defuse Elkan's critique and in particular to show that fuzzy logic was not the victim of the paradox evoked, by positing a cultural incommensurability between the approach taken in Elkan's demonstration and the approach attributed to fuzzy logic. In their messages, they presented fuzzy logic as indissociably linked to an Eastern mode of thought and to Asian civilization. The concepts of binary logic, especially what they referred to as "the"—supposedly unique and well-identified—notion of paradox, associated with Western civilization and a tradition of Aristotelian thought, would not make it possible to grasp the subtleties of fuzzy logic.

By adopting this culturalist position, some participants in *comp.ai.fuzzy* attempted to show that the contradictions Elkan evoked were relative to a mode of thought that valorized a particular notion of precision, one that cannot be retained in efforts to conceive of the coherence that is proper to fuzzy logic and the value of fuzziness. A new way of evaluating Elkan's demonstration was thus introduced to short-circuit the debates over the details of the proofs. This evaluation proceeded from a form of cultural relativism.[4] The following excerpt of a message by J. Wiegand illustrates this approach:

[4] This form of slippage from cultural relativism to logical relativism can be compared to the possible passage from cultural relativism to aesthetic relativism. On this subject, see Menger (1989).

> Last week at AAAI, a paper by Charles Elkan of UCSD was given
> with the above title. It concludes that a "standard version" of fuzzy
> logic collapses mathematically to binary logic.

I should hope so! Fuzzy logic is the generalization of binary logic. It sounds as if Mr. Elkan has slipped back into the trap of Aristotelian logic. Witness the title: "The Paradoxical Success ..." There are no such things as paradoxes, only semantic dead-ends that result from the Western mindset.[5]

The demonstrative schema used in this message is easy to identify. Rejecting the very notion of paradox, the author opposes Elkan's patient demonstrative efforts. The notion of paradox, mobilized by Elkan to stigmatize an impasse proper to the approach of fuzzy logic, is exhibited here as an impasse inherent in the Western mode of thought. To organize this shift of focus, the writer identifies and then disparages as a whole the tradition of thought of Western civilization in general. Wiegand is not simply adhering to cultural relativism here; he moves on to establish a hierarchy among the cultures he identifies. The exhibition of fuzzy logic as a generalization of binary logic (also labeled Aristotelian) leads him to consecrate an Eastern mode of thought as triumphing over the failures (the "impasses") of the Western mode of thought. Asserting the opposite of the text he is citing, the writer thus offers a new evaluation of Elkan's demonstration, which is reduced to an absolute misunderstanding of fuzzy logic that originates in a fundamental cultural limitation.

In the face of such a positioning of the debate, several participants intervene to denounce an excessive culturalism. We can analyze some of the demonstrative strategies used to confront this new evaluation by studying excerpts from a few messages. I will begin by examining a response formulated by Elkan himself to this type of critique:

PHILOSOPHICAL AND MATHEMATICAL ISSUES The claimed binary opposition between a Western discrete mindset and a continuous Eastern mindset is ridiculous. In the history of Western thought, there have always been competing holistic, continuous and reductionist, discrete points of view. It makes reasonable sense to talk of "Western" thought because there has been continuity in philosophy from the Greeks to the present. I know almost nothing about non-Western philosophy. However I am willing to conjecture that "Eastern" thought is much less of a unitary tradition, and that in any particular division of Eastern thought (for example Confucianism) there are also parallel strands of continuous and discrete points of view.[6]

To defend his demonstration and bring back into view contradictions proper to fuzzy logic, Elkan adopts a strategy consisting of calling the nature and

[5] J. Wiegand, message 767. [6] C. Elkan, message 794.

scope of the dichotomy between Eastern and Western modes of thought back into question. To this end, he sets himself up as a historian of civilizations and philosophy: he challenges the distinctions made between two modes of thought by evoking the diversity of their respective tendencies, their hybridization, even their possible unity; in the process, he formulates a thesis about the continuity of Western philosophical thought from the Greeks to our own day, and develops hypotheses on the history of the various tendencies of Confucianism.

We see here to what extent debates about an article in AI can consist in writing or rewriting the cultural history of philosophy and civilizations. The work of the actors, far from being limited to the drafting of symbolic proofs, also consists in producing reflections stemming from epistemology, from the history of science, or even from philosophy and history in general.

In fact, it may be noticed that the development of the exchanges on *comp.ai.fuzzy* is related to the introduction of an increasing volume of considerations, including competing viewpoints as to what it is acceptable and important to debate, and what can legitimately stem from a scientific exercise. In other words, the disagreements spread to the definition of borders between what may or may not constitute an acceptable element for discussion, between what may or may not be considered as legitimately arising from a debate in AI, or between what does or does not constitute a good evaluation.

Elkan benefits here from the support of converging points of view. Several participants express decided and sometimes virulent opposition to the type of culturalist thesis examined above, as the next excerpt attests:

> I do object to the characterization of anyone who doesn't use or accept fuzzy logic as "Aristotelian" or "Western," as if fuzzy logic were the Holy Grail, and the entire universe has changed since its introduction.[7]

The attempt to neutralize Elkan's demonstration by recourse to culturalism hardly garners the support of all. It is met with vigorous hostility from contributors who deem this register unacceptable. In the message above, the writer contests a form of reductionism that consists in judging that the rejection of fuzzy logic by an individual would stem essentially from the latter's Western origins. He also simultaneously opposes any reconstruction of the history of civilizations around the discovery of fuzzy logic, targeting the praise of fuzzy logic and Eastern civilization simultaneously.

The shared rejection of cultural relativism paired with a form of orientalism gives Elkan an opportunity to rally actors around his cause. Antagonism

[7] W. Dwinnell, message 769.

constitutes the glue binding together several evaluations intended to neutralize a demonstrative register that threatened Elkan's demonstration, with at least indirect support for the latter as a consequence.

As another example, the various ways in which a matter of fashion in Japan is invoked should be noted: an infatuation with fuzzy systems on the national scale would explain the surge in their development. This thesis is suggested in the following passage:

> One possible reason for Fuzzy's popularity is
> that it fits iterative development so well—not
> because of some attraction for imprecision.
> Controlling subway cars doesn't have much in common
> with Zen archery.

Apparently not, it seems that in Japan, Fuzzy Set Theory is a fad and anything fuzzy is to the consumer a warm and . . . (wap!) . . . {sorry}[8]

This contributor rejects the culturalist arguments in his turn even as he gives a new orientation to the readers' shift of focus, a shift initiated by the author he cites. Instead of exhibiting a discourse celebrating fuzziness and explaining its emergence by invoking a structural equivalence between techniques for perfecting fuzzy systems and traditional working methods, he refers to a social phenomenon, a fashion affecting Japanese consumers. The Japanese are not shown here as determined in their behavior by a distinctive mode of thought. Their singular behavior is explained as a function of a local phenomenon, whose principle (fashion) is not itself associated with any eastern cultural specificity.

Here, the indirect defense of Elkan's demonstration in the face of orientalism thus passes through a reformatting of the Japanese mode of thought, evaluated according to a more universalist mode. This reformatting is pursued in other texts, moreover, through the invocation of linguistic particularities. In these messages, instead of invoking differences in mental structures, the writers put forward specific connotations of the term "fuzzy" in English and in Japanese in order to account for the differentiated infatuations with fuzzy logic in the United States and in Japan. The following messages are illustrative:

Fuzzy seems a very practical tool for us. Now if it only had a better name . . .[9]

- The fact that Japan leads the world in fuzzy patents
- and in fuzzy controllers has several plausible non-
- mystical explanations: . . .

[8] M. Aichlmayer, message 789. [9] H. Bonney, message 787.

> Here is another: Both Japan and America use the same word "fuzzy." In America, the word has many old negative connotations. In Japan, it has no old connotation, positive or negative. (And it is a wonderful sounding word.) Thus, "fuzzy camera" sounds high-tech in Japan, and goofy in the U.S.[10]

In these two texts, and more specifically in the second, particular attention is paid to the word "fuzzy" in order to account both for a certain hostility to fuzzy logic in the United States and for the attractiveness of fuzzy logic in Japan. The writers do not invoke the existence of distinct mental structures that would characterize Japanese people on the one hand, Americans on the other. They assess the dissimilar effects that the use of the same expression in two different spaces allegedly induces, in terms of linguistic idiosyncrasies. In other words, the reformatting of the Japanese mode of thought that we are considering here depends on the substitution of a sociolinguistic register for an orientalist register. It is no longer a question, in these messages, of a gap between two civilizations, but simply of language effects. The people concerned are no longer so foreign; they are endowed with identical cognitive and behavioral faculties.

Dissonance and After: The Emergence of a Collective Statement

We have seen how heterogeneous principles of evaluation were used on *comp.ai.fuzzy* to produce antagonistic judgments about the value and success of fuzzy logic, as well as the validity and value of Elkan's and other arguers' demonstrations. Multiple orders of worth of fuzzy logic and of everybody's arguments were invoked in the framework of the debate. They were based on logical, technological, economic, business, organizational, psychological, linguistic, sociological and cultural considerations. What came out of this dissonance, if not cacophony? Could the latter be transcended in one way or another and lead to an innovation? If so, what did this innovation consist of? In order to address these issues, it is now useful to study the dynamics of the debate and its outcome.

Although highly fragmented and contradictory points of view about Elkan's paper and fuzzy logic resulted from the discussions on *comp.ai.fuzzy*, some few points of view began to acquire more visibility than others after a few months: those of Elkan and those of fuzzy logic personalities working in computer science, such as Didier Dubois and Henri Prade from the CNRS (France) and Enrique Ruspini from SRI (United States). Before they

[10] C. Kadie, message 798.

could express themselves in journals, Dubois, Prade, and Ruspini had put in token appearances in the forum to counterbalance the proliferation of representations of fuzzy logic, and of Elkan's paper, that were too distant from what they considered to be reasonable or desirable. Some researchers adopted several mechanisms to make certain messages more visible than others: citations and frequent reappearance, with placement of some texts in a data base accessible via Internet, the electronic address of which was often displayed.

Mechanisms such as these promoted differentiated visibility of messages; the texts displayed in the forum did not have the same impact on the shaping of representations of Elkan's paper. The capacity for conviction that could be granted to these new demonstrations and counterdemonstrations was therefore not only limited; in the very frame of this material economy of access to texts, it was also variable.

Messages posted on *comp.ai.fuzzy* show that the shaping of representations of Elkan's paper in the context of the forum did not derive from a sum of individual homogeneous examinations obtained from attentive readings of an easily accessible text (Rosental 2008). Nor was this process reducible to an exchange of arguments, or even discussions, given the scriptural dimension of these interactions and the aforementioned material economy of access to texts. The process of dialectical evaluation to which Elkan's paper had been subjected by no means resulted in a clear and uniform collective view. Even the apparently simple question of the exact scope and significance of the theorem was the object of a multiplicity of representations. Now what were the consequences, from the point of view of representations of Elkan's demonstration, of the continuation of debate in other forums?

In the eyes of many leaders in fuzzy logic, the publication of Elkan's paper could have resulted in a substantial loss of credibility for their research domain. Several of these leaders rallied together to organize a counterattack on several fronts. In conjunction with their intervention in the electronic forum, they addressed a protest letter to the organizers of the AAAI, and made plans for the publication of several responses to Elkan's article.

Thus, several months after the conference, the center of debate shifted from the electronic forum to journals that specialized in artificial intelligence. This shift of exhibited interaction to other arenas was accompanied by a radical transformation in the time-scale of debates and a substantial rise in the barriers to be surmounted to stay in the game. Making a point of view public now required authors to produce polished texts and to subject them to editorial constraints. Some authors could reuse many arguments they had posted in the electronic forum, but not all of them, and some of them needed to be reformulated. Making a point of view public also required full investment in a milieu in which interindividual relations and reputations were essential in the processes of selecting (and often commissioning) articles, as Davis and Hersh (1987) have noted for the field of mathematics.

Some of the leading researchers in fuzzy logic already had extensive experience in facing criticisms. The reactions of this group to Elkan's article were thus based on already well-established know-how and wide experience with methods of counterattack. Endowed with these specific competencies, several of the defenders of fuzzy logic united to formulate a response that would stand in deliberate contrast to the dissonance, if not cacophony of messages in the electronic forum. The repetition, the constant attack on the same elements of counterproof, was a source of discredit to Elkan's theses but was at the same time likely to give the impression that all researchers in fuzzy logic stood against Elkan in a knee-jerk fashion. No fewer than eight major researchers in fuzzy logic, including Dubois and Prade, cosigned an article whose content was close to that of a message in the electronic forum, thus subjecting Elkan's point of view to the test of numbers in case their approach was not enough to marginalize it (Berenji et al. 1994).

The multiplicity of counterdemonstrations formulated in the electronic forum were thus succeeded by only a small number of interventions by recognized spokespersons for fuzzy logic organized around united points of view. "The" theorem of Elkan (as if its formulation was unique and clearly identified) was presented, in particular, as a result that was in fact very simple when reformulated appropriately, known for a long time and without any effect on the foundations of fuzzy logic. The image of two-sided confrontation between Elkan's viewpoint and that of researchers in fuzzy logic was progressively built up through additional approaches such as publication in journals (and shortly afterwards in the electronic forum) of abstracts of papers, in which the standpoints of participants in the debate were simplified and frozen in a conflict reduced to its simplest expression.

As mentioned above, considering editorial constraints, which required a heavier investment in contributions to the debate, the possibilities for publicly expressing a point of view were substantially reduced for some. The journal *IEEE Expert*, by eventually devoting a special issue in August 1994 to the debate on Elkan's article, had nevertheless opened an exceptional forum for the question. But even this apparent opening remained relatively limited: The fuzzy logic authors it included were "stars" of the field, such as Didier Dubois, Henri Prade, Enrique Ruspini, and the computer scientist Lotfi Zadeh from UC-Berkeley, who was generally reputed to have invented fuzzy logic.

The number of participants in the debate was very small, and most of those who earlier had participated in the interaction in the forum disappeared. This did not mean, however, that they had been convinced by the very few emerging points of view and had adhered to them. The sudden unity stemmed from the fact that some had managed to remain in the debate and to enhance its resonance, while others had kept silent because they were unable or unwilling to overcome the obstacles to asserting their opinions. This singular dynamic worked powerfully toward stabilizing the debate. What about Elkan's approach and that of his partisans? How did they react to this counter-attack?

The series of viewpoints researchers in fuzzy logic expressed in the forum and in private letters sent to Elkan, as well as in personal conversations with him, had allowed him an opportunity to clearly perceive the divisions that existed between researchers, especially regarding the definition of their own research objects and their representations of his arguments. The existence of these evolving rifts gave Elkan the opportunity to multiply reformulations and personalized approaches. This, in turn, afforded him the chance to quiet antagonistic reactions and to solidify his results—or, more precisely, what he could re-present in each case as the nature of his results. This was a considerable resource for Elkan, for it emerged in circumstances wherein little support for his theses had been publicly expressed during the months immediately following publication of his article.

Elkan thus had the opportunity to elaborate and test several different reformulations on diverse interlocutors. He adapted his talks to suit his interlocutors and the forums at which he presented, adjusting his presentations in a differentiated, evolving, and sometimes personalized way. To a certain extent, Elkan was more involved in dramatic than in communicative acting (Goffman 1959; Habermas 1984). His talks could possibly help to curb the virulence of the counterattack by fuzzy logic researchers without, for all that, deceiving researchers in so-called classical AI.

For example, Elkan formulated the following private answer to criticism of his supposedly implicit introduction into his proof of an axiom (equivalent to the law of excluded middle)[11] rejected by fuzzy logic, in order to obtain the theorem: The average user of fuzzy logic could fail to be aware of the impossibility of mobilizing this type of axiom, despite it being so usual. Although, according to one of Elkan's interlocutors, the author affirmed on the day he delivered his paper at the AAAI conference that he did not see why he could not introduce the law of the excluded middle, Elkan subsequently claimed, in another context, that he had used an equivalent axiom in his proof by attributing an essentially educational value to his theorem.

Such nuances underline the primacy of public expression for grasping acts of enunciation (Quéré 1990). Because his article, a singular material device launched in the world, eventually proved somewhat ineffective in countering

[11] The law of the excluded middle is generally defined as, "every statement is either true or false."

criticism, Elkan added to it by producing new texts and new speeches. He thus provided new instructions for his text's interpretation and general comprehension, thereby forging new tools for changing readers' relationship to his original text.[12] Such adjustments also helped to stabilize debate because they limited disagreements by making them appear, retrospectively, and at least partly, as misunderstandings (which differed, of course, depending on the interlocutors and the publics).

Elkan thus undertook a fundamentally different exercise than did his opponents. While their idea was to organize their debate as a unified and coordinated counterattack, Elkan proved to be a highly mobile target and was a talented mediator capable of producing different and evolving responses. He had to be more skilled at improvisation than orchestration. For him this exercise appeared to be no routine exercise but rather a new, ad hoc approach.

Yet after a few months Elkan had forged tools that enabled him to limit his production of differentiated answers that had become consuming, both in time and energy. In a new version of his article, published in the journal *IEEE Expert* and available on the Internet at the end of 1993, he perfected reformulations that he could present to widely diverse publics. Considering their polysemic nature, their effect was twofold. After the different readers or listeners had read or heard these reformulations, they could adopt radically divergent points of view on the nature and meaning of his demonstration. But at the same time they could agree to grant it a form of validity and value that was obviously variable.

For example, in the new version of his text, Elkan stated that the four axioms he had used to prove his theorem offered an "apparently reasonable" description of fuzzy logic. This expression could be interpreted in different ways by its readers. For some researchers in fuzzy logic who were relatively well-informed as to current debate on the subject, the use of this expression might have signified that Elkan had taken into account criticism on the limited scope of his theorem, and that the misunderstanding was over. However, a reader who had little knowledge of the debates and fuzzy logic literature could attribute a far more general scope to his theorem.

Elkan could easily use this type of text in discussions with proponents of so-called classical AI to assert that he had stood his ground and shown the limits of fuzzy logic. But he could also affirm to fuzzy logic researchers, supposedly without any contradiction, that he had essentially formulated constructive criticism and had delivered an educational message, to the point of being able to consider collaboration with them in the future. Indeed, after a

[12] For an illustration over a longer period of the fully historical nature of readings in the field of mathematics, see Goldstein (1995) on the case of Fermat's theorem.

long discussion with one of the leading figures in fuzzy logic, Elkan and his interlocutor had considered co-authoring a paper.

The polysemic nature of Elkan's text thus enabled him to assert its validity and value in his private interaction. He could also rely on a single reformulation of his article for all his answers, thus avoiding the need to multiply adjustments to suit the forums in which he had to express himself.

Yet the stabilization of Elkan's formulations caused the debate to run out of steam because different forms of agreement became possible. After *IEEE Expert* had devoted a special issue in August 1994 to reactions to Elkan's article, in which a few papers by important representatives of both fuzzy logic and so-called classical AI were published side-by-side, there seemed to be no fundamentally new elements left to advance. Although these debates finally appeared to be drawing to a close, however, no real consensus had been reached. Partial agreement that seemed to have formed around the validity and value of Elkan's demonstration was simply apparent and must be understood in relation to the distribution of the distinct, and often antagonistic, points of view to which its expression corresponded. It was a matter not of a univocal statement overcoming or sublimating temporary divergences, but of a *collective statement* in Alain Boureau's (1989, 1992) sense. The medievalist Boureau coined the term to refer to a "verbal or iconic fragment that creates around itself a certain convergence of languages and thoughts, through the play of a structural fuzziness allowing to capture a still implicit thematic and to welcome the most diverse projections and appropriations" (Boureau 1992: 1072).

Here, "the validity and value of Elkan's demonstration" may be seen as a collective statement as it was appropriated by each participant in the debate into its own specific mode of agreement, while at the same time the statement managed to serve as a point of coordination for various points of views regarding fuzzy logic. Thus, although it became necessary to talk about the relative recognition of Elkan's demonstration, such recognition was possible only as a distribution of partly united representations in a void produced by the increasing scarcity of publicly expressed points of view.

Details of arguments and the way they were handled, no less than social positions and relations, all played a role in the dynamics of the debate under study and its outcome. A reductionist analysis that would have put exclusive attention to one of these dimensions would not have allowed us to grasp the social process at stake. More particularly, the changing shape and general evolution of representations of Elkan's demonstration—like the reduction of a set of points of view expressed publicly in a small number of contributions to journals—did not occur as a simple formulation of more or less convincing univocal arguments. Other factors also played a decisive role in this dynamic—especially polysemy, certain actors' capacities, depending on the case, to coordinate with, to consult, and to co-opt one another, to materially manage the visibility of their texts, to show flexibility by producing evolving and differentiated discourses, or to accompany their claims by reformulating them as much as necessary. Such competencies were all instrumental in stabilizing debate and played a key role in the emergence of a demonstration lastingly endowed with multiple significations.

Indeed, I found traces of comments on Elkan's article that supplemented citations in *comp.ai.fuzzy* and in specialized journals since 1994. A few articles have been devoted to continuing the debate,[13] but on the whole the controversy has clearly run out of steam since 1994. The arguments put forward have not been really new and have triggered no new outbursts of reactions. An analysis of the context of citations of Elkan's article provides some evidence that the representations of Elkan's paper have hardly evolved since summer 1994.[14] To date, the outcome of the process, as described here, appears to have lasted.

Lasting Dissonances and Misunderstandings

To conclude, one may say that the dissonant evaluations that were produced during the debate have led to a scientific innovation, namely the emergence of a collective statement—"the validity and value of Elkan's demonstration." This statement has become part of the world and served as a point of reference for various judgments and actions (Rosental 2008). To a limited extent, the scientific community has eventually recognized and valorized the last version of Elkan's demonstration that was produced after the first months of the debate. But this recognition and valorization has not been based on a simple process of conviction or on the clear victory of one side over others. Only partial agreement on the validity and value of the demonstration has been reached, largely based on misunderstanding. Besides, only a few contradictory points of view have finally become more visible than others. But it does not mean that those who became silent in public shared the same viewpoints. Conflicts between evaluative principles and what could be called different scientific cultures could not be overcome here. Most participants did not change positions once the debate was over: they stuck to their guns on different modes of evaluation of fuzzy logic and of demonstrations and kept contradictory views on the validity and value of Elkan's arguments.

[13] For references, see Rosental 2008: 237–238.

[14] The analysis could be conducted thanks to Nec Research Institute's Citeseer Search Engine (citeseerx.ist.psu.edu) and the Science Citation Index.

This case shows how misunderstanding, like dissonance, may play a decisive role in the dynamics of the scientific field, no less than in other fields and social spaces. Such is the case of the artistic field in which experts produce a large set of dissonant evaluative principles (Quemin 2005, Velthuis and Kharchenkova in this volume). In other spaces, Proust (1927: 565–568) demonstrates, through his characters, how individuals have diverging viewpoints and very different ways of apprehending signs that might seem most likely to support a single interpretation and a programmed effect. Besides, people are often confronted with polysemic formulations that are intended to convey distinctly different messages to different actors. But such communicative aims regularly fail too. As a result, misunderstandings and dissonances represent significant aspects of ordinary life.

This case also illustrates how conflicts or dissonances between different evaluative principles and value systems may lead to an innovation—a collective statement here—although they may not be surmounted or transcended in a straightforward way. Dissonance is not necessarily resolved and may last, even in science. Scientific statements like technologies may have a "permanently beta" status (Neff and Stark 2004). But a stable status granted to a given scientific statement may be based in fact on permanent disagreement and misunderstanding. Contradictory viewpoints may co-exist until scientists die (Bloor 1978). Even under conditions that may seem favorable, such as when science is taught, misunderstandings and disagreements (between students and professors) may not be resolved (Rosental 2009).

Since antagonistic viewpoints produced during a scientific debate are not inevitably transcended or sublimated in a clear-cut way, reaching transcendence or something equivalent to it appears to require very specific conditions of possibility that are not always met. This raises an important issue: Under what conditions can dissonances be transcended and in what sense? Hopefully, future research will help answer this question and help us better understand how innovations are socially produced, and what they are made of.[15]

References

Berenji, H. R., Bezdek, J. C., Bonissone, P. P., Dubois, D., Kruse, R., Prade, H., Smets, P., and Yager, R. R. (1994). A Reply to the Paradoxical Success of Fuzzy Logic. *AI Magazine,* Spring: 6–8.

[15] I would like to thank the volume editors and the participants to the "Valorizing Dissonance" Conference held at WZB, Berlin, in June 2013, for their insightful comments on an earlier draft.

Bloor, D. (1976). *Knowledge and Social Imagery.* London: Routledge.
Bloor, D. (1978). Polyhedra and the Abominations of Leviticus. *British Journal for the History of Science* 2(39): 245–272.
Boltanski, L. and Thévenot, L. (2006). *On Justification: The Economies of Worth.* Princeton, NJ: Princeton University Press.
Boureau, A. (1989). Propositions pour une histoire restreinte des mentalités. [Propositions for a Limited History of Mentalities]. *Annales ESC* 6: 1491–1504.
Boureau, A. (1992). L'adage *vox populi, vox dei* et l'invention de la nation anglaise (VIIIe-XIIe siècle). [The Proverb "The Voice of the People is the Voice of God" and the Invention of the English Nation, 8–12th Century]. *Annales ESC* 4–5: 1071–1089.
Collins, H. M. (1985). *Changing Order: Replication and Induction in Scientific Practice.* London: Sage.
Collins, H. M. (1990). *Artificial Experts: Social Knowledge and Intelligent Machines.* Cambridge, MA: MIT Press.
Davis, P. J. and Hersh, R. (1987). Rhetoric and Mathematics. In J. S. Nelson, A. Megill, and D. N. McCloskey (eds.), *The Rhetoric of the Human Sciences: Language and Argument in Scholarship and Public Affairs.* Madison, WI: University of Wisconsin Press, 53–68.
Elkan, C. (1993). The Paradoxical Success of Fuzzy Logic. In *Proceedings of the Eleventh National Conference on Artificial Intelligence.* Menlo Park, CA: AAAI Press. Cambridge: MIT Press, 698–703.
Goffman, E. (1959). *The Presentation of Self in Everyday Life.* Garden City, NY: Doubleday.
Goldstein, C. (1995). *Un théorème de Fermat et ses lecteurs.* [One of Fermat's Theorems and its Readers]. Saint-Denis: Presses Universitaires de Vincennes.
Guice, J. (1998). Controversy and the State: Lord ARPA and Intelligent Computing. *Social Studies of Science* 28: 103–138.
Habermas, J. (1984). *The Theory of Communicative Action.* Boston, MA: Beacon Press.
Hutter, M. and Throsby, D. (eds.) (2008). *Beyond Price. Value in Culture, Economics and the Arts.* Cambridge: Cambridge University Press.
Lamont, M. (2009). *How Professors Think: Inside the Curious World of Academic Judgment.* Cambridge, MA: Harvard University Press.
Latour, B. (1987). *Science in Action: How to Follow Scientists and Engineers through Society.* Cambridge, MA: Harvard University Press.
Menger, P. M. (1989). *Les Laboratoires de la Création Musicale. Acteurs, Organisations et Politique de la Recherche Musicale.* [The Laboratories of Musical Creation]. Paris: La Documentation Française.
Neff, G. and Stark, D. (2004). Permanently Beta. Responsive Organization in the Internet Era. In P. N. Howard and S. Jones (eds.), *Society Online: The Internet in Context.* London: Sage, 173–188.
Proust, M. (1927). The Captive. In *Remembrance of Things Past.* Vol. 2. New York, NY: Random House, 453–788.
Quemin, A. (2005). L'art plus fort que la science? L'affrontement entre expertise stylistique et expertise scientifique dans une querelle sur l'authenticité d'une

œuvre d'art: l'affaire Sésostris III. [Art Stronger than Science?]. *Sociedade e Estado* 20(2): 403–424.

Quéré, L. (1990). Agir dans l'espace public: L'intentionnalité des actions comme phénomène social. [Acting in Public Space]. In P. Pharo and L. Quéré (eds.), *Les formes de l'action*. Paris: Editions de l'EHESS, 85–112.

Rosental, C. (2003). Certifying Knowledge: The Sociology of a Logical Theorem in Artificial Intelligence. *American Sociological Review* 68: 623–644.

Rosental, C. (2004). Fuzzyfying the World: Social Practices of Showing the Properties of Fuzzy Logic. In M. N. Wise (ed.), *Growing Explanations: Historical Perspectives on Recent Science*. Durham, NC: Duke University Press, 279–319.

Rosental, C. (2008). *Weaving Self-Evidence: A Sociology of Logic*. Princeton, NJ: Princeton University Press.

Rosental, C. (2009). L'erreur en logique. [Error in Logic]. In C. Chauviré, A. Ogien, and L. Quéré (eds.), *Dynamiques de l'erreur.* Paris, Éditions de l'EHESS (« Raisons Pratiques » N° 19), 285–307.

Rosental, C. (2010). Social Studies of Evaluation. *Social Studies of Science* 40(3): 481–484.

Stark, D. (2009). *The Sense of Dissonance: Accounts of Worth in Economic Life.* Princeton, NJ: Princeton University Press.

Zadeh, L. A. (1965). Fuzzy Sets. *Information and Control* 8: 338–353.

11

Accounting and the Plasticity of Valuation

Andrea Mennicken and Michael Power

Introduction

In the mid 1970s a new kind of accounting statement appeared—value added accounting. Promoted by an accounting professional body in 1975 which sought to reengineer the corporate report (Accounting Standards Steering Committee [ASSC] 1975), a number of companies engaged experimentally in producing value added accounts (Burchell et al. 1985). Conventional financial statements were rearranged so that "'value added' appeared as an indicator of the value created by the activities of an enterprise in a number of different sites (private companies, newspapers, government bodies, trade unions, employer associations, professional accountancy bodies, etc.)" (Burchell et al. 1985: 385). Statements of value added were created which were aimed at "showing how the benefits of the efforts of an enterprise are shared between employees, providers of capital, the state and reinvestment" (1985: 386). As Burchell et al. (1985: 387) highlight, the accounting category of value added was not only promoted as a vehicle for information disclosure. It was also put forward as a basis for determining rewards at the level of the enterprise, and it appeared on several occasions in the context of policy discussions concerned with the performance of British industry.

This accounting innovation had its conditions of possibility in a distinctive alignment of three key arenas: professional concerns with the nature of accounting; macro-economic concerns aimed at making productivity issues more visible; and the political concerns of a leftist government and the trade unions with "industrial democracy" (Müller-Jentsch 2008). Despite the great variability in value-added accounts, and their technical ambiguity, they could appeal simultaneously to a number of different actors with different agendas. In so doing, value added accounting functioned as a kind of "boundary object" (Star and Griesemer 1989). As Star and Griesemer (1989: 393) would put it, value added accounts were plastic enough to adapt to the different local needs and constraints of the parties employing them. Yet, at the same time, they were robust enough to maintain a common identity across sites (1989). However, the value added accounting experiment was short lived, and the alignment of the different arenas began to break up—a process completed with the election of Margaret Thatcher in 1979.[1]

The value added accounting case perfectly illustrates the theme of this volume and is an exemplification of both newness and dissonance. The temporary stability of value added accounting was itself conditioned by a fragile alignment of interests, which hung together for a while without any explicit agreement or consensus. In this case one might say that mutual misunderstanding created coordination benefits (Stark 2009)—at least for a while. Value added accounting statements eventually disappeared and their rise and fall can be regarded as a *valuation event* in a specific time and place. The event reveals both the temporal nature of valuation, its dynamics and contingent path over time, and also its spatial localization. The value added accounting event took place at a specific site, in a particular social space marking out a distinctive field of operations. In Burchell et al.'s (1985: 400) words, "the necessity of talking about value added in a particular way arose as a result of the conditions that made the value added event possible"—conditions which Burchell et al. seek to encapsulate within the notion of an "accounting constellation" comprising a network of intersecting valuation practices, processes and institutions (1985). The constellation concept resembles that of *assemblage* (see the introductory chapter by Hutter and Stark in this volume),[2] which also refers to a temporarily stabilized set of linkages and practices formed between diverse actors, agencies, activities, technologies and ideas.

Value added accounting was above all an attempt to represent a new kind of object in accounting terms, namely the firm as a cooperative endeavor. This effort to re-value the nature of the firm involved the use of value-added statements not simply as a mirror of activity, but as a mechanism for the production of an alternative reality, albeit within the existing frame of

[1] To put it in Burchell et al.'s (1985: 405) terms, "with the election of a new Conservative Government in 1979 the three arenas of the value added constellation were suddenly ruptured and transformed." Value added accounting was no longer seen as a tool for the management of the national economy. "Market pressures in an increasingly high unemployment economy were seen to offer more effective means for income control" (ibid.).

[2] See also Miller (1991), Miller and O'Leary (1994), and Muniesa et al. (2007). In his study of the rise of discounted cash flow accounting, Miller (1991: 737), for example, uses the notion of assemblage to unravel how a complex interplay of new bodies of knowledge, novel regulatory attempts, changing theoretical conceptions of governance and a new accounting rhetoric contributed to the rise of discounted cash flow techniques.

standardized corporate financial reporting. That reality failed to materialize, but the case shows how accounts can be regarded as constrained ways of producing fictions. Value added accounts needed to be grounded in a transactional reality—they were not "made up" and needed to pay attention to a certain institutional obligation. But they were also fictional in the sense of projecting a new model of economic cooperation in the production of surplus. In this sense we can think of accounting valuation as having a quality of *plasticity*, namely being both simultaneously stable and unstable. It is the nature of this *plasticity* and its dual properties which is the subject of this chapter.

Carruthers and Kim (2011: 253) argue that "the plasticity of valuation is [...] apparent with every accounting restatement, but such episodes do not simply reflect valuation-gone-wrong. Rather, they reveal how much value is a contested and provisional judgment whose complexity lies buried beneath a surface of numbers and quantification." In this chapter, we seek to track specific conflicts and tensions within accounting valuation as they bear on wider questions of value. Indeed, we suggest that while valuation practices may eventuate in single figures for further consumption and processing (Miller 1992; Vollmer 2007), they are sustained by an apparatus which becomes visible in times of crisis and whose components are inherently malleable and contestable.[3] In short, "economic valuation processes are eminently contingent" (Fourcade 2011).

The chapter is structured as follows. In the next section we discuss the role of accounting in terms of its *economizing* (Çalışkan and Callon 2009, 2010) effects on entities and actors. We distinguish between *economizing* as a general mode of valuing in economic terms and *financialization* as a distinctive mode of economizing, drawing on the ideas and practices of financial economics. This distinction enables us to say that accounting is plastic in the sense that it is both a powerful and stable vehicle of economization, valorizing economic constructions of actors and things, and yet is also itself a practice which has been significantly destabilized by processes of financialization.

Building on our discussion of the economizing role of accounting, we explore three historical moments of dissonance and innovation in accounting valuation. First, we examine the event of brand accounting in the 1980s and early 1990s in the United Kingdom and investigate the struggles involved in the creation of a new object of accounting valuation. Brand accounting was an attempt to transpose values from the marketing world into the pre-existing frame of corporate financial statements, making brands, as

[3] For example, the contingent nature of the discrete accounting entity assumption underlying corporate financial statements became visible and problematic in the financial crisis of 2008 as risk and value migrated from fragile corporate entities to states and to citizens.

"intangible" sources of value, financially re-cognizable in accounts. Second, we discuss the more recent case of fair value accounting and, third, related efforts to standardize and stabilize the practice of asset impairment testing. The case of fair value accounting gives insight into the dynamics of an aspiration to move from a (historical) cost-based valuation system to an accounting system based on (current) market price. While fair value accounting aims at the systemic financialization of financial accounting, we analyze how this process runs up against limits at the level of impairment testing of asset values, where new frictions are created which reveal unstable accounts of worth.

Taking these three cases with the opening discussion of value added provides four historical moments which demonstrate the plasticity of accounting, both as a powerful force for economizing but also as a vehicle for different and often frictional conceptions of value. This chapter shows that accounting research can add to the sociology of finance and valuation studies more broadly by jointly navigating both the general role of accounting as a source of economization and also delineating the complex and contested nature of specific accounting valuation practices as sub-frictions within these general economization processes. In what follows, we seek to reveal the productive frictions involved in accounting valuation as different calculative frames are used, compared, put in relation to each other (Stark 2009, 2011), and how this uncertainty is only ever temporarily effaced.

Accounting and Economization

The opening example of value added accounting demonstrates that accounting statements are less stable and durable than may be commonly imagined. There is no institutional logic of accounting as such. Accounting has no essence, which would tempt us to seek such a logic. As Hopwood puts it, "accounting is not imbued with purpose, but it can be made purposeful" (Hopwood 1992). Yet accounting provides often powerful narratives of economic life (Hines 1988)—perhaps the most powerful—which expand and spill into many other domains and organizations, changing them in the process and challenging existing configurations of value. In this respect accounting is both a valorizing practice (Vatin 2013) which promotes and reproduces a conception of worth, as much as one which contains valuations. Regarding the former, it is a force for the *economization* (see also Çalışkan and Callon 2009, 2010) of entities and activities. Accounting makes it possible to render the representation of entities and activities in economic terms, even though it does this in many different ways (Miller and Power 2013). It is a mechanism by which the economization of organizational life becomes elaborated and institutionalized, and which increasingly provides the dominant narrative of market rationality at an organizational and at a societal level (Carruthers and Espeland 1991; Weber 1978), particularly since the collapse of the Soviet Union.[4] Accounting practices operationalize and materialize *fictions* of efficiency, economy, "value for money" and "value-added" (Hines 1988; Miller 2001).

Specific accounting valuations of assets and liabilities are intertwined with this more general constitutive role for accounting. The roles that accounting plays within organizations and society co-emerge with the economized social relations that in turn provide its rationales and that shape the organization as an accounting entity (Hopwood 1986; Miller and Power 2013). While "new modes of financing and their representations have brought attention to accounting's dependency on legal presumptions of the boundary of the enterprise" (Hopwood 1992: 127), accounting practices also enable an entity to be observed and constituted in economic terms, particularly in the public sector (Kurunmäki 1999; Mennicken 2013).

That accounting entities are fictional and network-effacing superimpositions on complex organizations is suggested by competing conceptualizations of heritage assets in museums (Barton 2000; Carnegie and Wolnizer 1996). If museums are regarded as separate accounting and therefore economic entities, heritage artefacts are "assets" on the entity's balance sheet, which can be valued in terms of their economic performance (e.g., paying visitors). But if museums are regarded as holding such heritage assets in trust for society and future generations, then they are also something that the organization owes, in other words a liability. In short, on which side of the balance sheet something appears is not a natural or obvious fact—it must be constructed as an organizational fact of a specific kind (Hines 1988) which in turn depends on the entity construct. In the case of heritage assets there is a persistent tension about what kind of accounting object they are, and this tension can be traced to fundamental differences in orders of worth (Boltanski and Thévenot 2006) implicit in entities.

Accounting bestows apparent precision and operational reality on economic categories of value and thereby constitutes a practical realm of the economic. As Hopwood (1992: 142) puts it, "a practical economy needs to be positively forged rather than merely revealed." Through this process organizational agents even acquire new preferences as they attend to accounting

[4] This is not to say that there were no attempts to "economize" organizations without appeal to the notion of markets—planned economies provide a good case in point here, and accounting played an important role in the organizing of state-led, planned economies. However, since the collapse of the Soviet Union, economization, predominantly, goes hand in hand with the marketization of much of social life.

constructs of economic value. Fourcade (2011: 1766) makes a similar point about the context of economic valuation practice: "Through this institutionalization process, it is not impossible that a new set of 'individual preferences' were actually made endogenous to the techniques supposed to reveal them."

This power of accounting to constitute and realize economic conceptions of entities and the performance of organizational agents, thereby shaping rather than reflecting preferences, does not imply in any way that accounting is a simple transmission mechanism for economic ideas. Indeed, this power paradoxically reveals the autonomy of accounting and its unstable relationship with economics: "The present practice of the accounting craft cannot be deduced from economic conceptions of it and economic ideas for its change and reform, although often articulated, find it difficult to become entangled with a craft that seems to have independence from what are seen as its essential roles" (Hopwood 1992: 130). As we show below, the financialization of accounting—the introduction of financial economics into accounting and the increasing intertwining of accounting with the governing of capital markets—involves a program of stabilizing this relationship and making accounting practice much less independent of economic ideas. And yet, despite this program, there has always been something permanently problematic and dissonant about the economizing role of accounting. Accounting gives life to economics but is not a part of economics, despite many efforts within the academy to position it as such.

Nowhere is this dissonant relationship more apparent than in asset accounting, which we discuss further in the cases presented below. Balance sheets are highly stable and institutionalized accounting forms. Yet they are also imperfect representations of value and have been a battleground for the representation of novel forms of asset, such as brands, and liabilities, such as future decommissioning costs for nuclear facilities. To pass the test of accounting recognition and make it into the balance sheet "valuable assets" like brands must be capable of being credibly measured. But even if something does make it onto the balance sheet at an entry cost to the entity, history shows that this may be a poor on-going reflection of economic value. Balance sheet values have periodically been challenged by alternative market-based values (which we discuss further below), thereby undermining the economizing role of financial accounting. Within these periodic crises of value representation and measurement, the apparent objectivity of accounting—its auditability (Power 1996, 1997)—has played a constraining role in the valuation deliberations of analytical communities, ensuring that not "anything goes." And yet, auditability is itself plastic and adaptable and networks of reliability can shift, and have shifted, over time.

All this makes the relationship between accounting and presumed communities of users a very complex one since accounting constitutes both objects and subjects of value, the latter being "accounting users" as they are commonly called (Young 2006). The power of accounting resides in the demarcation of "calculating selves and calculable spaces" (Miller 1992). Presuming, and bringing into existence, users who value accounting is one of the most powerful accomplishments of accounting infrastructure. Yet, exactly how accounting values enter the valuation processes of market actors is poorly understood even though some effects can be observed (Power 2012; Vollmer et al. 2009). There can be little doubt that accounting plays an important systemic role for capital markets; it is a force for the economization of organizations and societies even if the valuation of assets remains a battleground for its practitioners. For example, the experience of transition in Russia and Eastern Europe demonstrated the centrality of accounting to the creation of new capital markets and economies, even as Western accounting was known to be riddled with difficulties (Mennicken 2008, 2010). This also suggests that imperfect accounts of worth may be highly functional for the constitution of markets and certainly preferable to nothing (Dambrin and Robson 2011).

Finally, accounting involves the specialist co-optation and economization of ordinary language notions of "appreciation," "depreciation," "recognition," "realization" and "goodwill" to name a few key concepts. Accounting is a continuous project of trying to fix a closed technical meaning for these terms, which always have the potential to overflow and create connections to many other ways of accounting and to other accounts. Accounting has enormous power; indeed the power of the balance sheet is tantamount to an *aestheticization* of economic value, namely a way of providing a persuasive "spectacle of value" which, in the case of financial institutions, as the financial crisis has shown, became also a spectacle of value destruction.[5] Through balance sheet focused accounting, organizations which may be ordinarily valued as non-economic in nature, such as prisons and hospitals, can be made into economic entities whose "asset-performance" can be tested, compared and ranked (Espeland and Sauder 2007; Kurunmäki 1999; Mennicken 2013). Private sector-oriented accounting instruments, tools of financial reporting, performance and cost management, have been introduced from the private sector to the public sector to define and operationalize "value for money," to better understand the link between resources and outputs, and to redefine public sector entities as competitive units responsible for their own performance management. Ironically, the balance sheet can make such newly economized public entities visible as failing and up for closure, while the failure of private entities, such as banks, and their interconnected vulnerabilities, is

[5] We owe this point to Kimberly Chong.

not made visible (Miller 2008). This means that the aesthetics of the balance sheet, its source of power, and its projection of the autonomously performing entity, is also highly selective, particularly in terms of the network properties which it systematically leaves out of account. This is shown by the case of brand accounting, to which we now turn.

New Accounting Objects: Brands in the Late 1980s

> A simple mind could hardly entertain the notion of intangible assets. In a child's tale, wealth is castles, land, flocks, gold—i.e. physical things. It is a long step forward to realise that the essence of wealth is the prospect of benefits, not their physical source. The accountant is thus showing his maturity if he ceases to use the physical test for deciding whether outlays fall under the heading of "asset" or "expense." The correct test, he then argues, is whether or not the outlay improves the firm's prospects (when the outlook at the year's end is compared with that at its start). Tangibility has nothing to do with this test; he applies it alike to the cost of a lathe and an advertising campaign. (Baxter 1984: 218; quoted in Napier and Power 1992: 85)

Brand accounting in the UK in the late 1980s and early 1990s is an illustrative case of accounting instability caused by new objects and a collision between two managerial "orders of worth" (Boltanski and Thévenot 2006)—marketing and accounting (Napier and Power 1992; Power 1992). At the time the accounting authorities sought to prohibit the recognition and valuation of brands as assets in accounting statements. Their general argument was that the valuation method was "unreliable" and too subjective. Furthermore, brand values were not observable in liquid markets; brands could not be sold as separate items distinct from the products to which they were attached. For regulators, accounting assets needed to be *separable* in order to be represented on a balance sheet. There also needed to be a clearly measurable entry cost. Brands did not seem to pass either of these accounting tests.

Many UK companies with economically valuable brands were opposed to this view. How could prominent brands with considerable economic value for companies simply be left out of account? What is the role and value of a balance sheet when it omits such significant assets? A few companies decided to oppose the accounting regulator and, using valuation consultants, they proceeded to value both their acquired and self-developed brands as separable assets on the balance sheet. The external auditors of these companies supported this strategy, and regulators—in particular the British Accounting

Standards Committee (which later became the Accounting Standards Board)—found themselves increasingly isolated.

The brand valuation methods themselves were varied and contested. Some involved the development of analogies with more liquid assets, such as patents. Others involved efforts to capitalize a stream of hypothesized premium profits attached to the brand. Yet below the surface of most of these methods there was usually some kind of implicit discounted cash flow model and a technical challenge to identify the separable cash flows in "joint assets" (Napier and Power 1992). In many ways, the methods were unsurprising, yet two different groups—accounting preparers and regulators—could come to opposed views about their reliability. The temporary alliance between marketing-based knowledge and the pro-brand capitalizers won the day and the effect was to reinforce the solidity and reliability of brand values. In some sense brands existed as objects before they were accounted for. Yet they came to have increased institutional credibility *because* they were used for balance sheet purposes. By being part of accounting statements, brands became part of a new and powerful "order of worth" (Boltanski and Thévenot 2006).[6] This in turn supported the idea of a "market" for brands although they were rarely traded separately from the underlying assets.

The brand accounting case shows how the credibility and reliability of an economic valuation methodology is not an inherent feature of that methodology but depends on networks of social support (Power 1992) and configurations of attachment (Hennion 2007). From this point of view, valuation practice is embedded in an apparatus which involves the constitution of the objects themselves. Seemingly very technical processes of valuation and evaluation (auditing and other tests) co-produce both the account of the object and the object itself. It follows that disputes about techniques are often also disputes about the object being valued. This was evident in the case of brands where the contestation was only partly about the reliability of a valuation technique but was also about the nature of brands themselves. For accounting standard setters at the time brands had a flimsy reality. Yet for marketing specialists and within capital markets more generally they were far from flimsy—indeed they were a source of comparative advantage for the firms which owned them. It was argued that failing to reflect brands as assets on the balance sheet could lead to firms being undervalued and vulnerable to takeover—and this belief motivated their valuation and capitalization. The brands case shows how the plasticity of accounting—its contingent

[6] (Accounting) worth was no longer manifested in terms of (historical) cost. Attention was shifted from (historical) cost accounting and "tangibility" to the accounting for a firm's prospects and the ways in which assets, tangible and intangible ones contribute to the generation of a firm's future cash flows.

combination of stability and instability emerges from both the variable and contested nature of valuation technologies and also the different, and often conflictual, orders of worth and their respective objects of attachment that get registered (and de-registered) through these methodologies over time.[7] This plasticity is even more evident in the related cases of fair value accounting and impairment testing practice.

"Fair" Value and the Financialization of Accounting

It is tempting to imagine that the essence of accounting is measurement. Yet the question of measurement is also a source of instability (Power 2012). Measurement conventions for financial accounting have changed over time (see the brand and value added accounting cases discussed above) and different measurement conventions coexist in the same financial statements. The earliest and simplest measurement method was that of an "entry", or a historical, cost basis. This is not a valuation method as commonly understood but it is a basis for recording a value in the accounts based on the acquisition costs of an asset. In the case of assets other methods were also used, for example to reflect different possible managerial intentions (namely to replace, to use or to sell). History shows that a mixture of measurement methods has been tolerated for many years by accounting practitioners and regulators alike. Indeed, what needs to be highlighted is the eclectic nature of accounting measurement and valuations. Unpacking the eclectic nature of accounting concepts and methodologies contributes not only to a better understanding of how accounting works; "it also contributes more broadly to the sociology of economic valuation, by shedding light on the different forms of calculability engrained in the establishment of market-economic loss and value" (Mennicken and Millo 2013: 1).

"Fair" value measurement in accounting, at least on the surface, disturbed this pluralistic world (Laux and Leuz 2009). Having existed as a category for many years, over the first decade of the twenty-first century, there was an ambition to expand its scale and knowledge base. Defined as "the price that would be received to sell an asset or paid to transfer a liability in an orderly transaction between market participants at the measurement date" (International

[7] Today, brand valuation still represents an important area of valuation. Yet, most accounting standards, including the international accounting standards, do not allow firms to account for brands separately in their balance sheets due to the stringency of their intangible asset recognition tests. Yet, indirectly, brands are often subsumed under the accounting for goodwill, which arises when one company acquires another company. Such externally acquired goodwill is measured in terms of the difference between the price paid for the acquired assets and their fair value at the purchase date.

Accounting Standards Board [IASB] 2011: A491), fair value is based on an *exit* price concept. The notion of fair value moves attention away from an asset's "value in use," i.e., an asset's contribution to the creation of value within the organization, established via managerial judgment, to its "exchange value," i.e., its current market price, established via an actual or imagined market transaction.[8] In other words, the concept of fair value assumes an intention to sell, even where this is not the case empirically. Furthermore, the connotation of *fairness* implies freedom from bias and impartiality. In this respect, fair value is both a valuation method and also an *idea* about what asset prices *should* be. It is for this reason that critics (Bromwich 2007; Ronen 2008) suggest that fair values are often hypothetical and imaginative constructs of value rather than representations of actual market price. However, this fictionality is the very source of the power of fair value. Just as value added accounting in the 1970s was in essence a fiction intended to reconstitute the enterprise as a cooperative endeavor, so does fair value accounting seek to make firms more market-facing, regardless of whether the markets are real or imagined. For its advocates, fair value was an idea motivating the alignment of financial accounting with capital markets, and increasing its relevance for capital market investors (Badertscher et al. 2010; Barth et al. 2001). The advocates were powerful because financial assets, such as derivatives and other financial instruments, rather than traditional assets, such as machinery, increasingly comprised a larger proportion of many balance sheets. In other words, changes and tensions in the accounting model of value—from historical cost to fair value, from organizationally focused use value to market-based exchange value—were implicated in changes in the wider economy (Espeland and Hirsch 1990; Perry and Nölke 2006) and in economic discourse (Burchell et al. 1985; Hopwood 1992; Miller 1991).

The rise of fair value accounting was enabled by an increase in the cultural authority of financial economics (Whitley 1986) and by the problem of accounting for, and representing the risks of *derivatives* (e.g., options, swaps, convertible bonds) (Morley 2011; Power 2010; Young 1996). These new derivative financial instruments were a challenge to existing plural measurement conventions in accounting and were an obvious application for fair value methods. The successful application of fair values in this context gave a small group of standard setters confidence to expand their usage in other areas. From this point of view, fair value accounting was more than a method but also a stake for the professional identity of accounting standard setters and

[8] For the distinction between "value in use" and "value in exchange" see for example Adam Smith (1970), cited in Hutter (2010: 244). This distinction has been re-inscribed into accounting, and although fair value focuses on the "value in exchange" of an asset, "value in use" is still used as an important reference point in accounting valuations, as for instance in the case of impairment testing, which we discuss further below.

a challenge to existing forms of accounting knowledge because of the need for more engagement with corporate finance and financial economics (Power 2010; Young 1996). Today, the "fair" valuation of assets has become a specialized activity, and accountants have no necessary monopoly. New centers of valuation have been invented, such as the International Valuations Standards Committee (IVSC). Valuation has not become more scientific or less judgment laden as a result—rather accountants are no longer necessarily in control of value.[9]

The battleground for the fair value debate was, and remains, the balance sheet, itself a deeply institutionalized and compressed representation of an organization as noted above. The promotion of fair value measurement of assets and liabilities was in effect a challenge to the legal logic of the balance sheet and its grounding in specific jurisdictions. The fair value program represented a de-localization of the balance sheet and greater alignment with global capital markets. It promoted the financialization of accounting practices—the integration of valuation techniques derived from financial economics into accounting measurements and the rise of definitions of value derived from capital markets and capital market-based exchange (Dore 2008; Froud and Williams 2001; Krippner 2005; Perry and Nölke 2006).

Yet, the financialization of accounting through fair value did not take the form of a sweeping colonizing process resulting in a new measurement stability. Indeed it was a source of considerable dissonance within the accounting profession, and beyond.[10] The pressure to expand the application of fair values within accounting was more the vision of a small sub-set of accounting policy makers and academics rather than any failure of the market for accounting information (Bignon et al. 2009; Botzem 2012; Power 2010). Indeed, it can be read as contest *within* the accounting policy world between measurement *idealists* and *pragmatists* (Power 2010; Walton 2004; Whittington 2008). The early success by the idealists involved a new conception of accounting *reliability*. In place of transactional pragmatism grounded in the cash *realizability* of recorded profits, fair value enthusiasts appealed to the reliability of valuation models derived and adapted from financial economics. Yet this assemblage of actors and ideas which gave fair value accounting its original boost

[9] Although the IVSC was originally established independently from the accounting profession, it is worth noting the increasing intertwinement between the IVSC and the accounting profession. In 2012, for example, Sir David Tweedie, former Chairman of the IASB was appointed as Chairman of the IVSC's Board of Trustees.

[10] The uses, usefulness and measurement of fair value were, and still are, hotly debated amongst accounting academics, accounting standard setters and preparers (see e.g. Bromwich 2007; Macve 2010; Penman 2007; Walton 2004; Whittington 2008). Fair value is also not a new concept. As Walton (2007: 5) points out, variants of fair value defined in terms of market value can be found in much of the nineteenth-century company legislation in Europe (Georgiou and Jack 2011). However, the focusing of fair value on exit price is relatively recent.

has changed significantly since the financial crisis (Laux and Leuz 2009). Not only have key figures who promoted fair value in accounting standard setting, like Sir David Tweedie, former Chairman of the IASB, moved on, but the authority of financial markets was temporarily damaged. In 2014, fair value remains strong but less imperialistic than hitherto, and we have seen a return to more valuation pluralism. This is particularly evident in the testing of asset valuations for impairment which we consider next.

Testing Valuations: Asset Impairment[11]

Combining managerial and market-based valuation approaches, impairment tests position accounting, and economic valuation more broadly, at the interface between markets and organizations (Mennicken and Millo 2013; but see also Huikku et al. 2012). Impairment tests are accounting valuations that seek to establish whether or not an asset's recoverable amount has fallen below its carrying amount (i.e. book value) due to obsolescence, damage, or a fall in market value. Impairment tests involve both the application of market-based valuation approaches based on a concept of "value in exchange" (fair value) and also "value in use" calculations based on managerial estimates of the discounted future cash flows that an asset is expected to generate for its owner.

Impairment tests are contrived and structured de-stabilizations of value. Impairment calculations require managers (and accountants) to put previous valuations into question. In so doing, the plasticity of accounting valuations becomes apparent. Paraphrasing Carruthers and Kim (2011: 253), we can say that impairment losses do not simply reflect accounting valuations gone wrong. Rather, impairment tests "reveal how much value is a contested and provisional judgement whose complexity lies buried beneath a surface of numbers and quantification" (2011). Impairment of assets, in the form of the requirement to write down assets when their value to the business is lower than their book value, has been in existence for more than 100 years. Write-down requirements are engrained in the long-standing accounting principles of conservatism and prudence (Dicksee 1892). Yet, it was only in the late 1980s that such write-down requirements started to be redefined and regulated in terms of standardized impairment tests (Mennicken and Millo 2013). In short, an apparatus for impairment testing involving standards and regulation was constructed, but that very construction is founded on, and reveals, the latent dissonance contained in any accounting valuation.

[11] This section is largely based on Mennicken and Millo (2013).

Like the rise of fair value accounting, the newness of impairment testing did not involve the construction of new accounting objects, as in the case of value added or brands. Rather it involved the expansion in scale and scope of practices to assure valuations. One might put it like this in the context of this volume: the growth and elaboration of *tests* of accounting values simultaneously reveals that these values are a matter of contested collective taste or judgment. Following Hennion (2007 and in this volume), who draws our attention to the importance of a continuous elaboration of procedures that put taste to the test, we argue that accounting valuations, like wine tasting and artistic evaluation, are also complex, distributed collective endeavors, shaped not so much by the mechanical following of some algorithm, but by emotion, ritual and the reliance on others—their judgment, experience and preferences (see also Boedker and Chua 2013; Pentland 1993).

The conceptual underpinnings and valuation methodologies underlying standardized impairment tests are not singular and stable. Impairment testing rules embrace different calculative components and valuation methodologies, partly drawing on concepts and techniques derived from financial economics (e.g., the concept of market relevance, fair value and discounted cash flow analyses), partly drawing on concepts and techniques derived from more traditional accounting practice (e.g., the principle of cost-or-less, the notion of cash-generating unit, and firm-based, rather than market-oriented, notions of deprival value).[12] Impairment tests require managers (and accountants) to compare and contrast different valuation methodologies: the market-based fair value of an asset or a group of assets (i.e. the amount at which an asset could be disposed of, less any direct selling costs) and the "value in use" of an asset (i.e. its value—defined in terms of usefulness—for the owning organization, e.g., in terms of future cash flows that can be generated by the asset in question).[13]

Through impairment tests, managers are being made aware of the importance of markets and of the implication of market-based information in organizational impairment valuations. But at the same time, market-based information becomes more managerialized (Mennicken and Millo 2013). Although impairment tests, like fair value accounting, shift attention from a notion of reliability rooted in the objectivity of historical cost to a

[12] Deprival value is the lower of replacement cost and recoverable amount, where recoverable amount represents the higher of an asset's net realizable value and its present value (value in use) (Macve 2010: 113). Macve argues that deprival value "copes much more adequately and simply than FV [fair value, added] does with typical real-world market imperfections" (2010: 114).

[13] See e.g., International Accounting Standard (IAS) 36 "Impairment of Assets," which states that the impairment review should comprise a comparison of the carrying amount (book value) of the fixed asset or goodwill with its recoverable amount, where recoverable amount is defined as the higher of an asset's or cash-generating unit's fair value less costs of disposal and its value in use.

notion of reliability rooted in future-oriented managerial estimations and market-based valuation expertise (Power 2010), it would be wrong to view this process of financialization as a one-way street, in which finance and financial markets capture and colonize accounting valuations and organizations. Both managerial and market-based orders of worth are co-transformed in a process of disciplined dissonance between them.[14]

Furthermore, it is important to note that impairment tests are not easily put into practice (Huikku et al. 2012). Rather than contributing to the "doing of accounting," impairment tests ask managers, accountants and auditors to "undo accounting" and to get *uncomfortable* with the numbers they produce. As Mennicken and Millo (2013: 8) highlight, although presented in the form of a standardized algorithm, impairment tests are "anti-formulaic." The dissonant nature of impairment tests requires that valuation is multiplied and accounting representation destabilized (see also Huikku et al. 2012). In essence, impairment rules carry with them the potential for inducing more reflexivity and therefore novelty about the positioning of organizations vis-à-vis markets.

The Plasticity of Valuation

This chapter has progressed through four "moments" in the history of accounting and value which are clearly "sites of dissonance." We began in the late 1970s with the short-lived appearance of value added statements. Then we considered the brand accounting controversy in the late 1980s. Thirdly we discussed the controversial rise of fair value accounting and how this constituted the financialization of financial accounting. Fourthly, the case of impairment testing reveals the limits of this financialization process and the essentially pluralistic and plastic nature of accounting values. All four cases or accounting "moments" suggest that valuation practices within accounting are far from being the cool application of calculative technique. Such techniques have complex contingent histories of institutionalization, and they depend on a web of tacit non-opposition from a variety of actors whose orders of worth are far from being aligned. Even at the specific practice level we have shown how the production of particular valuations is the temporary outcome of a network of actors, and a form of distributed cognition with plural attachments, which is not necessarily fully coordinated. The limited stability of this network is quick to reveal itself at moments of innovation and related conflict and uncertainty.

[14] We speak of disciplined dissonance here to stress its orchestrated and organized nature.

In addition we have suggested that valuation is also *work* involving the multiplication and transportation of inscriptions (Latour and Woolgar 1979; Robson 1992) as well as acts of judgment and interpretation at crucial junctures. This work more or less hangs together, and more or less produces stability. In times of great instability, such as we saw in the financial crisis of 2008 onwards, the work of accounting valuation intensifies. For example, the valuation of financial assets with questionable liquidity during the height of the financial crisis shows how the extended elaboration of inscriptions is necessary to anchor judgments about value, essential fictions, in a material substrate, not least for the purposes of making such valuations auditable, traceable and comfortable again (Power 2013). Indeed, as valuations come to be recognized as essentially plastic and contestable we observe an intensification of evidentiary processes. Actors involved in valuation work increasingly record and defend what they do. Key value judgments must be articulated and defended. Forums exist, such as risk and audit committees inside organizations, where such judgments are challenged and must be justified. In this respect judgment, calculation and auditability are necessarily intertwined in the production of accounting values.

This plasticity of accounting valuation has broader implications for valuation studies and social studies of finance. Our arguments cut across finance and accounting (Power 2012; Vollmer et al. 2009), and we suggest an under-explored dynamic within broad arguments about financialization. We have argued that the economizing role of accounting is not invariant. Financial accounting as a deeply institutionalized practice within capital markets has undergone profound changes and challenges to its internal logic or orientation, becoming progressively financialized by stages. By prioritizing balance sheet elements over income statements in the early 1970s, the US-American Financial Accounting Standards Board (FASB) created a gateway for measurement practices informed by financial economics. As the latter grew in legitimacy and was validated, financial accounting was effectively "financialized" (Morley 2011), but this is less a story of inevitable progress in valuation methods so much as the temporary triumph of one value center—the capital markets—over another—the judgments of management.

It is in the playing out of value and valuation clashes *within* accounting that a moral economy of valuation in general becomes visible. Where is the ground of value? Whose value counts? Is it the work of multiple actors on the crowded virtual trading floor, whose screen-based technologies fascinate sociologists of finance? Or is it to the managers of firms with intimate knowledge of their products? This is a tension, which has been addressed by many scholars of economic life, but we have suggested that this is also a drama played out within the realm of market-oriented valuation practices themselves, and revealed by our four moments in accounting history.

To conclude: We have argued that an investigation of the multiple forms of calculability engrained in economic valuation and measurement technologies, such as brand accounting and impairment tests, helps to get to grips with the "plasticity of valuation" (Carruthers and Kim 2011) and "the productive frictions" (Stark 2009, 2011) involved as different calculative frames are used, compared, and put in relation to each other. The plasticity of accounting valuation underwrites the power of accounting and should not be read as weakness. Indeed, as sociologists we need to work harder to understand how the systemic reach of accounting as a practice of ordering and valorizing has never been greater at a time when core practices of measurement and valuation are most contested and plural, and ever vulnerable to the dissonant force of innovation.

References

Accounting Standards Steering Committee [ASSC] (1975). *The Corporate Report.* London: ASSC.

Badertscher, B., Burks, J., and Easton, P. (2010). Fair Value Accounting, Other-Than-Temporary-Impairments, and the Financial Crisis. *The Acccounting Review* 87(1): 59–90.

Barth, M. E., Beaver, W., and Landsman, W. (2001). The Relevance of the Value Relevance Literature for Financial Accounting Standard Setting: Another View. *Journal of Accounting and Economics* 31(1–3): 77–104.

Barton, A. D. (2000). Accounting for Public Heritage Facilities—Assets or Liabilities of the Government? *Accounting, Auditing and Accountability Journal* 13(2): 219–236.

Baxter, W. T. (1984). Inflation Accounting. Deddington: Philip Allan.

Bignon, V., Biondi, Y., and Ragot, X. (2009). *An Economic Analysis of Fair Value: Accounting as a Vector of the Crisis:* Prisme, 15, Cournot Centre for Economic Studies. www.centre-cournot.org/?wpfb_dl=15 (accessed February 5, 2014)

Boedker, C. and Chua, W. F. (2013). Accounting as an Affective Technology: A Study of Circulation, Agency and Entrancement. *Accounting, Organizations and Society* 38(4): 245–267.

Boltanski, L. and Thévenot, L. (2006). *On Justification. Economies of Worth.* Princeton and Oxford: Princeton University Press.

Botzem, S. (2012). *The Politics of Accounting Regulation: Organizing Transnational Standard Setting in Financial Reporting.* Cheltenham: Edward Elgar.

Bromwich, M. (2007). Fair Values, Imaginary Prices and Mystical Markets. In P. Walton (ed.), *The Routledge Companion to Fair Value and Financial Reporting.* London: Routledge, 46–68.

Burchell, S., Clubb, C., and Hopwood, A. G. (1985). Accounting in its Social Context: Towards a History of Value Added in the United Kingdom. *Accounting, Organizations and Society* 10(4): 381–413.

Çalışkan, K. and Callon, M. (2009). Economization, part 1: Shifting Attention from the Economy Towards Processes of Economization. *Economy and Society* 38(3): 369–398.
Çalışkan, K. and Callon, M. (2010). Economization, part 2: A Research Programme for the Study of Markets. *Economy and Society* 39(1): 1–32.
Carnegie, G. D. and Wolnizer, P. W. (1996). Enabling Accountability in Museums. *Accounting, Auditing and Accountability Journal* 9(5): 84–99.
Carruthers, B. G. and Espeland, W. N. (1991). Accounting for Rationality: Double-Entry Bookkeeping and the Rhetoric of Economic Rationality. *American Journal of Sociology* 97(1): 31–69.
Carruthers, B. G. and Kim, J.-C. (2011). The Sociology of Finance. *Annual Review of Sociology* 37: 239–259.
Dambrin, C. and Robson, K. (2011). Tracing Performance in the Pharmaceutical Industry: Ambivalence, Opacity and the Performativity of Flawed Measures. *Accounting, Organizations and Society* 36(7): 428–455.
Dicksee, L. R. (1892). *Auditing: A Practical Manual for Auditors*. London: Gee.
Dore, R. (2008). Financialization of the Global Economy. *Industrial and Corporate Change* 17(6): 1097–1112.
Espeland, W. N. and Hirsch, P. (1990). Ownership Changes, Accounting Practices and the Redefinition of the Corporation. *Accounting, Organizations and Society* 15(1–2): 77–96.
Espeland, W. N. and Sauder, M. (2007). Rankings and Reactivity: How Public Measures Recreate Social Worlds. *American Journal of Sociology* 113(1): 1–40.
Fourcade, M. (2011). Cents and Sensibility: Economic Valuation and the Nature of "Nature." *American Journal of Sociology* 116(6): 1721–1777.
Froud, J. and Williams, K. (2001). A New Business Model? The Capital Market and the New Economy. *Economy and Society* 30(4): 467–503.
Georgiou, O. and Jack, L. (2011). In Pursuit of Legitimacy: A History Behind Fair Value Accounting. *The British Accounting Review* 43(4): 311–323.
Hennion, A. (2007). Those Things That Hold Us Together: Taste and Sociology. *Cultural Sociology* 1(1): 97–114.
Hines, R. D. (1988). Financial Accounting: In Communicating Reality, We Construct Reality. *Accounting, Organizations and Society* 13(2): 251–262.
Hopwood, A. G. (1986). Management Accounting and Organizational Action: An Introduction. In M. Bromwich and A. G. Hopwood (eds.), *Research and Current Issues in Management Accounting*. London: Pitman, 9–30.
Hopwood, A. G. (1992). Accounting Calculation and the Shifting Sphere of the Economic. *European Accounting Review* 1(1): 125–143.
Huikku, J., Mouritsen, J., and Silvola, H. (2012). The Production of Numbers about the Future: The Case of Impairment Testing of Goodwill, *IPA Conference Cardiff, July 11–13, 2012*.
Hutter, M. (2010). Wertwechselstrom: Zum Verhältnis zwischen ökonomischem und ästhetischem Wert [Alternating Currents in the Creation of Value: On the Relationship between Economic and Aesthetic Value]. In M. Hutter (ed.),

Wertwechselstrom: Texte zu Kunst und Wissenschaft. Hamburg: Fundus Philo Fine Arts, 241–278.

International Accounting Standards Board (IASB) (2011). *International Financial Reporting Standard 13: Fair Value Measurement*. London: IASB.

Krippner, G. R. (2005). The Financialization of the American Economy. *Socio-Economic Review* 3(2): 173–208.

Kurunmäki, L. (1999). Making an Accounting Entity: The Case of the Hospital in Finnish Health Care Reforms. *European Accounting Review* 8(2): 219–237.

Latour, B. and Woolgar, S. (1979). *Laboratory Life: The Construction of Scientific Facts*. Beverly Hills: Sage.

Laux, C. and Leuz, C. (2009). The Crisis of Fair Value Accounting: Making Sense of the Recent Debate. *Accounting, Organizations and Society* 34(6–7): 826–834.

Macve, R. (2010). The Case for Deprival Value. *Abacus* 46(1): 111–119.

Mennicken, A. (2008). Connecting Worlds: The Translation of International Auditing Standards into Post-Soviet Audit Practice. *Accounting, Organizations and Society* 33(4–5): 384–414.

Mennicken, A. (2010). From Inspection to Auditing: Audit and Markets as Linked Ecologies. *Accounting, Organizations and Society* 35(3): 334–359.

Mennicken, A. (2013). "Too Big to Fail and Too Big to Succeed": Accounting and Privatisation in the Prison Service of England and Wales. *Financial Accountability and Management* 29(2): 206–226.

Mennicken, A. and Millo, Y. (2013). *Testing Values: Financialization, Organization and the Emergence of Impairment Rules*. Working Paper. London School of Economics and Political Science.

Miller, P. (1991). Accounting Innovation Beyond the Enterprise: Problematizing Investment Decisions and Programming Economic Growth in the U.K. in the 1960s. *Accounting, Organizations and Society* 16(8): 733–762.

Miller, P. (1992). Accounting and Objectivity: The Invention of Calculating Selves and Calculable Spaces. *Annals of Scholarship* 9(1–2): 61–86.

Miller, P. (2001). Governing by Numbers: Why Calculative Practices Matter. *Social Research* 68(2): 379–396.

Miller, P. (2008). When Markets and Models Fail: Rethinking Risk, Regulation and the State. *Risk and Regulation. Special Issue on Financial Crisis* December: 6–7.

Miller, P. and O'Leary, T. (1994). The Factory as Laboratory. *Science in Context* 7(3): 469–496.

Miller, P. and Power, M. (2013). Accounting, Organizing, and Economizing: Connecting Accounting Research and Organization Theory. *Academy of Management Annals* 7(1): 557–605.

Morley, J. (2011). *Sequences of Change in Financial Reporting: The Influence of Financial Economics in Financial Reporting Practice*. PhD Thesis, London School of Economics and Political Science, London.

Müller-Jentsch, W. (2008). Industrial Democracy: Historical Development and Current Challenges. *Management Revue* 19(4): 260–273.

Muniesa, F., Millo, Y., and Callon, M. (2007). An Introduction to Market Devices. In M. Callon, Y. Millo, and F. Muniesa (eds.), *Market Devices*. Oxford: Blackwell, 1–12.

Napier, C. and Power, M. (1992). Professional Research, Lobbying and Intangibles: A Review Essay. *Accounting and Business Research* 23(89): 85–95.
Penman, S. (2007). Is Fair Value a Plus or Minus? *Accounting and Business Research* 37(4): 33–44.
Pentland, B. T. (1993). Getting Comfortable with the Numbers: The Micro Production of Macro Order. *Accounting, Organizations and Society* 18(7–8): 605–620.
Perry, J. and Nölke, A. (2006). The Political Economy of International Accounting Standards. *Review of International Political Economy* 13(4): 559–586.
Power, M. (1992). The Politics of Brand Accounting in the United Kingdom. *European Accounting Review* 1(1): 39–68.
Power, M. (1996). Making Things Auditable. *Accounting, Organizations and Society* 21(2–3): 289–315.
Power, M. (1997). *The Audit Society: Rituals of Verification*. Oxford: Oxford University Press.
Power, M. (2010). Fair Value Accounting, Financial Economics and the Transformation of Reliability. *Accounting and Business Research* 40(3): 197–210.
Power, M. (2012). Accounting and Finance. In K. Knorr-Cetina and A. Preda (eds.), *The Oxford Handbook of the Sociology of Finance*. Oxford: Oxford University Press, 293–314.
Power, M. (2013). *Organizations and Audit Trails*. Working Paper: London School of Economics and Political Science.
Robson, K. (1992). Accounting Numbers as "Inscription": Action at a Distance and the Development of Accounting. *Accounting, Organizations and Society* 17(7): 685–708.
Ronen, J. (2008). To Fair Value or Not to Fair Value: A Broader Perspective. *Abacus* 44(2): 181–208.
Smith, A. (1776/1970). *An Inquiry into the Nature and Causes of the Wealth of Nations*. Harmondsworth: Penguin.
Star, S. L. and Griesemer, J. R. (1989). Institutional Ecology, "Translations" and Boundary Objects: Amateurs and Professionals in Berkeley's Museum of Vertebrate Zoology, 1907–39. *Social Studies of Science* 19(3): 387–420.
Stark, D. (2009). *The Sense of Dissonance: Accounts of Worth in Economic Life*. Princeton, NJ: Princeton University Press.
Stark, D. (2011). What's Valuable? In J. Beckert and P. Aspers (eds.), *The Worth of Goods. Valuation and Pricing in the Economy*. Oxford: Oxford University Press, 319–338.
Vatin, F. (2013). Valuation as Evaluating and Valorizing. *Valuation Studies* 1(1): 31–50.
Vollmer, H. (2007). How to Do More With Numbers: Elementary Stakes, Framing, Keying, and the Three-dimensional Character of Numerical Signs. *Accounting, Organizations and Society* 32(6): 577–600.
Vollmer, H., Mennicken, A., and Preda, A. (2009). Tracking the Numbers: Across Accounting and Finance, Organizations and Markets. *Accounting, Organizations and Society* 34(5): 619–637.
Walton, P. (2004). IAS 39: Where Different Accounting Models Collide. *Accounting in Europe* 1(1): 5–16.

Walton, P. (2007). The Nature of Fair Value. In P. Walton (ed.), *The Routledge Companion to Fair Value and Financial Reporting*. London: Routledge, 3–8.
Weber, M. (1922/1978). *Economy and Society*. Berkeley, CA: University of California Press.
Whitley, R. (1986). The Transformation of Business Finance into Financial Economics: The Roles of Academic Expansion and Changes in U.S. Capital Markets. *Accounting, Organizations and Society* 11(2): 171–192.
Whittington, G. (2008). Fair Value and the IASB/FASB Conceptual Framework Project: An Alternative View. *Abacus* 44(2): 139–168.
Young, J. J. (1996). Institutional Thinking: The Case of Financial Instruments. *Accounting, Organizations and Society* 21(5): 487–512.
Young, J. J. (2006). Making Up Users. *Accounting, Organizations and Society* 31(6): 579–600.

12

Clean and Profitable: Entangling Valuations in Environmental Entrepreneurship

Liliana Doganova and Peter Karnøe[1]

Waste and Worth: Valorizing Slurry

What does a farm produce? Certainly, a pig farm for example, produces meat. But in addition to the trucks full of animals heading for the slaughterhouse, a pig farm lets out a number of other entities which economists would call "externalities." Externalities take different forms. For neighbors, they are to be found in the unpleasant smell which dwells in the farm's surroundings. For environmentalists and governments, they lie in the pollution that animal waste and the use of pesticides and fertilizers generate into the air, soil and water. While such undesirable effects have long been accepted as part of the production process that transforms animals into food, they are turning into matters of concern (Latour 2004) which can stir up existing practices of valuation. Consumers are increasingly aware of the invisible costs at which food is made available to them. Governments have implemented more stringent regulation to change farming practices and curb the pollution they cause. In this move towards the "cleaning" of agriculture, entrepreneurs have discerned opportunities for developing new technologies and making profits. This chapter examines such attempts by governments and entrepreneurs at transforming polluting substances and cleaning technologies into *matters of worth*.

[1] This research was supported by the project "The pragmatics of valuation in clean-tech innovation" funded by Grundfos and the Danish Industry Foundation, and by the project Performabusiness funded by the European Research Council (ERC Starting Grant 263529). We are grateful to Zeinab Al-Asfoor and Roman Graber for their help with data collection.

One of the main sources of externalities in animal farming is slurry, that is, the liquid manure formed by animal feces and urine. Slurry holds a central place in the farm. As animals are fed, they produce slurry which goes down the slatted floors on which animals stand in the stables and accumulates in the underneath pits. From there, slurry is pumped into outside storage tanks, covered with a lid, and kept there until the time comes for it to be spread in the fields where it can serve as a natural fertilizer. As slurry moves from the stables through the storage tanks into the fields, it generates two kinds of externalities. One is positive: when farmers spread slurry into the fields, they get fertilizer—and an organic one, for that matter—for free. The other one is negative: the nitrogen that slurry contains tends to spread into the air, soil and water in the form of ammonia emissions and nitrate leaching, causing severe environmental damage.

Thus, slurry has a dual nature: it is both a good and a bad, a nutrient for the crops growing on farms' fields and a pollutant for the surrounding air, water and soil because of the ammonia emissions and nitrate leaching that it provokes. This peculiar characteristic has rendered slurry difficult to regulate, as the boundary between fertilizing and polluting substance is not stabilized and there has been doubt, both in legislation and in farming practices, as to whether slurry should be treated as waste to get rid of or a useful input to production activities (Borraz 2008; Bourblanc and Brives 2009).

But this also makes slurry an interesting object for the sociology of valuation. Following Latour (2004), we can depict slurry as a hybrid object—one that has no clear boundaries, and confounds nature and culture or politics. We will show that valorizing slurry means meddling with its nitrogen content, the fields of the farm, the surrounding air, water and soil, neighbors, animals, environmental regulation, and even carbon markets. We will have to abandon the dichotomy of natural vs. "socially constructed" objects, and study the transformations induced by competing attempts to make slurry valuable. Slurry is also a hybrid object in a second—related, and yet slightly different—sense. Because the nitrogen that it contains is both a pollutant and a fertilizer, slurry lends itself to multiple evaluation principles (Stark 2009), pertaining to the environment and the economy. Its valuation becomes all the more complicated as these principles are often thought of as not merely different, but antagonist. Scoring positive on one side and negative on the other, slurry is both an environmental bad and an economic good; an object, could we say, that carries a double—plus and minus—value charge.

This chapter examines regulatory and entrepreneurial experiments with "cleaning" agriculture by reducing the ammonia emissions resulting from animal slurry in farms. We focus on the case of Denmark—a country with both acute environmental conscience and intensive animal farming. Denmark has, besides cattle and poultry, the highest annual pig production per capita in the world, with almost five pigs per person; about 90% of the production of pigs and pig meat is exported, and constitutes a major source of income for the country. While regulation aiming at reducing nitrogen discharges from agriculture in the aquatic environment had been implemented since the late 1980s, a report published in 1999 by the Ministry of the Environment stated that "ammonia [was] the most important air pollution problem outside cities" in the country, and that 98% of ammonia emissions [stemmed] from agriculture (Jørgensen and Tybirk 1999: 5). This and other publications stimulated intense debates between farmers' spokespersons, politicians, NGOs and journalists on the state of ammonia emissions and the competitiveness of Danish agriculture. Following two European directives (IPPC in 1996 and NEC in 2001) which included ammonia emissions in their scope, new regulations were implemented in Denmark throughout the 2000s.

We analyze two initiatives carried out in Denmark: the implementation of the EU Integrated Pollution Prevention and Control (IPPC) directive for livestock farming, and the commercialization of a new environmental technology by a start-up firm which we will call CleanFarm[2]. What these initiatives have in common is the hope that the development of new technologies will help reconcile environmental and economic value. In the European IPPC directive, the attempt to reconcile environmental and economic value appears through the notion of "best available techniques" (BAT). Instead of setting a target for emissions reduction, the directive stated that "emission limit values [...] should be based on the best available techniques," that is, the techniques that are both "most effective in achieving a high general level of protection of the environment as a whole" and "developed on a scale which allows implementation [...] under economically and technically viable conditions, taking into consideration the costs and advantages" (European Council 1996). This definition pegged environmental to economic value: techniques were to be assessed not only according to how well they protected air, water and land, but also according to how much they cost.[3]

[2] Our analysis draws on two sources of data: interviews and documents. Interviews were conducted with the start-up's managers and investors, as well as various stakeholders (including the Ministry of the Environment and the farmers' professional association). We were given access to a wide range of internal corporate documents relating to the CleanFarm project (including business plans and PowerPoint presentations), and we collected public data relating to the political, legal and technical aspects of ammonia emissions reduction (e.g., legislation, scientific papers, reports). The quotations used in the remainder of the chapter come from these interviews and documents. For confidentiality reasons, the name of the start-up has been replaced by a fictitious name.

[3] This is even more visible in the term "best available techniques not entailing an excessive cost" (BATNEEC), which was used in a previous EU directive on air pollution from 1984, and from which "BAT" was derived.

Similarly, the attempt to reconcile environmental and economic value lies at the heart of the "clean-tech" industry in which start-ups such as CleanFarm position themselves—an industry described as a range of "new technologies and related business models that offer competitive returns for investors and customers while providing solutions to global challenges."[4] In their reconciliation aspirations, clean-tech entrepreneurs reached even further than regulators. Economic value, when blended with environmental value, would not be reduced to a definition based on costs, be they lower or at least "not excessive." A clean technology that would alleviate the negative externalities of agricultural production would also bear the potential to generate economic returns for those who used and developed them as well. In other words, technology could transform slurry from a matter of concern into a matter of worth, and make the farm both clean and profitable.

Regulatory and entrepreneurial attempts at cleaning agriculture are closely related. For regulators, the development of new environmental technologies is a key means towards achieving targets for emissions reduction. For entrepreneurs, the implementation of environmental regulation is a key driver for the sales of their technology. In the case that we examine here, the entrepreneurs have not only sought to exploit the opportunities arising from stricter environmental regulation, but they have contributed to bringing this regulation into existence and giving voice to the issue of ammonia emissions in the public debate. In what follows, we describe the valuation work performed by regulators and entrepreneurs on animal slurry, its ammonia emissions and the technologies developed to "clean" them. We explore how they proceeded in reconciling environmental and economic value, the chains of scientific measurement and economic calculation on which they relied, and the characteristics of the objects upon which they acted.

Framing Valuation

Which Emissions? What Farm?

In 2007, the Danish government passed the Environmental Approval Act for Livestock Holdings which conditioned the building of new (and the expansion of existing) livestock farming installations upon the issue of an environmental permit by local municipalities. To obtain an environmental permit, the farmer has to demonstrate that the new (or expanded) installation will meet certain emission levels. The proof of emission levels does not rely on

[4] This definition is taken from the website of the Cleantech Group (http://www.cleantech.com/about-cleantech-group/what-is-cleantech/, accessed in June 2013).

direct measurement of the emissions produced by a given farm, but on a list of technologies (hereinafter, the "technology list") managed by the Danish Environmental Protection Agency (EPA).[5] In line with the EU IPPC directive and its BAT concept, to get on the list, a technology has to be qualified as the "best available." Since the best available techniques are defined as those that achieve target emission levels "without entailing excessive cost," the EPA engaged in producing a series of "standardized BAT-assessments" to provide measures of the "environmental advantages" and "economic costs" of different technologies.

The production of BAT assessments raised a number of questions about what counts and how it counts. Answering these questions involved a threefold framing (Callon 1998). First, when studying the "environmental advantages" of different technologies, regulators drew a boundary between the emissions to be taken into account and those to be left unconsidered. The EPA followed the BAT Reference Documents (BREF) produced by the EC, which included ammonia emissions, phosphorous emissions, nitrate leaching, odor, noise and dust as relevant effects to be addressed—and excluded other emissions, such as greenhouse gasses. Emissions of odor, noise and dust were then excluded in turn, because they were considered as better dealt with at the local, rather than the national, level.

Left with ammonia emissions, phosphorous emissions and nitrate leaching, the EPA drew a second boundary, which separated the parts of the farm that fell into the scope of the new regulation from those that were to be governed by other means. The livestock farm was split in two parts: the animal houses and storage tanks were taken in charge by the EPA and the 2007 Environmental Approval Act for Livestock Holdings, while the farm's field remained under the preceding Danish legislation. As a result, "the only environmental effects left for quantitative BAT-assessment [were] the ammonia and phosphorous emissions from the farm, defined as the combined housing and storage emissions" (EPA).

Economic Costs

In their investigation of the "economic costs" of the environmental technologies applying for the EPA's list, regulators retained two evaluation criteria. The first one was the ratio between the cost of a technology (i.e., how much it costs the farmer to purchase the technology and operate it) and its environmental benefit. This ratio, explains the methodology, is the "cost of emission reduction." It expresses "the cost effectiveness of the technology for

[5] The Environmental Protection Agency is part of the Ministry of the Environment.

Table 12.1. Example of a best available techniques (BAT) assessment

1.	2. Total additional cost per kilogram of milk produced, excluding the value of saved nitrogen and sulphur		3. Total additional cost per kilogram of nitrogen reduced, excluding the value of saved nitrogen and sulphur	4. Total additional cost per kilogram of milk produced, including the value of saved nitrogen and sulphur		5. Total additional cost per kilogram of nitrogen reduced, including the value of saved nitrogen and sulphur
(animal units)	DKK (Danish krona)	(%)	DKK	DKK	(%)	
75	0.15	7	100	0.13	6	81
...	...	...	...	...	...	...
950	0.04	2	25	0.01	0	7

Adapted from Miljøministeriet, Miljøstyrelsens BAT-blade, 2. Udgave Kvæg—Malkekvæg Revideret: 19.05.2009 [Ministry of Environmental Affairs, The Environmental protection Agency, Best Available Technologies, 2. ed., dairy cattle, revised 19.05.2009], Table 1, p. 5 http://mst.dk/media/mst/Attachments/MaSt01forsuringkvgBATbladudenkorr.pdf

society as a whole" and is measured in Danish krona per kilogram of reduced nitrogen emissions. The second one was the ratio between the cost of a technology and the production of the farm where it is implemented. It represents "increased production costs," expresses "how a certain technology affects the economy of the agricultural sector" and is measured in Danish krona per kilogram of e.g., meat or milk produced. As one of the entrepreneurs we interviewed summed up, the first one is the "environmental cost"—a number about which "the farmer does not care, but the authorities care"—and the second one is "the cost to the farmer."

These calculations were made available in the "technology sheets" that accompanied the technology list and provided "short descriptions of the technologies' environmental and economic performance." Their results were summed up in tables; for example Table 12.1 illustrates the estimation of the "economic impact" of CleanFarm's technology for the smallest and the largest farm sizes.

Take the numbers in the first row of this table. For a cattle farm that has 75 animal units[6] (column 1), implementing this ammonia emissions reduction technology has a "production cost" of 0.15 krona per kilogram of milk

[6] The "animal unit" (AU) is a standard figure constructed out of a complicated set of assumptions and calculations based upon the nitrogen (N) content of slurry for different types of animals. For example, one lightweight milk-cow counts as 0.75 AU, one heavyweight milk-cow counts as 0.88 AU, and one pig for slaughterhouse counts as much as 34 AU. The animal unit establishes a relation between the number of animals, how much nitrogen they produce, and how much nitrate can be spread on the field and absorbed by the soil.

yield (column 2) and an "environmental cost" of 100 krona per kilogram of nitrogen reduced (column 3). These costs are lower if the farm is larger and if the savings on fertilizer enabled by this particular technology are taken into account[7] (columns 4 and 5). In any case "cleaning" a cattle farm has a cost—at best, a zero cost (this is the case for the largest farms and if the savings on fertilizer are taken into account). In other words, the framing of the calculation positions the cleaning technology as a product that has a negative, or at best, no economic value for its user.

Such numbers play a crucial role in this peculiar form of regulation that combines government intervention, through the setting of sustainability-oriented targets, with market mechanisms, through which competing products (here, the different best available techniques) are offered by manufacturers (e.g., CleanFarm) and selected by consumers (i.e., farmers). The technology list and its tables draw the calculative space (Callon and Muniesa 2005) within which different ammonia emissions reducing technologies are evaluated. They are used by farmers and their advisors in order to compare and choose among the best available techniques that regulation compels them to implement. They also instrument regulators' decision-making, as to whether a given technology is part of the "best available" and whether it is "reasonable" and not too "expensive" to enforce regulation. An informal rule, one respondent explained to us, is that regulation should not be enforced if the cost of the environmental technologies that it would compel farmers to purchase goes beyond 1 or 2% of the production cost. How such thresholds were determined is beyond our knowledge; what we would like to emphasize here—and this has been observed for other "golden numbers" in environmental issues (Reich 1983)—is the contrast between the far-reaching effects that the numbers in the technology list induce in markets and politics, and the fragile framing of the calculations from which they result.

This framing is threefold: it draws boundaries between different types of emissions, between different parts of the farm, but also between the domain of the environment and that of the economy. What we can see at play in the tables of the technology list is a process of commensuration through "the transformation of different qualities into a common metric" (Espeland and Stevens 1998: 314), which is Danish krona in this case. Ammonia emissions are given an economic cost: for example, when a cattle farm with 75 animal units implements an environmental technology, society pays 100 krona for a kilogram of nitrogen reduced. (Reduced) nitrogen and (produced) milk are thus made commensurate through the measuring rod of money (Fourcade

[7] CleanFarm's technology uses acidification to reduce ammonia emissions from slurry. A result of acidification is that the treated slurry has a higher nutrient content and, when spread in the farm's field, requires less additional fertilizer—hence the savings.

2011). However, they are still kept separate, as they are qualified by reference to different evaluation principles—the environment and the economy—and different actors—society and the farmer. The coexistence of an "environmental" and a "production" cost in the technology list tables illustrates a particular mode of reconciling environmental and economic value, which proceeds by *juxtaposition*. Here, the environment is not subsumed into the economic, but they are both kept in play—in dissonance (Stark 2009).

Categories

Boundaries between the environment and the economy are also drawn when it comes to categorizing best available techniques. Currently, five techniques have been approved to be the best available in Denmark. Two of them, floor design and air cleaning, focus on the emanations of slurry. Their logic is the following: if the problem is the ammonia that evaporates into the air, it can be solved by changing the design of floors (i.e., shrink the openings in the slatted floors which cover the slurry pits beneath the animals) or by cleaning the air (i.e., install air scrubbers) in the stable. Another technique, feeding, goes upstream the slurry production chain to consider how animals are fed: if the problem is the nitrogen contained in slurry, it can be solved by reducing the amount of nitrogen in the slurry that animals produce (i.e., change the diet of animals). The last two techniques, cooling and acidification, differ from the former in so far as they operate on the slurry itself: if the problem is that ammonia evaporates from the slurry, it can be solved by better fixing nitrogen in the slurry to impede its evaporation; this, in turn, can be done by lowering the temperature in the stable or by adding acid in the slurry (this latter technique, called "acidification," is the one developed by CleanFarm).

There are different ways to distinguish these best available techniques. We have just described them by focusing on how they treat slurry, that is, whether they operate on the emanations of slurry (as floor design and air cleaning do) or on slurry itself (as cooling, acidification and, albeit indirectly, feeding do). However, from the standpoint of regulation, this—material, could we say—criterion is not the most relevant one. There is another kind of categorical division that comes into play here: that between "environmental techniques" and "production techniques." The latter refer to activities like designing the floors of a stable, controlling the temperature and feeding the animals which are part and parcel of the "normal" running of a livestock farm as a production unit. Conversely, installing environmental techniques such as air scrubbers or acidification means adding something new to the "normal" farm with the unique purpose to make it "cleaner."

Historically, it is the Ministry of Agriculture that has dealt with production-related issues, including ammonia emissions from the field in so far as they have been envisaged as a loss of fertilizer (for nitrogen evaporates in the air instead of nurturing the crops). A technique like floor design, for example, does not fall under the scope of the Ministry of the Environment, because, as a respondent put it, "it has nothing to do with the environment, [...] it is something about basically agricultural business." Hence, as explained by one respondent from the Ministry of the Environment, the production of numbers on the environmental impact of floor design:

> has been driven not only by environmental needs, but also because it is important for livestock producers, to help to increase efficiency, that they can follow the input and output. What are the emissions—because what they lose into ammonia emissions, they lose from the nitrogen that goes to the slurry tank and goes to the field. So it is a part, it has been a part historically, of an efficiency and optimization strategy from the farmers. [...] For that reason, it has been in the Ministry of Agriculture for a long time. And since 10 years, we [the Ministry of the Environment] started to focus on more classical environmental technologies, like we had in the industrial area, like air scrubbers and so on.

This categorical division between environmental and production techniques has further implications as to the calculative apparatus through which different technologies are valued. While data on newly developed environmental technologies such as acidification has been produced through experimentation in test farms, the evaluation of production technologies such as floor design relies on standard data. Hence the following complaint of a manager from a company that markets an environmental technology competing with floor design:

> [Floor design] is regulated with some kind of standard figures for production: how much slurry does a cow produce etc. [...]; so, for this floor design, you could expect that you will have this amount of ammonia in the slurry. [...] They have actually issued a new catalogue with a lot of new floor designs. They have not been tested technology- or environmentally-wise, but the Danish environmental ministry is just taking the figures from the agricultural ministry, and the figures have not been made for environmental purposes but have been made more for standard figures to explain how much ammonia could get into the field.

An interviewee from the Ministry of the Environment stresses the differences in the procedures used by the two ministries in gathering data from farmers:

> We have encountered some problems with the procedures, because our demands for documentation and so on are set relatively high, for [the technologies whose efficiency] we are responsible for evaluating—the real environmental technologies. We have found that in some cases the demands that the [Ministry of Agriculture] sets for documentation for some of the efficiency rates of other strategies and basic emission rates from the animals are not always as tight, as

good as we would like them to be. So we are having a dialogue now with [...] the Ministry of Agriculture to kind of harmonize the documentation demand on a higher level.

Our point here is not to say that one method is better than the other, but to stress that categories matter. They are part of what Beunza and Garud (2007: 14) called "calculative frames...internally consistent associations between categorizations, analogies and key metrics". They are one of the intertwined nominal, ordinal and numerical scales of valuation (Guyer 2004). That floor design, for example, is categorized as a production technology has consequences on how it is valued: it becomes part of farmers' efficiency and optimization strategy, falls under the jurisdiction of the Ministry of Agriculture, and is measured through standard figures. Conversely, acidification is categorized as an environmental technology, becomes part of a collective effort to reduce ammonia emissions, falls under the jurisdiction of the Ministry of the Environment, and is measured through experiments. The framing that draws boundaries between the environment and the economy is thus inscribed not only in calculative devices such as the tables of the technology list, in which environmental and production costs are juxtaposed, but also in categories and institutional arrangements that delegate the handling of environmental and economic matters to different technologies and actors.

Multiplying Valuation

"Value Streams, Instead of Problem Streams"

One of the environmental technologies that feature on the technology list is acidification. CleanFarm—the start-up that developed and commercialized it—describes the technology as something like a *"washing machine"* for slurry. Its basic principle is the following: sulfuric acid is injected in the slurry accumulated in animal houses; this decreases its pH level and hence the proportion of ammonia (NH3) relative to ammonium (NH4+). A full scale experiment over a whole year in four farms, conducted in collaboration with universities and the Danish farmers' association, showed that the resulting "acidified slurry" emits less ammonia in the air and has a richer nutrient content. More precisely, the study showed that the acidification technology reduced ammonia emissions from the slurry in animal houses, in storage tanks and in the field by about 70%, while increasing its mineral fertilizer equivalent (i.e., the mineral fertilizer required to achieve the same yield as 100 kg manure) by 40% (Kai et al. 2008). This enabled putting CleanFarm's product on the technology list and assessing its "environmental cost," including and excluding the savings on fertilizer enabled by the richer nutrient content of slurry that has been acidified (see table 12.1).

For the entrepreneurs, tying "slurry washing"—this is how we will refer to CleanFarm's acidification technology hereinafter—to regulation and inscribing it on the technology list meant potentially higher demand for their product. However, that farmers *had* to buy a technology to reduce their ammonia emissions did not mean that they *would* buy CleanFarm's product, nor that this product would have any value to them—other than ensuring compliance with regulation. Even worse, in the calculative space of the technology list, slurry washing was no more than an additional cost—a zero cost, at best. The entrepreneurs thus found themselves in the uncomfortable marketing position of selling a product that had no value for its customers. As one of them complained to us:

> At the moment, the customer applying our system would much rather not buy. They are forced to buy. They feel forced to buy. [...] So we have a hard time explaining [the product's] value to the farmer, because the document [the technology list] clearly says that it is costing them money. [...] We cannot show that our system is actually beneficial to them. At least not in monetary terms.

The entrepreneurs hence engaged in a campaign to demonstrate that slurry washing was more than a necessary nuisance for farmers: as one of them put it, the technology had the potential to transform slurry into "value streams, instead of problem streams." They drew a picture of an envisioned world in which the "clean" farm became a profitable economic entity, industrialized livestock farming went hand in hand with tighter environmental conditions, and Denmark met its ammonia emissions reduction commitments while ensuring the prosperity of its agriculture. They tried to "plant [these] visions and ideas at central places," such as farmers' associations, the Danish parliament, the Ministry of the Environment and the European Commission.

Building on the documents that CleanFarm's managers fabricated and used to present the world they envisaged to various stakeholders, we provide an account of how slurry was made into an opportunity for generating "value streams" and how the "clean" farm was made into a profitable economic entity, at least on paper. What is at stake in these documents is the qualification of the acidification technology as a new product. The idea of qualification (Callon et al. 2002) emphasizes that the qualities of goods are not pre-given, but result from a process that implicates different metrologies as well as actors with sometimes competing strategies and definitions of value. Qualities may include both economic and non-economic parameters such as the price, the "fair trade" label for coffee, the expiration date for milk, or the CO_2-emissions for automobiles. An important part of the valuation work is to enact and stabilize such qualities, and it is here that documents, ranging from PowerPoint presentations and business plans to reports and scientific articles, play a crucial role (Doganova and Eyquem-Renault 2009; Stark and Paravel 2008).

What CleanFarm attempted to demonstrate in these documents is that slurry washing had qualities other than the reduction of ammonia emissions. This alternative calculative framing made slurry washing both environmentally and economically valuable. First, they claimed, slurry washing allows for higher savings from fertilizer and increased crop yield in the farm's field. Second, it helps fight climate change by reducing greenhouse gas emissions from slurry. Third, it is friendlier to the animals in the stable than other best available techniques such as feeding, cooling or floor design. Each of these qualities challenged the boundaries drawn by the technology list, by shedding light on what was left outside the frame: the farm's field, greenhouse gas emissions, animals.

How Can the Economic Be Made Tangible?

The peculiarity of slurry washing is that, unlike floor design or air scrubbers for example, acidification operates on the slurry itself, instead of on its emanations, and hence alters the chemical composition of slurry. Once washed, CleanFarm argued, slurry is no longer the same: it keeps more nitrogen in, which makes it more productive when it is spread in the field. Conversely, regulation would see slurry through the lens of a standardized metric which converts a number of animal units into an amount of slurry produced and hence into a certain level of pollutants emitted or fertilizer put in the fields. In other words, regulated slurry is a "normal" one: while its emanations may vary, its composition is supposed to remain fixed. Thus, a key issue that the entrepreneurs faced in demonstrating the qualities of slurry washing consisted in *not* "making things the same" (MacKenzie 2009), but in making them different, "re-cognizing" (Stark 2009) washed slurry as a new substance whose polluting and fertilizing power was to be reassessed.

Ambiguity remained, though, as to whether acidified slurry is the same or different from "normal" slurry. On the one hand, regulators included the savings on nitrogen enabled by slurry washing in the calculations of the technology list—thereby suggesting that acidified slurry behaves differently in the field and that acidification can be an environmental and production technology as well. On the other hand, the entrepreneurs took advantage of the omnipresence of "normal" slurry in regulation to claim another quality for their product: increased crop yield. Their argument was the following. According to a Danish legislation which attempts to reduce nitrate leaching from farms' fields, farmers can only use an amount of fertilizer than is lower than the "economic optimum." Hence, Danish farmers "under-fertilize" their fields, compared to what would be economically optimal. As a result, Danish farmers get a lower yield of cereal crop (used to feed the animals in the stable) and biomass (that can be sold to power plants). The limit on fertilizer is expressed in terms of the quantity of slurry that can be spread over the field, and here the law does not make any difference between normal and acidified slurry. But if the spread slurry is acidified, it contains more nitrogen; hence the crops could get more fertilizer, and the farmers could get closer to the economic optimum without infringing the law. As one of CleanFarm's managers explained to us, "The extra nitrogen [contained in the acidified slurry] will then not just be a replacement of fertilizer, but it would actually give more crops. [The farmers] will grow more crops. And the value of crops is higher than the value of the fertilizer."

But how can the farmers see that slurry washing "would actually give more crops" if this is not written in the calculations of the technology list? Making the economic value of slurry washing tangible proved difficult for CleanFarm, our respondent continued, for:

> [Increased yield] is not tangible at all, because you cannot measure it. You have to completely believe the surveys that have been done, the scientific documents which are there. Because from year to year, you always have a variation in the yield from the land, depending on the temperature, the rain, the sun, you know... So a farmer cannot really say: 'Oh well, now I am getting 10% more yield because I have acid in my slurry,' he doesn't really know.

Moreover, as long as farmers are not selling their crop but use it to feed their animals,

> they don't see the [benefit]. They see the cost: they have to buy [...] the sulfuric acid, they get the electricity bill, they get the maintenance bill [...]. But they do not get an invoice or a credit note saying 'Ok, I am getting this much money in the field'.

The entrepreneurs failed to demonstrate that slurry washing is not only an "economic cost" that farmers are to bear on behalf of society, as suggested by the calculations of the technology list, but also has economic value for the farmers because it increases crop yield. That slurry washing increases crop yield, and by how much it does so, is not directly observable by the farmers. Nor is it inscribed in scientific articles and regulatory documents, contrary to the "ammonia emissions reduction" quality of slurry washing. Nor is it made valuable through market transactions in which crops would be sold and converted into the currency of krona. Given this situation, the entrepreneurs resorted to yet another combination of scientific measurement and market mechanisms in order to demonstrate a quality of slurry washing that they had not envisaged initially: the reduction of greenhouse gas emissions.

Reconnecting the Farm to Government and Markets

In 2009, the Danish Environmental Protection Agency conducted a life cycle assessment which compared the environmental impact of different slurry management technologies, among which was CleanFarm's slurry washing (Wesnæs et al. 2009). The framing of emissions and of the farm here differed from that of the BAT-assessments. It included gaseous emissions caused not only by the handling of slurry in animal houses and storage tanks, but also by its application to the field. Moreover, it included not only ammonia, but also methane (CH4) and carbon dioxide (CO_2), and assessed how the implementation of the slurry management technologies considered would change the level of these emissions. The study concluded that the acidification of slurry reduces methane emissions, "probably due to that the biological activity is inhibited at the low pH" and thus leads to a reduction of the contribution to "global warming" by 10 to 36% (pp. 30–31).

While the study stressed the high uncertainty of the data and the need for more scientific research, it drew a link between slurry washing and global warming which opened new opportunities for valorization. Not only could the entrepreneurs claim a new environmental quality—"greenhouse gas emissions reduction"—for their product, but they could rely on an existing infrastructure—carbon markets—to transform environmental concerns into economic value. For example, in a PowerPoint presentation addressed to the Minister of food and agriculture after the publication of the life cycle assessment study, the valuation of CleanFarm's product was expressed as including an initial investment in the technology, operational costs (acid, electricity, etc.), and—a crucial new element—an "income from selling CO_2-quotas" on carbon markets, estimated at about €1 per ton of washed slurry. A new image of the farm could thus appear: encompassing animal houses, storage tanks and the field, and plugged into carbon markets, the slurry washing farm became clean and profitable as well.

A clean and profitable farm was an image attractive not only for farmers, but also for government who had launched a new initiative called Green Growth which aimed at "both reducing pollution and increasing growth opportunities" in Danish agriculture. Green Growth also included commitments to reduce ammonia and other polluters, increase biomass and biogas, and reduce CO_2 emissions with 700,000 tons CO_2-equivalents. CleanFarm proposed the following calculation. The 2000 biggest farms in Denmark have a total of 850,000 animal units. Based upon the standardized numbers, they produce 17 million tons of slurry annually, which corresponds to "reduction potential" of 1.14 million tons CO_2-equivalents—which is much more than the Green Growth commitment to 700,000 tons. If the government subsidized the purchase of 2000 acidification technologies, it would find a relatively cheap way to meet its commitments to CO_2 reduction.

Such calculations—translating a number of farms into an amount of slurry, and hence a number of slurry washing products into an amount of CO_2 emissions reduced—could be extended from the national to the European level, as indicated in the following excerpt from a position paper that CleanFarm circulated among EU policy makers:

> The potential greenhouse gas reduction with slurry washing is enormous. If 2,500 large pig farms and 7,500 large dairy productions were to be fitted with [slurry washing], this would save 5.2 million tons of CO_2-equivalent annually. This amounts to 25% of greenhouse emission form slurry production in EU. This is significantly more than alternative EU measures such as the implementation of environmental crops production methods, water and resource management and some soil technologies for carbon sequestration.

By making such demonstrations, the entrepreneurs assembled and made visible a hitherto non-existing entity: a farm, which stretched over stables, storage tanks and fields; washed the slurry that its animals produced; resonated with the concerns of farmers and governments; was connected to regulatory instruments and carbon markets; and thus became simultaneously clean and profitable. The calculations of costs and benefits, which were performed in the tables of the technology list and the slides of CleanFarm's PowerPoint presentations, played a crucial role in the constitution of the farm as a clean and profitable entity. Following Miller and Power (2013), we can describe that role as "territorializing": (re)drawing the boundaries of the farm as a polluting and productive entity, to create a calculative space in which slurry could be transformed from a stream of problems into a stream of value.

The constitutive role of entrepreneurial calculations does not imply that "interpretation" is an "unlimited source of value" (Aspers and Beckert 2011: 26). Valuation is inscribed in material devices and institutional arrangements which bear upon the scope and the mechanics of the qualification of goods. It can neither be detached from the materiality of the objects that are valued. That slurry is a hybrid object, and that slurry washing operates on its very composition by simultaneously altering its polluting and fertilizing properties, helps understand their "engagements in certain regimes of circulation and valuation rather than in others" (Çalişkan and Callon 2009: 389).

Animals Or The Environment?

What does it take to transform slurry from a matter of concern into a matter of worth? No less than altering its composition, setting up experiments and measuring emissions, redrawing the boundaries of the farm to constitute it as a polluting *and* productive entity, connecting it to carbon markets, challenging the framing imposed by regulatory instruments and the division of valuation labor between ministries, getting the government to subsidize slurry washing... And yet, it takes even more than that.

At the time of writing, the technology list still indicated that slurry washing had a cost (equal to zero, at best) and CleanFarm's efforts to make increased yield tangible and monetize the reduction of greenhouse gas emissions had not brought any fruits. Even worse, farmers' associations had challenged the official cost calculations, arguing for example that slurry washing incurred higher operating costs because it consumed more acid than expected. In a period of financial crisis, costs were a sensitive issue and farmers would rather turn to less expensive "low tech" ammonia reduction solutions, such as feeding or cooling. Nevertheless, CleanFarm was selling its product, and an increasing—albeit slightly—number of farms was starting to wash slurry. Why did farmers buy this product? A market study conducted by CleanFarm revealed that slurry washing had another quality that mattered to farmers: it was good for their animals. As a manager from CleanFarm explained to us, other best available techniques such as floor design have a negative impact on animal welfare:

> [With floor design] there is a trade-off between animal welfare and environmental [issues], because solid floors are not so good for the animals, because they are more dirty, there is more slurry on the floors. [...] For animal welfare, all farmers would like to have slatted floors, because a lot of slurry can quickly disappear and the surface that the animals are walking on is as clean as possible. But the environmental legislation forces them to make a choice.

A similar tradeoff appears for the techniques that consist in changing the diet of animals or lowering the temperature in the stable:

> The colder you can make the slurry, the less ammonia is evaporating. But again it is a tradeoff, especially with pigs: if you lower the temperature too much, the animals are sleeping here, they will start getting chilly and they will die [...] So again, it is a balance between animal welfare and environmental protection. This is also something that farmers complain a lot about: 'Ok, you want me to do something for the environment, but you know, what about my animals?'

Underlying these quotations is a definition of the environment that is not confined to the emission or leakage of polluting substances which are to be measured and traced, but embraces the animals that produce these emissions. This changes the issues at stake: the problem no longer lies in reconciling environmental and economic value, but in finding "a balance between animal welfare and environmental protection". The emerging issue of animal welfare thus creates a friction in what is valued—emissions or animals—but also in how it is valued. Regulation- and market-based arrangements to reduce ammonia emissions which rely on quantification and the scientific measurement of environmental benefits and economic costs appear in contrast to situated moments of mundane valuation which are embedded in farming practices (Jasanoff 2010).

Could animal welfare, as a new form of environmental value emerging from the comparison, in practice, between different techniques for reducing ammonia emissions, be reconciled with economic value? For the entrepreneurs, this could be made possible if animal welfare was translated in monetary terms. As one of them explained,

> We have the animal welfare, which eventually also is money, right. [...] So if the animals are more sick they have more diseases, they will produce less milk, they will grow less pork, and [the farmer] has to give them medication more often. He needs the veterinary more often, right. But again, how much is this? We have no numbers. [...] Our original [...] investigations on ammonia emission showed that [...] the pigs in our group, compared to the control group, were growing faster. They had better food conversion and [...] lower mortality than the control group—but not statistically significant.

In process here is a mode of reconciling environmental and economic value which is distinct from the one performed by the regulators' technology list. Its underlying principle is *translation*, rather than juxtaposition. Like carbon markets for greenhouse gas emissions, experimental data, statistics and accounting technologies translate animal welfare into money. When sick animals come to mean less milk and more veterinary bills, that is, less production and more costs, environmental value is subsumed into economic value, and farmers are placed back in the position of economic agents whose valuation repertoire is confined to the comparison of economic benefits and costs.

Conclusion

In public debates on climate change, environmental and economic value are often seen as standing in contradiction: environmental benefits are typically poised against economic costs, as if a choice has to be made between protecting nature and ensuring competitiveness and growth, as if less emissions could only mean more costs, and environmental goods could only be economic "bads." Faced with this dilemma, attempts at reconciling environmental and economic value have emerged. We have examined two such attempts, contrasting the initiatives carried out by policy makers and entrepreneurs in order to reduce ammonia emissions from animal slurry in Danish agriculture. They both rely on a combination of regulation, technological innovation and market mechanisms, but they experiment with different modes of reconciling environmental and economic value.

Regulators' efforts have focused on *framing* value: fixing the qualities of slurry as a standard object, and determining the environmental advantages and economic costs of the "best available techniques" for cleaning slurry. We have shown that, in the categories in which cleaning technologies are positioned and in the calculative device in which their performance is assessed, environmental and economic value are *juxtaposed*: they are both kept in play—in dissonance. Conversely, entrepreneurs have aimed at *multiplying* value, by challenging a framing in which slurry is a standard object, the farm is modelled as a polluting entity, and the environment is confined to ammonia emissions and defined as distinct from the economy. In their demonstrations, environmental value is *translated* into economic value, as entrepreneurs mobilize market and accounting technologies to transform matters of concerns, such as greenhouse gas emissions or animal welfare, into matters of worth.

In relation to the field of valuation studies, our approach and findings make a distinct mark on how the materiality of objects shapes the process of valuation. In the case examined here, slurry behaves in different ways dependent on it being acidified or not, and "re-cognized" as such or not. The valuation claims that different actors make on slurry are attached to its composition and to the empirical assessments made possible by the metrologies in place. Our analysis pertains to a "realist attitude" (Latour 2004) which differs from a strict social construction perspective on valuation in which human "interpretation" is imposed on passive objects and seen as an "unlimited source of value" (Aspers and Beckert 2011: 26). In line with other chapters in this book, our study calls for more research into the different ways in which the materiality of objects plays itself out in valuation processes.

References

Aspers, P. and Beckert, J. (2011). Value in Markets. In J. Beckert and P. Aspers (eds.), *The Worth of Goods: Valuation and Pricing in the Economy*. Oxford: Oxford University Press, 3–38.

Beunza, D. and Garud, R. (2007). Calculators, Lemmings, or Frame-makers? The Intermediary Role of Securities Analysts. In M. Callon, Y. Millo, and F. Muniesa (eds.), *Market Devices*. London: Blackwell, 13–39.

Borraz, O. (2008). *Les Politiques du Risque*. [The Politics of Risk]. Paris: Presses de Sciences Po.

Bourblanc, M. and Brives, H. (2009). La Construction du Caractère « Diffus » des Pollutions Agricoles. [The Construction of the « Diffuse » Nature of Agricultural Pollution]. *Etudes rurales 183*: 161–176.

Çalişkan, K. and Callon, M. (2009). Economization, Part 1: Shifting Attention from the Economy Towards Processes of Economization. *Economy and Society* 38(3): 369–398.

Callon, M. (1998). An Essay on Framing and Overflowing: Economic Externalities Revisited by Sociology. In M. Callon (ed.), *The Laws of the Markets*. Oxford: Blackwell, 244–269.

Callon, M. and Muniesa, F. (2005). Economic Markets as Calculative Collective Devices. *Organization Studies* 26(8): 1229–1250.

Callon, M., Méadel, C., and Rabeharisoa, V. (2002). The Economy of Qualities. *Economy and Society* 31(2): 194–217.

Doganova, L. and Eyquem-Renault, M. (2009). What Do Business Models Do? Innovation Devices in Technology Entrepreneurship. *Research Policy* 38(10): 1559–1570.

Environmental Protection Agency (n.d.). *Assessment of BAT in Livestock Regulation in Denmark—From Adjusted System Analysis to BAT Conclusions in Practice*. http://eng.mst.dk/media/mst/69150/BAT-Systemanalysis%20and%20Economic%20Considerations%20in%20Practice.pdf (accessed March 11, 2014).

Espeland, W. N. and Stevens, M. L. (1998). Commensuration as a Social Process. *Annual Review of Sociology* 24: 313–343.

European Council (1996). Council Directive 96/61/EC of 24 September 1996 Concerning Integrated Pollution Prevention and Control.

Fourcade, M. (2011). Cents and Sensibility: Economic Valuation and the Nature of "Nature." *American Journal of Sociology* 116 (6): 1721–1777.

Guyer, J. (2004). *Marginal Gains. Monetary Transactions in Atlantic Africa*. Chicago, IL: The University of Chicago Press.

Jasanoff, S. (2010). A New Climate for Society. *Theory, Culture & Society* 27(2–3): 233–253.

Jørgensen, V. and Tybirk, K. (1999). Ammoniak i Landbrug og Natur. [Ammonia in Farming and Nature]. *Danmarks Miljøundersøgelser and Danmarks Jordbrugs Forskning, Ministeriet for Fødevarer, Landbrug og Fiskeri*.

Kai, P., Pedersen, P., Jensen, J. E., Hansen, M. N., and Sommer, S. G. (2008). A Whole-farm Assessment of the Efficacy of Slurry Acidification in Reducing Ammonia Emissions. *European Journal of Agronomy* 28(2): 148–154.

Latour, B. (2004). *Politics of Nature: How to Bring the Sciences into Democracy*. Cambridge, MA: Harvard University Press.

MacKenzie, D. (2009). Making Things the Same: Gases, Emission Rights and The Politics of Carbon Markets. *Accounting, Organizations and Society* 34(3–4): 440–455.

Miller, P. and Power, M. (2013). Accounting, Organizing, and Economizing: Connecting Accounting Research and Organization Theory. *The Academy of Management Annals* 7(1): 557–605.

Reich, M. R. (1983). Environmental Politics and Science: The Case of PBB Contamination in Michigan. *American Journal of Public Health* 73(3): 302–313.
Stark, D. (2009). *The Sense of Dissonance: Accounts of Worth in Economic Life*. Princeton, NJ: Princeton University Press.
Stark, D. and Paravel, V. (2008). PowerPoint in Public: Digital Technologies and the New Morphology of Demonstration. *Theory, Culture & Society* 25(5): 30–55.
Wesnæs, M., Wenzel, H., and Petersen, B. M. (2009). *Life Cycle Assessment of Slurry Management Technologies*. Environmental Protection Agency, Danish Ministry of the Environment.

13

Reframing Expertise: The Rise of Behavioral Insights and Interventions in Public Policy

Holger Strassheim, Arlena Jung, and Rebecca-Lea Korinek

Introduction

In November 2013, the 8th Innovative Delivery Award, posted to "promote innovation and creativity across government"[1] went to Her Majesty's Revenue and Customs (HMRC) and the Behavioural Insights Team (BIT) at the Cabinet Office for launching a series of experimental tax trials. These trials involved writing tax reminder letters based on behavioral economics principles, and testing these against randomized control groups. Residents either received a message informing them that the vast majority of people in their area had already paid their tax or a control letter. HMRC and the BIT were able to show that using these and other social norms in tax reminder letters could result in an increase in tax revenue of around 210 million British pounds per year (BIT 2012).

In Whitehall's overcrowded "ecosystem of expertise" (Doubleday and Wilsdon 2013: 10) the BIT, often called the "Nudge Unit," has received remarkable attention. Established in 2010 by the Coalition Government, its objective "is to transform how government thinks about the behavioural aspects of public policy, making it easier for citizens to make better choices for themselves" (BIT 2010b: 4). The BIT tests and applies insights and interventions from behavioral economics and psychology focusing on policy areas such as food safety, energy policy, employment, health, and crime prevention.

[1] http://my.civilservice.gov.uk/reform/civil-service-awards-2013 (accessed December 28, 2013).

It develops more or less nonintrusive instruments of behavioral intervention based on forms, commitments, labels, or visual signals to alter choices and decision-making. The BIT also aims at expanding "public knowledge of policy relevant behavioural effects" (2010b: 4).

Using the well-known term coined by Richard Thaler and Cass Sunstein (2008), the *New York Times* recently baptized the BIT as "Britain's Ministry of Nudges" (Bennhold 2013). While the origins of behavioral economics and its application lie in the United States, the BIT has now become a place of pilgrimage for public administrators and advisors from Australia, Ireland, Denmark, the European Commission and the United States (Lunn 2012; Sunstein 2013).

In many ways, the rise of behavioral insights and interventions as a novel mode of public policy applied by organizations such as the BIT is a puzzling case. The basic principles of behavioral economics such as bounded rationality and biased judgment were already formulated in the 1950s (Simon 1957). Moreover, Tversky and Kahneman's (1982b) seminal studies on judgment under uncertainty identified three central heuristic principles—representativeness, availability, anchoring—that form the basis of behavioral economics until today. According to their research, people trying to predict the future intuitively rely on similarities and stereotypes (representativeness); the probability of risks is assessed by the ease with which instances or occurrences can be brought to mind (availability); in many situations, estimates are biased by an initial value or a starting point that works as "default" (anchoring).

From the beginning behavioral economics has always been systematically related to policy issues, using its key heuristics as a source for creating behavior changing interventions: Behavioral economists have extensively suggested corrective procedures, modes of removing biases and instruments of behavioral intervention to improve collective judgment heuristics and risk perceptions as a core challenge for public management (Kahneman et al. 1982; APSC 2007). When Kahneman received the Nobel Prize in Economics in 2002, behavioral economics and related fields of research were already well established in academia and known to a wider audience (Kahneman and Tversky 2000; Kahneman 2011). Although behavioral economics is continually gaining importance in many OECD countries, it is especially in Great Britain that it is enjoying a prize-winning reputation and a high governmental institutionalization.

This raises two simple questions: Why is it that in both Britain and Germany, among many other OECD countries, the moment of behavioral insights and interventions in public policy seems to have come? More importantly: Why is it that behavioral expertise, its remarkable degree of institutionalization within the British government and its public sector application come to be positioned as prize-winning innovations, while in other countries, such as Germany, it has received considerably less attention and has not become institutionalized?

To explain both aspects of the sketched puzzle we are proposing three main arguments:

First, on a general level, we claim that the relevance of behavioral economics for policy makers can be explained by fundamental characteristics of this knowledge form: Behavioral expertise gains politico-epistemic authority by using easily demonstrable forms of experimental evidence (*object dimension*), by positioning behavioral experts as intermediaries between science, politics and citizens on the one hand while empowering policy makers with the expertise to act as "choice architects" (Thaler and Sunstein 2008: 253) on the other (*social dimension*) and by formalizing experimental policy procedures through guidelines and other regulatory documents (*temporal dimension*). This research field thus combines the persuasiveness of experimental insights with the political practicality of easily understandable heuristics and biases.

Second, to explain, why the moment of behavioral economics has only come recently, we focus on current shifts in reflexive debates on the science-policy-politics interface. We show that behavioral approaches promised to re-define the relationship between science, politics and citizens during a period of increasing public scepticism towards both new public management as well as evidence-based policy (Straßheim and Kettunen 2014). Under the heading of "libertarian paternalism" (Sunstein and Thaler 2003), diverging political positions on individual autonomy are reconciled, while policy makers are given the capacity as choice architects to create situations where citizens make better choices to their own benefit.

Third, to explain differences in the reputation and institutionalization of behavioral economics in Great Britain and Germany, we argue that domestic "cultures of expertise" and trajectories of governance narratives in both countries have to be taken into account. By invoking the concept of "cultures of expertise" we refer to the standards of legitimacy and accountability that profoundly shape the relationship between epistemic and political authority, i.e. the more or less unquestioned ways in which knowledge claims are validated and understood as politically relevant at the science-policy-politics-nexus, (Jasanoff 2005; Jung et al. 2014). As will be shown there is a considerable difference between what counts as politically and socially legitimate evidence in Britain and in Germany.

This chapter presents selected findings from an ongoing research project comparing the changes in expertise and evidence across three different countries (Great Britain, Germany and the US) and two policy fields (labor market policy and food safety).[2] It is structured as follows: In section 2 we briefly illustrate our research puzzle, i.e. differences in the institutionalization of

[2] "Studying the Changing Orders of Political Expertise" (SCOPE), carried out at the WZB Social Science Center in cooperation with the Humboldt-University and the University of Darmstadt, funded by the Volkswagen Foundation.

behavioral economics as a politically relevant source of expertise in Britain and Germany. In section 3 we focus on the behavioral discourse in general and how it generates politico-epistemic authority at the science-policy-public nexus. To provide an adequate understanding of the exceptional influence of behavioral economics in Britain we analyze the timing and specific situation in which the behavioral discourse gained attention (section 4). In section 5 we show how cultural configurations of expertise and governance shape the varying valuation and institutionalization of behavioral economics in both countries. We conclude in section 6 by summarizing our arguments; we also discuss how the analysis of cultures of expertise might contribute to recent endeavors of developing a "comparative sociology of valuation and evaluation" considering "how evaluative practices are moulded by radically different political institutions and cultures" (Lamont 2012: 15; see also Lamont and Thévenot 2000).

States of Mind? The Institutionalization of Behavioral Expertise in Britain and Germany

The institutionalization of behavioral science within the British government began more than a decade ago. Initially, it was part of a general agenda of developing empirically informed theories of human behavioral change, particularly psychological approaches that go beyond the theoretical construct of the homo economicus. With New Labour's second term in office following the 2001 election, psychologically trained civil servants began entering the government. From the very first days of its establishment in 2002, the Cabinet Office Strategy Unit (COSU) within the New Labour administration had sought to identify the departments' existing capabilities in behavioral expertise. The COSU has also published the first comprehensive overview of behavioral theories, titled the "Personal Responsibility and Changing Behaviour" report, highlighting the utility of behavioral science for developing policies (Jones et al. 2013: 28–33). Applying behavioral approaches to policy development, this report argued, would help to establish the appropriate division of responsibility between the individual, the community and the state. It would empower citizens to make decisions that are in their own interest, while improving the cost-effectiveness of government interventions (COSU 2004). As we shall see later in the paper, this discursive framing is constitutive for the rise of behavioral economics in British public policy.

In addition to issuing such a sensitizing overview report, the Strategy Unit also started to work on a more 'joined-up' approach to behavioral science across Whitehall. To this end, it worked together with individual departments and individual government officials who were known for being behaviorally inclined. The Department for Environment, Food and Rural Affairs, for instance, had already set up the cross-Whitehall Centre of Expertise on Influencing Behaviours. Its main task was to identify possible applications for behavioral economics and behavioral changes across government departments (Jones et al. 2013: 32–33). A further milestone in the rise of the behavior change agenda in British government was the issuing of "MINDSPACE: Influencing Behaviour through Public Policy" (Cabinet Office and Institute for Government 2010).[3] The report differed from previously published documents in that it set out standards and guidelines for the systematic public sector application of behavioral approaches instead of simply summarizing the literature.

It was, however, not until after the election of the Conservative-Liberal Coalition Government in 2010, that behavioral economics became an official government priority subjecting the entire British public sector to the sustained application of behavior change principles. In July 2010, Prime Minister Cameron set up the BIT or "Nudge Unit," as it is often called, as part of the Cabinet Office. It comprises experts with backgrounds in behavioral sciences, experimental methodology and psychology. The BIT's prime task is to push for the public sector wide application of behavioral science in public policy. To this end, the BIT works together with almost every government department, local authorities, NGOs and even with foreign governments. It conducts research in diverse areas such as smoking, organ donation and food hygiene (BIT 2010a), reduction of energy consumption (BIT 2011), consumer empowerment (BIT and BIS 2011) and charitable payroll giving (BIT 2013). Government agencies from various policy fields have implemented behavior change measures, e.g., Britain's Food Standards Agency (FSA) with its National Food Hygiene Rating Scheme (FHRS) providing hygiene information about food businesses by voluntarily displayed stickers in the windows of restaurants and bars (SSRC 2012). The BIT doubled in size within the first four years of its existence, employing 13 behavior change experts in 2014. The success of the BIT can also be seen in the fact that the unit has a waiting list of government departments eager to work with it. In addition, in 2012, the government announced that behavioral science is to be included in the required curriculum for civil servants (Thaler 2012: 4). Every British civil servant is now being trained in behavioral expertise.[4]

[3] The mnemonic "MINDSPACE" refers to central influences changing behavior (messenger, incentive, norms, defaults, salience, priming, affect, commitments, ego).

[4] This is not to say that the behavior change agenda was undisputed: See the 2011 report by the House of Lords' Science and Technology Select Committee (House of Lords 2011: 69; for a critical discussion see SSRC 2012).

In sum, behavioral insights and interventions in Britain have risen sharply, from a small initiative to a highly praised policy instrument. For countries such as Australia, Ireland, Denmark, New Zealand or even the United States, Britain has become a role model, leading to an ever-growing literature on the possible applications of behavior changes for the public and private sector (for an overview see Lunn 2012).

In Germany, in contrast, the career of behavioral expertise has taken a completely different course. Behavioral expertise has not become institutionalized within government. Its proponents are actors outside government: academic policy experts, interest groups and, most recently, think tanks and foundations. In 2008, the authors of the first German report on the subject, prepared on behalf of the Federation of the German Consumer Organizations (vzbv), concluded that behavioral economics was nearly invisible in both consumer policy as well as consumer research (Oehler and Reisch 2008: 71). In a further study, again carried out on behalf of the vzbv three years later, the opportunities and limits of behavioral economics in consumer policy were again considered. The authors came to the conclusion that a "realistic" and evidence-based understanding of consumer behavior was still "overdue" (Hagen et al. 2011: 10). The call for a "realistic idea of consumer behavior, its complexity and diversity" has also been picked up by the Scientific Advisory Board on Consumer and Food Policies advising the Federal Ministry of Consumer Protection, Food and Agriculture (Strünck et al. 2012: 14). In a similar vein, policy experts have asked for a more "behavior sensitive" employment policy in Germany (Schmid 2011). In short, the German debate on behavioral economics has started with actors outside the government advocating it as a "scientifically based" alternative to dominating ideal models (*leitbilder*) of the rational consumer and citizen (Hagen et al. 2013). Only recently the German chancellery has begun to build up its own working group on behavioral insights.

Of course, there may be good reasons to be sceptical about the implications and effects of "nudging" on social welfare and consumer protection and to resist the "siren call" (Rose 1991) of behavioral expertise in politics. Some authors strictly advise against "nudge" tactics as being incompatible with political ideals such as empowerment, freedom and fairness (Saint-Paul 2011; Goodwin 2012). Some argue convincingly that the rise of behavioral approaches marks a shift "at the nexus of power and knowledge" in a Foucauldian sense, leading to a general spread of "behavioural governance" (Jones et al. 2013: 185). The aim of this chapter, however, is somewhat different. Bracketing out these normative issues, we address the fundamental characteristics of behavioral economics that explain its epistemic-political authority and we discuss the socially and historically contingent mechanisms of valuation that account for the different degrees of prestige and influence in British and German politics.

Gaining Politico-Epistemic Authority: The Discourse on Behavioral Expertise

To understand how behavioral expertise in general gains epistemic authority and political relevance, we distinguish between three different dimensions. Its persuasive power is based on (1) the demonstration of relevance for politically salient issues using easily understandable experimental evidence, that appeals to common sense reason, while at the same time being linked to scientific norms and standards (*object dimension*), (2) the positioning of behavioral experts as intermediaries between science, politics and citizens on the one hand and the empowerment of policy makers with the expertise to act themselves as "choice architects" on the other (*social dimension*) and (3) the fixation of experimental policy procedures in guidelines and other regulatory documents (*temporal dimension*). In the following we briefly describe these three characteristics (see also Lunn 2012; Jones et al. 2013):

(1) On the object dimension the epistemic authority of behavioral economics can be attributed to the demonstration of immediate practical relevance to politically salient issues such as efficiency and applicability. For instance, pension schemes such as the "Save More Tomorrow" plan developed by Thaler and Benartzi (Thaler and Sunstein 2008: 125) show how the simple application of behavioral principles can lead to a massive increase in participation rates, nudging people to save higher rates of their income for retirement. The promise to deliver effective and efficient solutions to contemporary political problems is, moreover, based on evidence forms that appeal to common sense and can be understood without specialized knowledge. Cognitive rules such as "loss aversion" or "overconfidence" seem to be taken directly from everyday experience (Tversky and Kahneman 1982a). Studies based on control group trials show that the prediction of the future is largely based on convenient, emotionally agreeable perceptions of the present: "As they stare into the crystal ball, politicians, futurologists and laypersons alike seek an image of the future that best represents their model of the dynamics of the present" (1982: 98).

While appealing to common sense, the epistemic authority of behavioral economics is, however, also safely nested in the academic recognition of peers. According to behavioral economists, for many decades orthodox economics has been a deductive endeavor, while behavioral economics rest on an inductive and experimental approach to produce easily demonstrable evidence and facts that are understood by every lay person. The heuristics and biases clouding the assessment of probabilities in everyday behavior are identified and demonstrated in numerous psychological experiments (Kahneman et al. 1982; Kahneman 2011). A constitutive aspect of the politico-epistemic authority that is attributed to behavioral expertise is the dual approach of referring to the scientific "gold standard" of experiments while at the same time presenting evidence that is easily graspable. In short, the "persuasive power" (Lunn 2012: 432) of behavior insights on the object dimension can be attributed to the way scientific norms and standards are linked with politically applicable "rules of thumb" (Thaler and Sunstein 2008).

(2) Essential to understanding the persuasive power of behavioral economics on the social dimension is that both judgmental and decision-making competence is attributed to behavioral experts and economists. In Thaler and Sunstein's perspective policy makers in government agencies need to be aware of the "misconception [...] that it is possible to avoid influencing people's choices" (Thaler and Sunstein 2008: 10–11). Since they are changing the 'choice architecture' with every decision, policy makers are supposed to rely on behavioral experts—both for the development of effective instruments of behavioral intervention *and* the use of randomized control trials (RCTs) to test and improve these instruments: "Nudgers will be able to make good guesses when they have much more expertise at their disposal" (2008: 247–248). Policy makers and behavioral experts cooperate to become what Thaler and Sunstein call "choice architects": "Choice architects can preserve freedom of choice while also nudging people in directions that will improve their lives" (2008: 253). The collaboration with behavioral economists is portrayed as enabling policy makers to become behavioral experts themselves, combining both the political competence to decide upon new instruments and the interpretive competence to judge whether these instruments are effective.

This collaborative image of behavioral experts and policy makers working together is embedded in the institutionalization of behavioral policy networks and behavior change forums connecting researchers, policy makers and governmental organizations (APSC 2007). It is, however, also cultivated by narratives such as the story of Cass Sunstein, one of the co-authors of the influential book "Nudge" describing his own career as a behavioral expert. As a young lawyer working in the Justice administration, Sunstein was also an advisor to the White House Office of Information and Regulatory Affairs (OIRA). After having moved from OIRA into academia, Sunstein dreamed of returning to what he calls the "cockpit of the regulatory state" (Sunstein 2013: 1). Thus, behavioral experts are framed as both reputable scientists and political realists, socialized at the science-policy interface.

(3) In the temporal dimension, behavioral experts have developed schemes for structuring the practices and sequences of decision-making. These include both guidance on how to ensure that "empirically informed regulation" gains real-time access to the rulemaking docket and checklists on how to use techniques to "quantify anticipated present and future benefits as accurately as possible" (Sunstein 2011: 1388). By defining the procedures and the temporal sorting of "behavioural policies" (Shafir 2013), the incorporation of behavioral expertise becomes a routine and a necessity in public policy making. Furthermore, the whole process of policy making is modelled following experimental designs. Agencies are advised to analyze in advance the effects of alternative policy designs and to adapt policy instruments in accordance with the results of experimental or quasi-experimental studies (Sunstein 2011: 1371). In a recent report prepared for the European Commission, the authors argue that the incorporation of behavioral sciences into the policy making process changes the very stages and time horizons of agenda setting, decision-making, implementing and monitoring (Bavel et al. 2013). It is argued that "good policy-making" is based on the right timing of identifying and studying behavioral elements (2013: 3).

With the amalgamation of experimental evidence, professional expertise and procedural guidelines at the science-policy nexus, behavioral economics has gained authority across a rising number of different countries and sectors. The next two sections, however, show that the actual valuation of behavioral insights and interventions as public policy innovation is dependent on moments and cultural contexts in which "opportunities for innovations" arise (Stark 2011: 334).

An Idea Whose Time Has Come?

It was John Kingdon who observed that change sometimes resembles an "irresistible movement that sweeps over our politics and our society, pushing aside everything that stands in its path" (Kingdon 1984: 1). In such moments, politicians as well as experts are absorbed by ideas whose time seems to have come. More often than not, pressing policy problems and public criticism are seen as "opportunity windows" (1984: 1) to implement new measures and promising instruments. Why, however, have behavior insights and interventions become so irresistible in Britain at this particular point in time? Behavioral economics has, as described above, received attention and stimulated policy debates long before the election of the

Conservative Party in 2010 and even before Cameron's encounters with the authors of *Nudge* (Jones et al. 2013: 24–33). Already in 2004, a discussion paper written by the Strategy Unit embraced "new insights on behaviour change and personal responsibilities" (COSU 2004: 3) that were regarded as particularly useful in the political strategy of "gradually establishing new social norms" (2004: 17). This strategy paper was written in a situation when the unintended side effects of radical public sector privatization and the paradoxes of public sector modernization became visible and officials were seeking new ways to gain legitimacy (Margetts et al. 2010). Transparency, notably through performance indicators, became a new catchword and a doctrine of good governance. At the same time, however, this commitment to transparency seemed to trigger an administrative reflex that has been described as "blame avoidance" (Hood 2007). To escape public scrutiny, officeholders developed all kinds of strategies—from the dispersion of responsibilities to the shaping of public impressions through "spinning" (Hood 2010). Criticism of these "blame games" (Hood 2007) was accompanied by a growing dissatisfaction with evidence-based policy strategies introduced by New Labour in its famous "White Paper on Modernizing Government" in 1999. The evidenced-based policy making movement was not only portrayed as "the latest form of inertia and control" (Healy 2002). Opponents claimed that it was also apt to having de-democratizing effects, becoming "a means for policy elites [to] increase their strategic role over what constitutes knowledge about social problems in a way that devalues tacit forms of knowledge, practice based wisdom, professional judgement, and the voices of the ordinary citizens" (Marston and Watts 2003: 158). The public criticism culminated under the label of 'policy-based evidence making' (Sanderson 2006; Straßheim and Kettunen 2014).

In this context it was the promise of behavioral approaches—as a renewed form of evidence-based policy making—to re-define the relationship between the state, science and the estranged citizens without taking back the modernization agenda of the earlier public policy reforms that was so persuasive. Referring to the behavioral agenda critics of evidence-based policy making and public sector modernization could be countered effectively as the following statement by the Prime Minister's Strategy Unit illustrates: "Policy tailored around a more realistic understanding of how people really do make choices and engage in society—twinned with sustained dialogue over the implications for the citizen's and state's responsibilities—lead not only to more effective policy. It should also enable citizens to feel more in control of their own lives"(COSU 2004: 68).

A key aspect of the resonance to the behavioral approaches in the British public sector was its tight embedment in the political narrative of libertarian paternalism. Like Blair's "Third Way" rhetoric, libertarian paternalism presents itself as bridging the two apparently contradicting ideas of "libertarianism" i.e. embracing freedom of choice and "paternalism" or moving people in "welfare-promoting" directions (Sunstein and Thaler 2003: 1159). In a much discussed paper from 2003 "Libertarian Paternalism is not an Oxymoron" Thaler and Sunstein argued "for self-conscious efforts, by private and public institutions, to steer people's choices in directions that will improve the chooser's own welfare [...]" (2003: 1162). Emphasizing the essential role of scientific authority, they framed behavioral experts as a means of helping the "Nudgers to make good guesses about what is best for the Nudgees" (Thaler and Sunstein 2008: 247).

To summarize, in a moment in which the unintended side effects of the previous public sector modernization agenda became visible, the narrative of libertarian paternalism successfully re-framed the public criticism of New Public Management and evidence-based policy making in favor of a new behavioral agenda. Embedded within the libertarian paternalism paradigm, behavioral approaches promised a renewed state-science-citizen relationship in which scientific experts and political actors would become collaborators, enabling citizens to make welfare-promoting choices, thereby improving public policies in a cost-effective way. It is exactly this combination of discursive authority and perfect timing that made behavioral economics so irresistible for British politics.

Valuation in Cultures of Expertise

Despite similar developments in public sector modernization, behavioral approaches are by far not as prominent in German politics. This is all the more puzzling since Britain served as a reference model during previous public sector reforms, notably for "activating" reforms in employment policy (Fleckenstein 2008). In the past decade, in both Britain and in Germany the science-policy-public nexus in consumer and in employment policy has undergone fundamental transformations that show remarkable similarities. The New Deal in Britain and the "Hartz" reforms in Germany, for instance, resulted in a fundamental transformation of employment policy, putting an emphasis on improving employability with instruments that are to be scientifically evaluated in terms of their effectiveness and efficiency. In both countries, the Bovine spongiform encephalopathy (BSE) crises lead to major organizational and regulatory reforms, granting food safety agencies more independence in assessing and publically communicating food risks (Korinek 2014). Yet, German politics has been largely immune to the promise of empowerment as choice architects.

Differences in the epistemic-political authority of this research field can in part be explained by different cultures of expertise. In each society, processes of valuation of expertise are structured by the more or less unquestioned ways knowledge claims are validated and understood as politically relevant at the science-policy-politics-nexus (for a conceptual outline see Jasanoff 2005, 2011; Jung et al. 2014; Strassheim 2010). Cultures of expertise entail standards of objectivity and validity as well as legitimacy and accountability that profoundly shape the relationship between epistemic and political authority. They are habitual and largely unquestioned modes of separating the publicly accepted procedures of policy-relevant knowledge production from unacceptable and illegitimate ones. As boundary-crossing arrangements they coordinate the production and circulation of politically relevant knowledge, regulate its legitimacy and validity, organize the credibility of experts and establish epistemic conventions and hierarchies of different forms of knowledge. They also determine the extent to which a material or symbolic artefact is identified as something new and relevant—an improvement, worth preserving, valuable and thus innovative. Cultures of expertise at the science-policy nexus mirror a country's wider politico-administrative traditions (Jasanoff 2005; Korinek 2014). Britain's "communitarian civic epistemology"—a specific culture of expertise linked to Anglo-Saxon "public interest and management" culture—is highly different from Germany's "consensus-seeking" epistemology with its more corporatist policy traditions and Weberian "Rechtsstaat" culture (Jasanoff 2005: 249–271).

Experimental results that appeal to common sense and that are linked with simple rules of thumb applying to the solution of contemporary politically salient issues are particularly valuable in the British culture of expertise. The British empiricist tradition (Shapin and Schaffer 1985) is characterized by its "commitment to empirical proofs, which, in principle, should be evident to everyone" (Jasanoff 2013: 141). Thus, there is a tight fit between this constitutive aspect of the British culture of expertise and the behavioral easily graspable demonstration of both scientific soundness and political relevance. This amalgamation of concepts of objectivity, however, stands in stark contrast to Germany's culture of expertise. In Germany, objectivity, i.e., the "appearance of a view from nowhere," as Jasanoff (2005: 267) puts it, "is achieved by resolutely embracing views from everywhere." Yet, in contrast to Britain where this view from everywhere is dressed in the "common sense vision" (Jasanoff 2011: 314), Germany's "view from everywhere" comprises all institutional perspectives that are deemed relevant for the issue at stake.

The behavioral discourse is additionally well compatible with another aspect of Britain's culture of expertise: It emphasizes the double socialization of behavioral experts as choice architects trained by rigorous methodical curricula, while at the same time educated as political realists with hands-on knowledge of both the functioning of the machinery of government and the salience of contemporary complex policy problems. In Britain, the attribution of expert competence is tied to an individual "who achieves standing not only through knowledge and competence, but through a demonstrated record of service to society" (Jasanoff 2005: 268). This pattern of attribution of expertise is closely linked to Anglo-Saxon public interest culture where the government or government departments (rather than the state) draw their authority from obtaining "the public's consent (or, at least, acquiescence) for measures devised in the public (general, national) interest" (Pollitt and Bouckaert 2004: 53). In Germany's "organic" state tradition, by contrast, state and civil society are part of an essentially organic whole—a citizen is not "an atomistic individual but rather a member of an essentially organic society that exists in more or less formalized relationship with the state" (Loughlin and Peters 1997: 48). In this context expert credibility draws on institutional affiliation as sufficient testimony to the quality of the ensuing judgments (Jasanoff 2005).

Moreover, the depth and breadth of institutionalization of behavioral expertise in British government—of which the recent incorporation of behavioral training into the curriculum for British civil servants is a primary example—is incompatible with the role of law in the German tradition of recruitment for public administrations. In contrast to Britain's generalist public servants whose profession is mostly about policy and pragmatic management, Germany's civil service is characterized by the prevalence of the so-called "Juristenmonopol." German civil servants, predominantly trained by lawyers, are not only public employees but also personifications of the legal state. Training them as nudgers would certainly collide with this still dominant tradition. In the UK, however, "the professional civil service provided politicians with both fearless advice, thanks to a security of tenure comparable to university professors" (Pollitt 2004: 103). This long-lasting influence of British administrative tradition is echoed in the evidence-based policy movement that accompanied New Labour's reform activities since the late 1990s but has never been a priority for the German civil service (Strassheim and Kettunen 2014).

Finally, it seems to be fitting that the behavioral agenda's procedural guidelines are valued within the British culture. In the British "stateless society" (Stillman 1991) the state has no legal basis and government is regarded as a "necessary evil, whose powers are to be no more than are absolutely necessary, and whose ministers and officials must constantly be held to public account" (Pollitt and Bouckaert 2004: 53). Accordingly, since all governmental authoritative claims require 'public witnessing' in Britain, British official science must allow the public to witness its scientific validity and objectivity. On the level of government agencies, this demand of public witnessing is increasingly met by publishing a dense web of science governance guidelines and meta guidelines (Korinek 2014). Again, there is an almost perfect fit between the behavioral economists' attempt to set standards and formalize them in guidelines visible to everybody and the general accountability relationships at the British science-policy-public nexus. In the German "Rechtsstaat" in contrast with its strong and all-encompassing body of public laws governing every administrative sphere, the state is a legal state vested with exceptional authority but constrained by its own laws. The Weberian accountability of public servants is ensured through constitutionally and legally mandated rules under which they carry out their functions. Accordingly, once the rules for a German advisory body have been formally set in place and once it has been constituted, science advice is carried out in a relatively "invisible sphere of expert decision-making" (Jasanoff 2005: 261) whose authority is not exercised by means of public witnessing as in Britain.

Summing up, these constitutive aspects of cultures of expertise can help to understand why the same characteristics of behavioral approaches explain their epistemic-political authority in England, while in fact hindering an equivalent valuation in Germany. It remains to be seen whether current efforts by the German government to incorporate behavioral insights will be successful.

Conclusions

To achieve a better understanding of what made behavioral insights and interventions appear to be so irresistible and innovative especially in the case of British civil service, we put forward three major arguments:

First, the increasing relevance of behavioral economics for policy makers across many countries is the result of a valuative discourse attributing both, epistemic as well as political authority in three dimensions of meaning: It combines persuasive power and political practicality by using easily demonstrable forms of experimental evidence (object dimension), positioning behavioral experts as intermediaries between science, politics and citizens (social dimension) and by formalizing experimental policy procedures through guidelines and other regulatory documents (temporal dimension). Thus, essential to the valuation of scientific expertise at the science-policy-politics interface is its ability to generate epistemic-political authority.

Second, the success of behavioral approaches especially in Britain is in part also the result of valuations in a specific moment: Behavioral approaches enter the political arena in a time of increasing public scepticism towards both, the new public management culture and the evidence-based policy movement. Under the heading of libertarian paternalism the behavioral agenda is presented as a solution to policy failures, promising to re-new the relationship between science, politics and citizens. In alliance with behavioral experts policy makers are empowered as "choice architects" supposed to preserve freedom while nudging people in directions that enhance their welfare. For a political sociology of valuation processes at the science-policy nexus it is vital to understand how particular discourses on policy failures and solutions devalorize established forms of political expertise, while other forms of expertise are valorized by construing them as the solution to these problems (Korinek 2013, 2014; Jung 2014).

Third, we argue that the differing reputation of behavioral economics in Great Britain and Germany is related to domestic cultures of expertise. The valuation of expertise is structured by these more or less unquestioned ways of regulating how knowledge claims are validated and perceived as politically relevant at the science-policy-politics-nexus. Cultures of expertise mirror a country's wider politico-administrative traditions. By defining the extent to which a material or symbolic artefact is identified as something new and relevant, valuable and worth being preserved, they also influence the emergence of political innovations (Strassheim 2010). However, future research should pay additional attention to policy-specific factors shaping cultures of expertise (Hoppe 2010).

In light of recent efforts to establish a "comparative sociology of valuation and evaluation" (Lamont 2012) some further aspects of our study might be worth mentioning: Jasanoff (2005; 2011) has pioneered the development of a comparative approach to understanding expertise. Attention to and validation of politically relevant knowledge claims, as she argues, depends heavily on the "carefully orchestrated performances" (Jasanoff 2011: 311) that are embedded in cultural assumptions on legitimacy, accountability, objectivity and reputation. Some have criticized this concept as overly fixated on stable patterns and national constellations, unable to account for the political tensions and conflicting rationalities that interpenetrate and fragment the national context on different levels (see Ezrahi et al. 2008).

This argument, however, seems to be based on a misrepresentation of the concept of civic epistemologies and related approaches. Conflicts between different worldviews are by no means excluded, in fact "these cultural formations shape the very concepts and discursive terms in which citizens are authorized to frame their critique" (Jasanoff 2008: 194). As "political imaginaries" (Ezrahi 2012) they provide a common frame of reference for societal dissent to be communicable. Boltanski has recently made this performative and pragmatic understanding of order very clear (Boltanski 2011). The power of discursive and institutional orders does not lie in the homogenization but in the provision of common options for collective criticism. They are sustained not despite, but because of the multiple registers of critical verification and validation made available. However, Boltanski also argues that in certain contexts some more radical formats of these tests might be turned reflexively against the already existing registers of validation.

The discourse on behavioral insights and interventions can be understood as such a reflexive format. Its transformative power is based on both, the visibilization and the invisibilization of political valuations: On the one hand, Rawls' publicity principle—a rule that "bans government from selecting a policy that it would not be able or willing to defend publicly to its own citizens"—is proposed as a core value of behavioral policies (Thaler and Sunstein 2008: 244). Radical visibility of political methods and motives is seen as a major mechanism to constrain nudgers and to avoid manipulations. On the other hand, critics of the behavior agenda have argued that it invisibilizes the value decisions that necessarily have to be made when translating behavior insight into interventions. For example, according to proponents of asymmetric paternalism, policies should be designed "that help the least sophisticated people in society while imposing the smallest possible cost on the most sophisticated" (2008: 249). However, such a design implies multiple assessments, including decisions about the right and wrong objectives of help, the differential between the "most" and the "least" sophisticated people and the relationship between costs and benefits of interventions. In the case of the tax trials mentioned at the beginning of this chapter this may mean that the least sophisticated people were the ones who responded to the letter and paid their taxes. For this reason, some maintain that "nudge works best in the dark" (Hansen and Jespersen 2013: 9). It is exactly this discursive invisibilization of values and its politico-epistemic repercussions that might be of interest for a future comparative sociology of expertise.

References

APSC (2007). *Changing Behaviour. A Public Policy Perspective.* Canberra: Australian Public Service Commission.

Bavel, R. van, Herrmann, B., Esposito, G., and Proestakis, A. (2013). *Applying Behavioural Sciences to EU Policy-making (JRC Scientific and policy reports).* Luxembourg: European Commission.

Bennhold, K. (2013). Britain's Ministry of Nudges. *The New York Times.* December 7, 2013.

BIT (2010a). *Applying Behavioural Insight to Health.* London: Cabinet Office Behavioural Insights Team.

BIT (2010b). *Behavioural Insights Team. Annual update 2010-11.* London: Cabinet Office Behavioural Insights Team.

BIT (2011). *Behaviour Change and Energy Use.* London: Cabinet Office Behavioural Insights Team.

BIT (2012). *Applying Behavioural Insights to Reduce Fraud, Error and Debt.* London: Cabinet Office Behavioural Insights Team.

BIT (2013). *Applying Behavioural Insights to Charitable Giving.* London: Cabinet Office Behavioural Insights Team.

BIT and BIS (2011). *Better Choices: Better Deals. Consumers Powering Growth.* London: Cabinet Office Behavioural Insights Team/Department for Business, Innovation & Skills.

Boltanski, L. (2011). *On Critique: A Sociology of Emancipation.* Cambridge/Malden, MA: Polity Press.

Cabinet Office and Institute for Government (2010). *MINDSPACE: Influencing Behaviour through Public Policy.* London: Cabinet Office/Institute for Government.

COSU (2004). Personal Responsibility and Changing Behaviour: the State of Knowledge and its Implications for Public Policy (David Halpern and Clive Bates, Geoff Mulgan and Stephen Aldridge with Greg Beales and Adam Heathfield). London: Cabinet Office Strategy Unit.

Doubleday, R. and Wilsdon, J. (eds.) (2013). *Future Directions for Scientific Advice in Whitehall.* Cambridge/Sussex: University of Cambridge/University of Sussex et al.

Ezrahi, Y. (2012). *Imagined Democracies: Necessary Political Fictions.* Cambridge: Cambridge University Press.

Ezrahi, Y., Grove-Wright, R., Robins, R., and Jasanoff, S. (2008). Review Symposium—Controlling Biotechnology: Science, Democracy and "Civic Epistemology." *Metascience* 17: 177–198.

Fleckenstein, T. (2008). Restructuring Welfare for the Unemployed: The Hartz Legislation in Germany. *Journal of European Social Policy* 18(2): 177–188.

Goodwin, T. (2012). Why We Should Reject "Nudge." *Politics* 32(2): 85–92.

Hagen, K., Micklitz, H.-W., Oehler, A., Reisch, L. A., and Strünck, C. (2011). *Mehr empirische Evidenz, mehr Realitätssinn—Vorschlag für einen "Check Verbraucherpolitik und Verbraucherbeteiligung."* [More Evidence, more Sense of Reality—Proposal for an "Assessment of Consumer Policy and Consumer Involvement"]. Berlin: Verbraucherzentrale Bundesverband e.V.—vzbv.

Hagen, K., Micklitz, H.-W., Oehler, A., Reisch, L.A. and Strünck, C. (2013). "Check Verbraucherpolitik und Verbraucherbeteiligung"—Empfehlungen für eine evidenzbasierte Verbraucherpolitik. ["Assessment of Consumer Policy and Consumer Involvement"—Recommendations for an Evidence-based Consumer Policy]. *Journal für Verbraucherschutz und Lebensmittelsicherheit* 8(1–2): 61–66.

Hansen, P. G. and Jespersen, A. M. (2013). Nudge and the Manipulation of Choice. A Framework for the Responsible Use of the Nudge Approach to Behaviour Change in Public Policy. *European Journal of Risk Regulation* 1: 3–28.

Healy, A. (2002). Commentary: Evidence-based Policy—The Latest Form of Inertia and Control? *Planning Theory & Practice* 3(1): 97–98.

Hood, C. (2007). What Happens When Transparency Meets Blame Avoidance. *Public Management Review* 9(2): 191–210.

Hood, C. (2010). *Blame Game: Spin, Bureaucracy, and Self-Preservation in Government.* Princeton, NJ: Princeton University Press.

Hoppe, R. (2010). *The Governance of Problems: Puzzling, Powering and Participation.* Bristol: The Policy Press.

House of Lords (2011). *Behaviour Change (HL Paper 179)*. London: House of Lords (Science and Technology Select Committee).

Jasanoff, S. (2005). *Designs on Nature: Science and Democracy in Europe and the United States*. Princeton, NJ: Princeton University Press.

Jasanoff, S. (2008). Author's response. In Y. Ezrahi, R. Grove-Wright, R. Robins, and S. Jasanoff, *Review Symposium—Controlling Biotechnology: Science, Democracy and "Civic Epistemology" Metascience* 17(2): 191–198.

Jasanoff, S. (2011). The Practices of Objectivity in Regulatory Science. In C. Camid, N. Gross and M. Lamont (eds.), *Social Knowledge in the Making*. Chicago, IL/ London: University of Chicago Press, 307–338.

Jasanoff, S. (2013). Epistemic Subsidiarity—Coexistence, Cosmopolitanism, Constitutionalism. *European Journal of Risk Regulation* 2: 133–141.

Jones, R., Pykett, J., and Whitehead, M. (2013). *Changing Behaviours. On the Rise of the Psychological State*. Cheltenham, UK/Northampton, MA: Edward Elgar.

Jung, A. (forthcoming, 2014). Autonomieansprüche und Leistungserwartungen—(Dis)Kontinuitäten in der wissenschaftlichen Politikberatung der deutschen Arbeitsmarktpolitik. [Claims of Autonomy and Performance Expectations—(Dis) Continuities in Scientific Policy Advice in German Labor Market Policy]. In M. Franzen, A. Jung, D. Kaldewey, and J. Korte (eds.), *Zeitschrift für Theoretische Soziologie, Sonderheft "Autonomie revisited—Beiträge zu einem umstrittenen Grundbegriff für die Beobachtung von Wissenschaft, Kunst und Politik"*.

Jung, A., Korinek, R.-L., and Strassheim, H. (2014). Embedded Expertise: The Contextualization of Knowledge Orders. *Innovation: The European Journal of Social Sciences*. DOI 10.1080/13511610.2014.892425

Kahneman, D. (2011). *Thinking, Fast and Slow*. London: Penguin Books.

Kahneman, D. and Tversky, A. (2000). *Choices, Values, and Frames*. Cambridge: Cambridge University Press.

Kahneman, D., Slovic, P., and Tversky, A. (eds.) (1982). *Judgment under Uncertainty: Heuristics and Biases*. Cambridge: Cambridge University Press.

Kingdon, J. W. (1984). *Agendas, Alternatives, and Public Policies*. Boston, MA: Little Brown and Company.

Korinek, R.-L. (2013). The Re-framing of Food Policy: Studying the Implications for the Construction of Expert Authority of Arm's Length Agencies. Paper presented at the 8th International Interpretive Policy Analysis Conference (IPA) "Societies in Conflict: Experts, Publics and Democracy," University of Vienna, July 3–5, 2013.

Korinek, R.-L. (2014). Comparing the Construction of Expert Reputation of Food Safety Agencies in Britain and in Germany. Paper presented at the conference "What Makes Governmental Agencies Tick?", IPSA Research Committee "Structure and Organization of Government," The Hebrew University of Jerusalem, January 5–7, 2014.

Lamont, M. (2012). Toward a Comparative Sociology of Valuation and Evaluation. *Annual Review of Sociology* 38: 201–221.

Lamont, M. and Thévenot, L. (2000). *Rethinking Comparative Cultural Sociology: Repertoires of Evaluation in France and the United States*. Cambridge: Cambridge University Press.

Loughlin, J. and Peters, B. G. (1997). State Traditions, Administrative Reform and Regionalization. In M. Keating and J. Loughlin (eds.), *The Political Economy of Regionalism*. London: Frank Cass & Co, 41–62.
Lunn, P. L. (2012). Behavioural Economics and Policymaking: Learning from the Early Adopters. *The Economics and Social Review* 43(3): 423–449.
Margetts, H. and Hood, C. (2010). *Paradoxes of Modernization. Unintended Consequences of Public Policy Reform*. Oxford: Oxford University Press.
Marston, G. and Watts, R. (2003). Tampering with the Evidence: A Critical Appraisal of Evidence-Based Policy-Making. *The Drawing Board: An Australian Review of Public Affairs* 3(3): 143–163.
Oehler, A. and Reisch, L. A. (2008). Behavioral Economics—eine neue Grundlage für Verbraucherpolitik? Eine Studie im Auftrag des Verbraucherzentrale Bundesverbandes. [Behavioral Economics—a new Foundation for Consumer Policy? A Study on Behalf of the Federal Consumer Association]. Berlin: Verbraucherzentrale Bundesverband e.V.—vzbv.
Pollitt, C. (2004). United Kingdom. In C. Pollitt, C. Talbot, J. Caulfield, and J. Smullen (eds.), *Agencies. How Governments do Things Through Semi-Autonomous Organizations*. Basingstoke/New York, NY: Palgrave Macmillan, 97–114.
Pollitt, C. and Bouckaert, G. (2004). *Public Management Reform. A Comparative Analysis*. Second Edition. Oxford: Oxford University Press.
Rose, R. (1991). What is Lesson-Drawing? *Journal of Public Policy* 11(1): 3–30.
Saint-Paul, G. (2011). *The Tyranny of Utility. Behavioral Science and the Rise of Paternalism*. Princeton, NJ: Princeton University Press.
Sanderson, I. (2006). Complexity, "Practical Rationality" and Evidence-Based Policy Making. *Policy & Politics* 34(1): 115–132.
Schmid, G. (2011). Stärkung der Autonomie durch verhaltenssensible Arbeitsmarktpolitik. [Strengthening of Autonomy through behavior-sensitive Labor Market Policy]. *WSI-Mitteilungen* 4: 188–194.
Shafir, E. (2013). *The Behavioral Foundations of Public Policy*. Princeton, NJ: Princeton University Press.
Shapin, S. and Schaffer, S. (1985). *Leviathan and the Air-Pump. Hobbes, Boyle, and the Experimental Life*. Princeton, NJ: Princeton University Press.
Simon, H. A. (1957). *Models of Man: Social and Rational—Mathematical Essays on Rational Human Behavior in a Social Setting*. New York, NY: Wiley.
SSRC (2012). *Introducing Behavioural Economics and Potential Applications for the FSA (SSRC 12/2/2)*. London: Food Standards Agency (Social Science Research Committee).
Stark, D. (2011). What's Valuable? In P. Aspers and J. Beckert (eds.), *The Worth of Goods. Valuation and Pricing in the Economy*. Oxford: Oxford University Press, 319–338.
Stillman, R. J. (1991). *A Preface to Public Administration: A Search for Themes and Direction*. New York: St. Martin's.
Strassheim, H. (2010). Innovation and Valuation: The Governance of Public Epistemologies. Paper presented at the First Berlin Forum on Innovation in Governance, Technical University Berlin, May 20–21, 2010.

Strassheim, H. and Kettunen, P. (2014). When Does Evidence-based Policy Turn into Policy-based Evidence? Configurations, Contexts and Mechanisms. *Evidence & Policy: A Journal of Research, Debate, and Practice* 10(2): 259–277.

Strünck, C., Arens-Azevêdo, U., Brönneke, T., Hagen, K., Jaquemoth, M., Kenning, P., Liedtke, C., Oehler, A., Schrader, U., and Tamm, M. (2012). *The Maturity of Consumers: A Myth? Towards Realistic Consumer Policy. (Statement by the Scientific Advisory Board on Consumer and Food Policies at the Federal Ministry of Consumer Protection, Food and Agriculture).* Berlin: Bundesministerium für Ernährung, Landwirtschaft und Verbraucherschutz.

Sunstein, C. R. (2011). Empirically Informed Regulation. *The University of Chicago Law Review* 78(4): 1349–1429.

Sunstein, C. R. (2013). *Simpler. The Future of Government.* New York, NY: Simon & Schuster.

Sunstein, C. R. and Thaler, R. H. (2003). Libertarian Paternalism is not an Oxymoron. *The University of Chicago Law Review* 70(4): 1159–1202.

Thaler, R. H. (2012). Watching Behavior Before Writing the Rules. *The New York Times*. July 8, 2012.

Thaler, R. H. and Sunstein, C. R. (2008). *Nudge. Improving Decisions about Health, Wealth and Happiness*. London: Penguin Books.

Tversky, A. and Kahneman, D. (1974/1982a). Judgments of and by Representativeness. In D. Kahneman, P. Slovic, and A. Tversky (eds.), *Judgment under Uncertainty: Heuristics and Biases*. Cambridge: Cambridge University Press, 84–100.

Tversky, A. and Kahneman, D. (1974/1982b). Judgment under Uncertainty: Heuristics and Biases. In D. Kahneman, P. Slovic, and A. Tversky (eds.), *Judgment under Uncertainty: Heuristics and Biases*. Cambridge: Cambridge University Press, 3–20.

Part IV
Valuation Within Organizations

14

Epistemic Dissonance: Reconfiguring Valuation in Architectural Practice

Ignacio Farías

Introduction

Consider the following situation:

> It's 6pm. José has been working all day on the possible layout of a social housing project trying to fit as many housing units as possible in a slightly triangular slot, while following the design principles they attain at Elemental, which do not just involve leaving space for the houses to grow, but also arranging groups of families around common yards. José has just finished a layout he considers to be the best possible solution, even though the slot's triangular corner remains a bit underutilized. Tomorrow morning José is joining one of the partners in a meeting with a municipality's chief of urban development who commissioned the project. I ask what else he needs to prepare. He says he will probably stay late night trying out other possible layouts. He mentions possibilities that go against Elemental's design principles and others where he already sees some potential problems. I am puzzled and ask why on earth he would stay late trying out things that he already knows are not good. His answer is that you never know. I don't say anything, but lucubrate that presenting bad options to the client might be a strategy to avoid the client questioning the suggested best option. During the meeting, I realize José was right, you simply never know. It turned out that the client had actually quite different concerns, so that instead of discussing the different layouts he explains current municipal plans for building a consistorial house. In this new scenario, the layouts came to be seen and assessed from a completely new angle, leading the architects to suggest building the consistorial house in the otherwise underutilized triangular slot corner. Such a possibility would change the slot's capacity for a social housing project and accordingly what counts as a good and a bad layout.

Architectural design is a process deeply imbued in valuation practices involving projects' conditions and constraints, as well as projects' architectural forms. The excerpt above provides a good example of the omnipresence of valuation practices: José spent a good part of the day figuring out the best possible solution for a social housing project. During the meeting with the client, new conditions were revealed, forcing the architects to re-evaluate the commission's conditions and constraints and suggest a solution they couldn't have considered before. Indeed, not only the divergent evaluative criteria of architects, specialists (from structural engineers to light designers) and certainly clients, but also the shifting nature of a project's conditions and constraints lead to a constant redefinition of what are considered good and bad architectural answers. This constant flow through different valuation moments, this need to constantly reassess what the situation actually is and how designed architectural forms would simultaneously respond to its different elements, is the everyday condition of architectural work.

In this context, it becomes particularly fascinating to look in more detail at moments and situations like the one José engaged in the night before the meeting with the client, and in which valuation is put on hold, resisted as it were, in order to enable the exploration of options that otherwise might seem irrelevant if not simply unsatisfactory and unworthy of pursuing. One could speak here of anti-valuation moments in larger transvaluation processes, that is, situations, in which people restrain from assessing or giving value to certain entities or courses of action, so that first alternatives and eventually new values can emerge. Perhaps the most classic example of such anti-valuation moments in organizations are brainstorming sessions as proposed by Osborn in 1957: "don't criticize, quantity is wanted, combine and improve suggested ideas, and say all ideas that come to mind, no matter how wild" (cited in Sutton and Hargadon 1996: 685). It is important to emphasize that these moments do not involve definitive rejections or renunciations to valuation, but rather temporary configurations, in which it is deemed convenient to distrust current values, abstain from valuation and experiment in the hope that a proximate future will bring new values. As such, anti-valuation moments are constituents of dynamic valuation processes. By creating breaks, introducing hiatuses, making interjections, opening interstices through which the new and unexpected slips in, they are capable of reconfiguring valuation processes. In architectural practices, these moments can thus transform the valuation processes through which design decisions are made and unmade.

Decisions and Dissonance: From Evaluative to Epistemic Uncertainties

The key role of anti-valuation moments in architecture becomes particularly evident, when we consider that architectural operations fundamentally involve decision-making. Schumacher (2011) suggests understanding design as a practice entangled in a recursive chain of decisions. From this perspective, visual architectural representations, such as sketches, plans, diagrams, renderings, cardboard and polystyrene models, or 3D animations, appear as incomplete "epistemic objects" not just requesting knowledge and interpretations from involved actors (Ewenstein and Whyte 2009), but also, and crucially, calling out for architectural decision-making and unmaking. Visual representations allow going back to decisions that have already been made and trigger new decisions, rendering the design process recursive. This is crucial, as it allows architects to engage creatively with the indeterminacy produced by the existence of various design alternatives of uncertain value. Schumacher's description thus suggests that architectural practice is fundamentally imbued in a process of constant valuation of multiple alternatives, which go through multiple valuation trials before and after becoming a design proposal.

Yet such decision-oriented accounts of architectural design take as a given an important condition for decision-making, that is, the actual existence of design alternatives. As is well known from Derrida's discussions on decision-making and responsibility, for a decision to be considered the responsible action of an actor, it needs to be made under specific conditions of "undecidability:" A decision that did not go through the ordeal of the undecidable would not be a free decision, it would only be the programmable application or unfolding of a calculable process (Derrida 1992: 24). Accordingly, in order for any decision to be made, it is imperative that there are alternative courses of action among which it is not possible to choose following heuristic formulas, abstract models or imperative principles. Only then it is possible and necessary to actually decide. In the case of architectural practice, alternatives are however not given by default, but need to be brought into existence by means of sociomaterial practices. In every phase of the design process, alternatives need to be seen, articulated and visualized in order for architects and other involved actors to e/valuate them and eventually make a decision.

The key question is then how design alternatives come about in the first place. The hypothesis I'd like to explore in this chapter is that they often emerge or appear in anti-valuation moments characterized by what I shall call epistemic dissonance. Different from José's night shift explorations and from rule-based brainstorming sessions, these moments occur due to the variety of epistemic positions vis-à-vis ongoing projects held by collaborating architects. This epistemic diversity produces unplanned, surprising or even disrupting moments of epistemic dissonance; moments in which how architects see a project and what they know about it is unexpectedly challenged and transformed, thus opening interstices for hitherto unknown alternatives.

In order to specify the nature and reach of epistemic dissonance, I shall turn first to Stark's (2009) work on the sense of dissonance in organizations. His various case studies demonstrate that uncertainty and undecidability regarding what is good and valuable are not given by default, but an organizational achievement resulting from specific collaborative dynamics and sociomaterial configurations. Hereby Stark reverses the traditional understanding of dissonance as an unpleasant, negative emotional state, or even a dysfunction, that arises rather unexpectedly when people have cognitions that are mutually irreconcilable (Cooper 2007; Festinger 1957). His fundamental contribution demonstrates that highly innovative organizations do not try to avoid, but actually cultivate and foster a shared sense of dissonance and uncertainty regarding what is considered valuable among its members. In an ethnographic study of a web design firm, Stark and Girard (2002) show that collaborating teams apply at least four different evaluation criteria when assessing websites. Web designers, marketing experts, computer programmers and content managers all have different ideas about what makes a "good" website, so that friction and dissonance are constant. Such evaluative dissonance, Stark (2009) argues, is however crucial for making the company more sensitive and open to new ideas and innovation.

Stark's understanding of evaluative dissonance can be very helpful for understanding how uncertainty regarding design decisions comes into play in architecture, as architects need to take account of the diverse evaluation criteria of clients and of a large number of specialists involved in an architectural project (see Cuff 1992). However, this does not yet provide a comprehensive explanation of how dissonance and design alternatives are cultivated within architectural offices. While clients and specialists shape the design process, bringing in their own evaluative criteria, they do so only in the context of meetings often held in separate rooms or even outside the office with partners or the project leader(s). Accordingly, they are not involved in the day-to-day spaces and dynamics in which architectural design processes are carried out. Indeed, most architectural practices, and certainly the ones I had the opportunity to observe, have mainly mono-disciplinary staffs of architects, so that evaluative dissonance emanating from specialists and clients do not shape everyday collaboration. Moreover, architects specifically emphasize the importance of developing and sharing common assessment criteria, as this would allow them to "play by heart," as one partner repeatedly emphasized in our conversations. Sunley and colleagues (2008) reported similar findings for the design sector, where senior designers expressed their preference for hiring younger designers who could be moulded, as well as their concerns about hiring more experienced workers who might hold differing design ideas—"You know we've developed our own way of working and our own view about what good design is" (partner cited by Sunley et al. 2008: 690).

In view of this situation, I shall argue that day-to-day collaboration in architectural offices is characterized by a different type of dissonance. Instead of the evaluative dissonance that emerges from the different professional cultures of interdisciplinary teams, I encountered a kind of epistemic dissonance, which capitalizes on the different perspectives and knowledge level architects have regarding a project in order to generate new questions, ideas, options and alternatives. Epistemic dissonance does not rely on differences regarding what is considered good or valuable in a project, but on differences with respect to what is known about a project. And what one knows depends, in turn, on how one sees a project, how closely or not one is working on it, and which visual representations are being produced and worked with. More specifically, I propose distinguishing between evaluative and epistemic dissonance as shaped by normative and cognitive expectations respectively (on this distinction see Luhmann 1995). Evaluative dissonance arises from disappointed normative expectations regarding what good design is for a particular actor. When a design proposal does not meet the actor's normative expectations, he or she does not abandon or revise his or her expectations. Quite the contrary, his or her expectations are maintained and become the source of dissonance. Epistemic dissonances, by contrast, are related to learning processes. An actor holds expectations regarding what an object actually is, how it behaves or which effects it produces, which when suddenly disappointed are called into question, leading to a thorough revision of the knowledge the actor held up to then.

The Practical Organization of Epistemic Dissonance

Ethnographic fieldwork carried out in three architectural practices in Santiago de Chile provided a range of examples of how epistemic dissonance is practised, organized and exploited on a day-to-day basis. This research took place between early September and mid-November 2009 in Klotz & Asociados, Sabbagh Arquitectos and Elemental[1]. All three are among today's

[1] Two research assistants carried out participant observation three days a week in the first two architectural offices. I did fieldwork at Elemental on a half-day basis and occasionally visited the

most successful Chilean practices. They all operate internationally, their work is regularly published in major international journals of architecture and their founding partners have received numerous honors, including the Marcus Corporation Foundation Prize, the Premio Borromini and prizes awarded at the Venice Biennale and the World Architecture Festival.

The architectural practice Klotz & Asociados is a home studio comprising around 70 square meters, where up to 12 architects and interns work together. In one of our first conversations, Mathias Klotz remarked that his studio was like an 18th century atelier. What he meant was that the craftsmanship, detail and care that characterize his working practice distinguish his studio from the more industrial approaches of other architects. I've come to think of his role as that of the classical master craftsman, supervising, controlling and deciding on every detail of the work, while at the same time training his assistants and interns in his understanding of what good architecture is.

Elemental is based in an iconic 1980s office building that also houses three other distinguished architectural studios. The premises consist of a large workspace with room for around 20 architects, separate offices for two partners, a model and storage room, and a conference room. Interestingly, the choice of location was not primarily based on the proximity to other architectural practices, but rather on the quest for a business-oriented context. Elemental was founded in 2003 as a state-funded innovation platform based at a prestigious university. When the start-up funding ran out, it became a for profit company with social interest, whose shareholders are the Universidad Católica de Chile, COPEC (Chilean oil company) and the partners. Elemental views itself not only as an architectural practice, but also as an innovation platform bringing forward architectural solutions to housing policy and urban planning challenges. It has defined itself for several years as a "do-tank."

Sabbagh Arquitectos is a two-storey architectural practice that employs about 40 people. The ground-floor area with two conference rooms hosts mainly meetings with clients and specialists. The production area in the basement includes four studios and offices separated by glass walls. The practice was founded in 1988 and was one of the first of many new architectural firms set up as the Chilean economy began a 15 years cycle of accelerated growth. It quickly became one of Chile's largest studios. Its founder was particularly concerned about fashioning the practice not as a master studio but as a professional service enterprise. This began first and foremost with the

other two offices. Towards the end of my research period, I interviewed a total of 12 partners and architects of these architectural practices. In addition, I conducted 18 interviews with other well-known Chilean architects focusing on design process and studio practice.

chosen name—not simply the owner's first name and surname, rather a more abstract name that reflected the image of a professional service company.

Despite these substantial differences in the profiles of the architectural practices studied, modelled as a master studio, an innovation platform and a service-oriented enterprise, I focus on practices and configurations that these offices share. In the following sections, I analyze the practical organization of epistemic dissonance with respect to three dimensions: socio-spatial conditions of casual engagement, socio-cognitive dynamics of review meetings and socio-technical manufacture of project mediators.

Casual Engagement

In the office of Elemental's main partner hangs a large board featuring a table with the development of around 20 ongoing projects. Each project is represented as a horizontal line and tagged with the initials of the architects involved. Architects are assigned to multiple projects. Some projects are represented with discontinuous lines as they are in a "cold phase"—waiting for a decision, a signature or a cut-off date. Thus, the board not only displays projects being worked on at present, but also those that must be continued at some point in time. The exact moment of reactivation is however unknown for most such projects. The architects' work load thus needs to be organized as flexibly as possible, so that the office could rapidly adapt to the reactivation of projects. Interestingly, when one leaves the partner's office and enters the large space in which up to 20 people sit and work, one encounters no signs of separate projects, responsibilities or team affiliations. Despite the coming and going of numerous projects, each with a different team structure, the workstations of most architects and draughtspersons do not change.[2] The spatial organization of the office thus remains steadfast in the face of the unpredictable progress of projects. This, however, poses the question as to how collaboration for specific projects actually takes place. In other words, how does the board hanging in the partner's office relates to the work dynamics that can be observed in the large room?

The fact that project collaboration does not occur within bubbles or subsystems of common activity with clear spatial boundaries does not mean that it does not occur. However, their structure is more similar to a rhizome which is open to interactions with all members of the practice and thus has no outside. When I first met Elemental's main partner, I mentioned that

[2] This is also the case in the other two architectural practices we did participant observation in, and also what many other architects reported in interviews. In this context, the spatial organization found by Yaneva (2009) at Rem Koolhaas' Office for Metropolitan Architecture (OMA), where architects are constantly changing work stations, seems to be a rather exceptional feature.

doing fieldwork in the office would also involve being present at meetings of project teams, as this would be very informative for understanding how people report on their own work and how decisions are made. He explained then that no such meetings take place. They were unnecessary because information about ongoing projects is always automatically shared thanks to the fact that everybody works in the same space. This co-presence, he went on, means that everybody is up to date about what is happening in other projects. He added that a few times a year an overview about new projects would be given, so that they could give some order to the constant flow of information that fills everyday office life. Later on, fieldwork confirmed that, contrary to what I had initially assumed, project teams were not the most suitable analytical unit for understanding work dynamics in architectural practices.

This, however, does not mean that the board hanging at the partner's office is just an artefact, in the sense of a faulty or false representation imposed by the very form of the table. The table is indeed quite reliable, as it clearly shows who is responsible for and who is working on which project. What the table, however, does not show is the overlapping between projects in the large workspace and the fact that in this manner the whole studio is party to every project. Indeed, those who are not working directly on a particular project are still involved in it in some way or another, even if this only means listening in to conversations, telephone calls or discussions. Every member of the office thus knows all projects in greater or lesser detail and is more or less informed about decisions, unforeseen problems, newly discovered constraints and sudden setbacks. This open, rhizomatic organization of project work not only means that they can adapt flexibly or undergo major transformations, for example when new projects arrive, current projects are suddenly suspended or old projects are reactivated. It also makes it possible for them to exploit the open workspace as a source of potential and unforeseeable connections. More importantly, this does not seem to be specific of Elemental. Despite the different office profiles described above—a master studio, an innovation platform and a service-oriented enterprise—we encountered in all three a similar socio-spatial organization: overlapping teams in relatively small spaces in which everybody effortlessly becomes engaged in every project and can interact with members of other teams. This appears to be an important principle leading to almost accidental exploration and exploitation of different epistemic positions and knowledge perspectives with respect to projects.

Daily proximity and spatial overlapping indeed provide important points of contact for casual engagement in a joint exploration of options, possibilities and alternatives. Even if everybody works at their own station, the architectural studio is never silent. Emotional reactions to the work can be heard constantly in the form of grumbling, sighing, groaning, cheering or expletives. These noises are generally easily interpretable communications that offer all the people in the vicinity the possibility of starting conversations and participating in their colleague's work by asking how it is going or walking over to look at the visual elements that provoked such responses. In this way, ideas often emerge from quite unexpected corners of the studio. For instance, an architect may be sitting at his computer and without removing his eyes from the monitor will make a comment. Or somebody will stand up to go to the plotter and on the way stop by a colleague, look briefly at his work and casually ask a question such as "why is the room like that?" or "what does this thick line mean?" Sometimes someone will make an association with another project, as in "it reminds me of that other commission, but the problem was solved differently there." Or there may be a direct suggestion—"I would do it this way."

This kind of interaction is especially evident when particular exhibits, such as a model for a consultation with clients or a poster for a competition, are being prepared in the office. We observed repeatedly in these situations that other members of the practice suddenly took on the role of the public. Often all it takes is for someone to stop briefly in front of a project sketch or a model in order to attract the attention of another architect and set off a joint exploration of the project, with its strengths and weaknesses, references, unresolved issues, alternative ideas, possibilities and options. These exchanges are particularly welcome, as architects preparing a presentation are very eager to anticipate how the work is perceived by an external observer. Of course, not all questions, comments, suggestions are equally significant, but such exchanges often take place with different types of colleagues, whether they are architects, draughtspersons, interns or even ethnographers like myself. What is also particularly interesting is that the aim of such conversations is not to arrive at an agreement or a decision regarding the project in question. Rather, they serve to create a space for sounding out associations triggered by an exhibit until they have been exhausted. It is less about interpersonal communication than a joint exposition of the impressions, ideas or doubts that a tangible exhibit provokes. The aim is thus to multiply, compare and contrast points of view and thereby enable the formulation of alternatives, not necessarily to carry out a dialogue geared towards a decision. Accordingly, the conversations do not end with a conclusion or with resolutions, rather they peter out and the contributors drift apart again.

It is important to emphasize that often these comments, associations, ideas and critiques are not necessarily of any help. In most cases, the architects responsible for the project have a good answer to the spontaneous inquiry of the engaged colleague. But sometimes they do not. Thus, these comments open up windows of opportunity. While they do not steer the project in a totally new direction, they reveal alternatives, and where there are alternatives, a decision must be made. The studio thus constitutes a diverse cognitive ecology in which different epistemic positions or knowledge perspectives adopted by different office members towards a particular project are collected in a casual way.

Review Meetings

Review meetings are extremely important in all the offices we observed. These are not scheduled appointments, rather recurring discussions usually held between the architect responsible for a project and one of the partners. The aim of such conversations is to fine-tune the project planning, discuss problems regarding the design, explore alternatives, and make decisions. Review meetings take place with varying frequency—several times a week or every two weeks, depending on the partner's schedule. In one practice, for example, one partner went on three business trips to India, Brazil and Germany during the observation period, so that his short presence in the practice was used intensively in order to run through all the projects, with half a day dedicated to each commission. The dynamic in another practice was that the main partner went around all the project studios in the office almost daily in order to discuss the work with the different teams, to effect corrections and to make decisions. Interestingly though, such corrections and decisions are not related to e/valuation practices, but to epistemic problems:

> Architectural projects [are] much better when more people are involved in them—without exaggerating the number—than when there is only one observer, because this observer does not see, unless it is somebody really brilliant—that happens, but it's not the norm for human beings. (partner#1, 1 November 2009)

Project architects also consider review meetings to be crucial, precisely for what is often called operational blindness. Architectural design is based on very meticulous and patient work with an immense number of details. Every line, drawing, floor plan, and section requires several hours, sometimes several days of work, which leads to a sort of absorption in a project. This is what makes review meetings essential:

> You've already been working on it for a long time and then they [the partner or some external person] come along and say 'there's something wrong here, and there's another problem there, and that doesn't work done that way.' It's the pits, [but] these reviews are really important because... Look at it the other way around: you're so involved in the stuff that you can't view it from a distance. (architect#1, 26 October 2009)

The ability to "view something from afar" is probably one of the most important sources of epistemic dissonance exploited in these review meetings. This is a paradoxical phenomenon. The person who seems to understand the overall project "best" is not the person who has been working on it for the longest time, but someone who is not involved in the project directly and is not so familiar with all the details. Moreover, such a perspective from afar enables the partner not only to see the project as a "whole," but also to set and consider it in relation to other projects being carried out in the office or that have been seen or discussed in another context.

> [The partner] maybe has this advantage [because] he's not up to his ears in things as much. He jumps from one problem to the next, from one project to another, and has the advantage of the freshness of the person who has just arrived [...]. It's like the tourist [...] who sees things that the locals don't [...]. You don't see the mistakes anymore because you basically don't want to see them, because you're exhausted and you say 'OK, it's good like that', and you don't have that distance to be able to say 'It's bad and it looks bad' [...]. [The partner] can maintain this objectivity all the time and for the whole office because he can stand back and because he's not so personally and deeply involved in any of the projects. (architect#2, 19 October 2009)

It is interesting to notice that in this context objectivity means, above all else, a certain degree of distance from the enormous amount of time put in and the personal and emotional investment made by the project architects. Objectivity is not meant as a form of superior knowledge, but rather as a personal detachment that allows the partner to say things like "this is wrong, that looks bad, do it again."

But this is only half of the truth of the review meetings. When following the development of these meetings more closely, and listening to the long and often very complex conversations, noticing the jumping back and forth between the different aspects and paying attention to the challenges and the restrictions involved in a project, it becomes clear that in the large majority of cases it is not just a question of mere corrections made by detached observers. Indeed, architects and partners agree that the downside of distance can be making bad decisions:

> The effect of not being deeply involved in the project is that you neither enthuse about it nor fall in love with it. And that allows you [on the one hand] to be very objective. But [at the same time] you can end up being not very objective at all, so you express only prejudices [...]. It's very easy to get carried away [...] because there are many solutions and many good solutions. At the end of the day, you always have to decide [...]. There is a moment when you have to say 'I'm going in that direction' and so you take a risk. (partner#2, 22 October 2009)

Indeed, review meetings are by no means only about eliciting a detached perspective. A crucial function is to enhance, transform and question the partner's visions for the project. Indeed, architects responsible for a project often have good reason to reject suggestions that are made on the basis of insufficient knowledge or too general views. Accordingly, review meetings are instances, in which project architects describe, explain and reflect upon each of the steps taken, the possibilities and constraints considered and decisions made by them. Not doing this is indeed problematic and makes review meetings ineffective:

> Many decisions are made after trying out lots of different things with a lot of clarity about the limitations. And often when I talk to [the partner] about the project. Well...you know, you forget something [...]. And [he says] 'OK, we'll do it like that, let's change that like this.' And then later you remember some reason why you can't do it like that. Yeah...all these decisions [...], it's like being in a maze. (architect#1, 26 October 2009)

Review meetings certainly involve eventually making decisions, but their primary aim seems to be re-viewing complex chains of decisions. Re-viewing, in the sense of viewing things from a new perspective, requires a balance between a dialogical reconstruction of decisions made and the exploration of dissonant observations and insights. Indeed, considerable time is invested in giving an account of the path that has been followed by the project architect through the decision maze, so that it becomes possible with the help of the distanced perspective of the partner to notice or look for ramifications, back roads or shortcuts.

Thus, review meetings enable the propagation of design alternatives. What is crucial is not the co-existence of different criteria of evaluation, which may well be shared by architects and partners, but the dissonances that emerge from their differing knowledge perspectives on a project. Better decisions are not arrived at by being closer to a project and having more detailed knowledge or being further afar and looking at the project as a whole, but through the dissonant combination of such perspectives. This epistemic dissonance is what is explored and exploited in these meetings.

Project Mediators

Moments of epistemic dissonance and the ensuing propagation of alternatives do not only result from the differing knowledge perspectives of architects in the studio. A crucial role is also played by project mediators, such as sketches, plans, renderings, 3D simulations or models, which are created using various media and with different representation techniques. As Hennion (1993, 1997) has demonstrated for music, these mediators do indeed more than provide simple materializations or representations of already existing mental or ideal forms. Architectural projects do not exist in an ideal space, but only through material mediators which enact the project in different ways. It is important to note that since there is no ideal form that would provide an ultimate reference point, the reality of a project is multiple and heterogeneous. Project mediators constitute thus an ontological multiplicity, with each giving body to the project in a different way, so that fractures, gaps and discontinuities abound. This is why epistemic dissonance cannot be traced back only to the different knowledge positions of the architects involved. Project mediators are painstakingly crafted by architects, precisely in order to gain insights into the reality of the project that may deviate from their own cognitive expectations. Accordingly, project mediators become independent components of the studio's distributed cognitive ecology and thus another source of epistemic dissonance. It is important to avoid two symmetrical errors—that of humanism, whose point of departure is the creative or brilliant individual, and that of sociologism, in which individual contributions are blanked out in the name of an emergent social dynamic. Instead this approach understands cognition as a distributed sociomaterial process. The cognitive sciences have long discussed how carrying out even simple and everyday cognitions requires collective action by multiple individuals and artefacts (Gieri and Moffat 2003; Hutchins 1995). Accordingly, cognitive processes involving inspiration, creativity and certainly dissonance must be viewed as occurring in empirical, concrete and local ecosystems and depending on specific capacities of people and things, of architects and project mediators.

Architectural plans, for example, enact a project in significantly different terms depending on the scale, the projection technique and the medium in which they are drawn. Scales are widely used for different types of drawings, such as floor plans (1:48), site plans (1:20 or 1:30), etc., as they enable architects to see different things. It is important to notice that architects don't simply work from larger to smaller scales, but rather switch between scales (Yaneva 2005). Apart from the scale, the media in which the plan is being worked on, whether it is printed on paper and intervened with a pencil or on a computer monitor and intervened with the mouse, radically alter what architects can see and do in a plan. When working on the computer, an architect explains, standing back, taking distance, looking at the project as a whole, simply becomes impossible:

> You plunge so quickly into the details that you end up dealing with every little door and every little screw [...]. And when you zoom out [on the computer] you can't see anything. You can't see because the screen is so small [...]. I find distance can be achieved with plans—you print the plans, you have time to look at them...yes, to look at the thing carefully [...]. Those are the two

> yardsticks: distance in the form of the plan, the printed paper, and the proximity is the screen. (architect#1, 26 October 2009)

One and the same set of geometrical lines that constitutes a plan is not or, rather, does not always do the same thing. What counts is the material medium in which plans are drawn and intervened. On paper or on the screen plans enact the architectural project differently and convey insights and understandings not easily exchangeable. A back and forth between media, between paper and monitor, is just as necessary and common, as moving up and down in the scale. At stake is the same type of cognitive dissonance between proximity and distance, between detail and whole, already described for the review meetings. Indeed, review meetings are often held at tables with various layers of printed plans, among which architects constantly switch, and in the proximity of a computer, where things can be consulted or observed in a closer scale.

While most architectural plans, such as elevations, floor plans or cross sections, are two dimensional, architects also produce three-dimensional representations of projects. There are, however, multiple projection techniques, each with different results. One example is isometric projection where, unlike central perspective, the scale along each coordinate axis of an object is always the same. The isometric projection method, which was commonly used by architects until the 1970s, offers a representation of the designed object that significantly deviates from human perception and from central perspective techniques, where the scale in which an object is represented varies depending on where it is placed in space. But, as we all know, objects do not change their scale, as they move closer to or further away from an observer:

> [Isometric projection] allows a more real representation of the object. This side here, for example, has the same length as that one. The way it is in reality. It's a mental representation, which allows things to be represented as they really are. And in this sense they are more real. (architect#2, 19 October 2009)

Architects often use isometric projection for showing the actual sizes of different sides of an object, something that is especially useful for drawing construction details, which will be used in the construction site to actually build things. Thus, while the reality enacted by the isometric representation is indifferent to the perspective of the human eye, to the point that sometimes is difficult to gauge depth or altitude of objects, it is extremely helpful to visualize objects in ways that are relevant for their actual manipulation.

Quite different is the case of architectural renderings, probably the most common three-dimensional representations, which are made by not just using central perspective, but also applying color and texture to surfaces and placing human and non-human objects in space, in order to create a photorealistic visualization of a building or its interiors. The reality enacted by renderings is thus neither about the spatial perception of the human eye, as in perspective drawings, nor about the construction or manipulation of the object itself, as in isometric projections, but about the photographic look of the constructed work. And, interestingly, architects make renderings not just for bids, client communication or PR purposes, but also "for the design process, to see how it [the building] functions" (architect#3, 9 September 2009). Renderings, architects suggest, are unique in their ability to anticipate the reality and the appearance of a constructed work and thus to bridge the insurmountable gap between abstract project representations, such as plans or perspectives, and the actual building. In one of the offices, I observed architects explaining to interns the need of adding "weight" to the renderings, of realistically conveying a sense for the actual materials and the actual physical weight of the finished construction. The problem, they would explain to me, is that one needs to counteract the weightlessness of abstract representations which wouldn't allow visualizing and grasping the built reality of the represented object. A rendering would gain in "weight" through the addition of hundreds of layers to represent textures, materials, shadows, lights, human beings, plants, etc. Importantly, the huge investments made in renderings don't simply aim at visualizing what has already been projected by the architects, but rather at providing new insights, generating new knowledge and eventually destabilizing architects' vision and understanding of a project. "Renderings," a partner explains "should be used to reveal problems" (partner#3, 7 September 2009).

Whereas renderings are a standard part of the day-to-day studio work, 3D animations that allow one to make a virtual tour through a building are normally made to convince an audience of the qualities of a project. In the observation period, two such animations were prepared in two different architectural practices: one of a Chilean university library destined primarily for presenting the project to university staff, another of a German winery as a special request of the client. Interestingly, even though 3D animations are not aimed to reveal problems, they become challenging tests, as they allow, for the first time, the perception in motion of a space from both the outside and inside and thus also lay bare the continuities and discontinuities between spaces, proportions and so forth. "You stick a camera into a project and see the interior and then you know what it actually looks like from the inside" (architect#2, 19 October 2009). When the 3D animation of the winery was seen for the first time by architects, key elements of the design reveal themselves as problematic. The 3D animation thus forced architects to generate design alternatives for aspects and configurations that had not previously been an issue, such as the number, form and distribution of a series of narrow, oblong windows on the winery's façade.

Project mediators play a key generative role of epistemic dissonance. As we have seen in previous sections, they are at the center of review meetings, what is re-viewed or viewed differently, and they are also what enables casual collaboration within the office, as their visual presence engages all office members without distinction. Project mediators thus can be seen differently according to the varying epistemic positions of involved architects. But, as these latter examples suggest, project mediators cannot just be viewed differently, they also enact the reality of a project in differing ways. In this sense, they produce insights and knowledge that can significantly depart from what had hitherto been visualized and known. Epistemic dissonance—in other words, the productive exploitation of alternative knowledge perspectives on an object that in the case of architecture does not yet exist—is thus not only an intersubjective phenomenon, but also a socio-technical phenomenon that critically depends on the capacity of visual mediators to redefine what the project is and to open up an exploration of alternatives.

The Occurrence of Dissonance

The organization of epistemic dissonance in architectural practices is sustained by multiple coexisting temporal regimes that allow for moments of dissonance to occur at different intervals and rhythms. Occurrence should be here understood in a double sense: moments of epistemic dissonance "occur" or take place in certain situational configurations in certain temporal intervals. At the same time, they also depend on "occurrences," on ideas, possibilities, alternatives that suddenly suggest themselves and come to mind.

This double sense is evident in socio-spatial dynamics of casual engagement, which occur on an everyday basis and potentially at any moment. Casual collaboration is indeed based on forms of mutual observation, feedback and dissonance that are constants in architectural practices as they are enabled by their very spatial arrangements. It is, as we have seen, the situation of co-presence among architects and mediators, humans and things that generates casual collaboration, unsought engagement and occurrences in the form of questions, comments, associations, etc. At the same time though, and precisely because casual engagement can occur at every moment, it is not expected at any particular moment. Thus, despite its recurrent occurrence, casual collaboration takes the form of the surprise, of the unexpected. Architects do not expect at any specific moment insights or comments from their colleagues, who in turn don't feel obliged to provide feedback or comment. In consequence, the moment a dissonant insight is communicated takes the form of a gratuitous act.

Project re-views occur in a quite different mode. This is not a spontaneous capacity of the practice's everyday work dynamics, but a carefully prepared and configured situation explicitly oriented at generating and exploiting dissonant views on a project. This requires coordination between architects and partners to find a time, print out plans, and sit together. Thus these review meetings occur in recurrent time intervals, marking a rhythm in the development of a project: every one, two, three weeks. Interestingly though, what can be carefully prepared are only the initial conditions for a joint exploration, a common time to sit together and go over plans and decisions, but not the kind of outcomes of the re-viewing. Apart from particular problematic issues that need to be discussed, what is expected is that through a joint revision of the state of a project, new options and unseen possibilities will occur. These occurrences do not result spontaneously in casual engagements with project mediators, but they are methodically sought for systematically questioning decisions. Occurrences thus don't take the form of surprises and gratuitous acts, but of findings obligatorily communicated. Accordingly, the great risk of re-viewing is not that dissonant occurrences might not be communicated, as in the case of casual engagements, but rather that they are forcedly arrived at.

Project mediators become unexpected and often unwanted sources of epistemic dissonance. They certainly play an enabling role for the unfolding of different perspectives both in casual engagements and review meetings, but it is in the very process of making them, of adding layers on a render, of running a virtual camera through a 3D model, that they bring along not just unexpected, but also unwanted dissonant enactments. Such dissonant insights do not occur in recurrent temporal intervals. Rather than moments in a larger temporal and rhythmic dynamic, they involve instances of sudden realization, both in the sense of making something real and of having an insight that cannot be subsumed in some overarching structures or contexts of office collaboration. Indeed, as mediators are carefully crafted over days and weeks, sociotechnical work is oriented at faithfully representing the projected reality of a building, especially considering the practices of future users and its emplacement in a larger environment. Dissonant occurrences occur then when the process of completing the reality of the project fails in spite of the efforts put into making it work, that is, in spite of the layers, the tweaks and filters added. Occurrences take the form here of neither welcome surprises nor expected findings, but rather of disappointments.

Finally, what these different modes of occurrence have in common is a positive reframing of dissonance as a source for variations and alternatives. Such positive reframing does not mean that disappointments and negative feelings are spared. Certainly, the casual remarks of colleagues who are not involved in the project and know only little about it, the corrections of the chief architect or sobering 3D animations are not always received with great joy. But the important thing is that this unpleasantness, this dissonance is necessary and even desirable for the proliferation of design alternatives.

The key question is then what enables organizations, such as these architectural practices, to turn dissonance into a source of new ideas. An initial answer can be found in the title of David Stark's book *The Sense of Dissonance*. The essential point here is the concept of "sense," which refers to a shared recognition and appreciation of dissonance. The shared culture of an organization could, accordingly, be described in relation to its more or less developed, more refined or more blunted sense of dissonance. But this does not explain where this shared sense of dissonance might come from.

Consider, this time to conclude, the following exchange with Rodrigo, a colleague of José:

> I am sitting next to Rodrigo looking at how he works on the computer. Without separating his eyes from the monitor, Rodrigo asks whether I could ever imagine working as an architect. I replied somewhat tersely that architecture is a pretty precarious profession. Rodrigo agreed, and then he added that architects do indeed often have only incomplete information at their disposal and have to make decisions lacking a watertight basis. I explained that I had really meant the precarious job security especially among freelance architects. Yes, replied Rodrigo, who had been in a permanent job for some years, this was also true. But by precarious he actually meant the methods and information used by architects.

Rodrigo's reflection about the precariousness of architectural practice can give us important clues about where a shared sense for dissonance might actually come from. The key, it seems, lies in the professionals' reflexive awareness of the need to wade through what Schön (1983) calls the swamps of professional practice, in which neither problems are well defined, nor univocal courses of action available, and the unavoidability of eventually going through what Derrida (1992) calls the ordeal of the undecidable. Allowing for dissonance to unfold is thus crucial to transform indeterminate situations into specific, defined problems and to open up alternatives. Herein lies the creative and constructive potential of dissonance as an anti-valuation moment calling for a transvaluation of values.

References

Cooper, J. (2007). *Cognitive Dissonance: Fifty Years of a Classic Theory*. Los Angeles, CA: Sage.

Cuff, D. (1992). *Architecture: The Story of Practice*. Cambridge, MA: MIT Press.

Derrida, J. (1992). Force of Law: The Mystical Foundation of Authority. In D. Cornell, M. Rosenfeld, and D. G. Carlson (eds.), *Deconstruction and the Possibility of Justice.* New York, NY: Routledge, 3–67.

Ewenstein, B. and Whyte, J. (2009). Knowledge Practices in Design: The Role of Visual Representations as "Epistemic Objects." *Organization Studies* 30(1): 7–30.

Festinger, L. (1957). *A Theory of Cognitive Dissonance.* Stanford, CA: Stanford University Press.

Gieri, R. D. and Moffat, B. (2003). Distributed Cognition: Where the Cognitive and the Social Merge. *Social Studies of Science* 33(2): 1–10.

Hennion, A. (1993). *La passion musicale.* [The Passion for Music]. Paris: Métailié.

Hennion, A. (1997). Baroque and Rock: Music, Mediators and Musical Taste. *Poetics* 24: 415–435.

Hutchins, E. (1995). *Cognition in the Wild.* Cambridge, MA: MIT Press.

Luhmann, N. (1995). *Social Systems.* Stanford, CA: Stanford University Press.

Schön, D. (1983). *The Reflective Practitioner: How Professionals Think in Action.* New York, NY: Basic Books.

Schumacher, P. (2011). *The Autopoiesis of Architecture. A New Framework for Architecture.* Vol. I. Chichester: John Wiley & Sons.

Stark, D. (2009). *The Sense of Dissonance: Accounts of Worth in Economic Life.* Princeton, NJ: Princeton University Press.

Stark, D. and Girard, M. (2002). Distributing Intelligence and Organizing Diversity in New-Media Projects. *Environment and Planning A* 34(11): 1927–1949.

Sunley, P., Pinch, S., Reimer, S., and Macmillen, J. (2008). Innovation in a Creative Production System: The Case of Design. *Journal of Economic Geography* 8(5): 675–698.

Sutton, R. I. and Hargadon, A. (1996). Brainstorming Groups in Context: Effectiveness in a Product Design Firm. *Administrative Science Quarterly* 41(4): 685–718.

Yaneva, A. (2005). Scaling Up and Down: Extraction Trials in Architectural Design. *Social Studies of Science* 35(6): 867–894.

Yaneva, A. (2009). *The Making of a Building. A Pragmatist Approach to Architecture.* Bern: Peter Lang.

15

Sources of Newness in Organizations: Sand, Oil, Energy, and Artists

Ariane Berthoin Antal[1]

Your company specializes in developing and implementing internet-based solutions for companies. To your surprise, management has invited an artists' collective to help the organization find a way to balance the interests of the end users and those of your clients. Now a choreographer and an actress want you to give bodily expression to the issues you need to address! Can their exercises really be of any use in resolving your technical problems?

You belong to a group of companies that promotes innovation systems in the areas of industrial, environmental and technological services. The group wants to change its business model in order to become an international brand, which also means remodelling the way it creates networks, and rethinking its own message. What kind of help can a person who describes herself as someone who "works with people and not canvases . . . a catalyst, a facilitator, a hostess, a service provider, a support structure, a manager, a producer—in short, an artist" possibly be?

Seeking newness in the form of products, processes, strategies, or organizational models, organizations from all sectors and of all sizes have been experimenting with artistic interventions. In other words, managers are bringing "people, products and practices from the world of the arts into the world of organizations" to intervene in established processes and mindsets with the hope of stimulating the development of new ways of seeing and doing

[1] I am grateful to *Conexiones improbables* for inviting me to study the artistic interventions they are producing in Spain; to the artists who responded to my many questions; and to Victoria Brattström, Gervaise Debucquet, Michael Eriksson, Victor Friedman, and Michael Hutter for their comments on earlier versions of this contribution.

things (Berthoin Antal 2009: 4).[2] Even the business weekly the *Economist* has encouraged managers to learn with the arts (2011). But, why bring in the arts, when, as Dewey pointed out, "to multitudes art seems to be an importation into experience from a foreign country" (2005: 11), and when artistic interventions are characterized as a process of crossing the "cultural divide" (Barry and Meisiek 2004: 5)? It is in fact the very foreignness of artists' codes and behaviors that managers see as a resource for stimulating the emergence of newness.

People who have worked with artists in organizations often use physical metaphors to describe this relatively recently discovered foreign resource for organizational learning. A process manager in an intermediary organization that produces artistic interventions recounted, "I tell them they should be the sand and the oil in the system," explaining that artists should serve both as an abrasive and as a lubricant. Employees and managers report that the artist "was the energy" in the process, helping them to overcome the inertia that impedes experimentation with new ideas and behaviors in established organizational cultures.

These physical, machine-related metaphors are intriguing because the application of such terminology to describe artists is (still?) jarring. Using terms associated with machines to describe organizations populated by people is a common metaphor (Morgan 1986) that seems quite acceptable, although it is a dehumanizing practice. Referring to artists as sand and oil to generate newness in organizations mixes two cultural codes with values that are generally considered incompatible. But the choice of these images reflects the belief expressed both by practitioners (Boyle and Ottensmeyer 2005; Seifter and Buswick 2005) and researchers (Barry and Meisiek 2010; Darsø 2004; Styhre and Eriksson 2008) that artists can stimulate change in organizations by increasing harmony or by generating friction. They may be invited in to help develop a sense of unity and rapport (e.g., by working on corporate identity or on communication problems), thereby enhancing the capacity for collective action. And they may be brought in to disrupt routines and taboos engrained in the organizational culture as steps towards developing new ideas and approaches. There is the implicit assumption that the dissonance between the different orders of worth of the world of the arts and the world of organizations will generate newness (Hutter 2013). Managers expect both the "harmony" and "friction" approaches to enhance organizational performance by stimulating innovative capacity, improving people's skills and motivation, as well as organizational processes and practices (Berthoin Antal 2009; Schiuma 2011).

[2] Some scholars introduce different terms. For example, Schiuma (2011) and Biehl-Missal (2011) prefer arts-based initiatives (ABIs).

The third element, "energy," is also physical: employees talk about "waking up," "sitting up," and about the very presence of artists as "energizing" the space (Berthoin Antal and Strauß 2013: 31). The physical aspect of energy is closely related to the emotional energy employees need in order to overcome scepticism to newness and organizational inertia. Two kinds of processes appear to be at work: artists can bring in fresh energy from their world of the arts into the organizational context, and they can release energy that has been pent-up in individuals or groups by restrictive rules, structures, norms and mindsets in the organizational culture.

Research on artistic interventions has been dominated by organizational, and particularly by managerial, perspectives. The artists' perspectives are missing. How do artists express what they value in engaging for a moment (which can last for a few hours, weeks, or months) with a world that is so foreign to their own? How does the experience of serving as sand, oil, and energy affect them? In this chapter I will try to enrich the discussion by drawing on the words of artists to attune the reader to the diversity of their experiences in the process of contributing to the emergence of newness in organizations.

Seeking out Artists' Voices: The Sample

Artistic interventions take many forms: they can be short or long, involve all kinds of arts, include all members of an organization, or just a few. There is therefore no "typical" artistic intervention because each one is unique, depending on the artists' approach, the issue they address, and the arrangement with the participating organization and its members. The artists[3] whose voices I present here participated in artistic interventions that the intermediary organization *Conexiones improbables* produced in the Spanish Basque country between 2011 and 2013. The intermediary undertakes numerous tasks, including identifying organizations that could benefit from working on an internal issue with an artist, matching each organization with an artist or an artists' collective, supporting the overall process, and disseminating insights from the experiences (Berthoin Antal 2012). *Conexiones improbables* produces artistic interventions in what they call "editions," whereby 6–10 organizations start and finish a project at the same time. They have an introductory methodological session together and share experiences in meetings during and at the end of their projects. This study draws on data from 62 artistic interventions, of which most (48) were short "Creative Pills,"

[3] I received 54 responses from artists to the pre-experience survey and 55 to the post-experience questions.

as the intermediary calls its 4-month format, and 14 (eponymously called "Improbable Connections") lasted 10 months.[4] The participating organizations were diverse in size and in sector, including, for example, neighborhood hair salons, local bakeries and grocery stores as well as a supermarket chain, several hospitals, IT-companies, and some rural development associations.

The issues to be addressed in the artistic interventions were also diverse, including exploring new ideas for products and services, trying out new modes of communication within the organization or with external stakeholders, and working on organizational identity. The artists come from a wide range of disciplines including the theater, conceptual art, and graphic design, but many resist using traditional art categories to describe themselves, preferring instead to use terms like "social artist" and "facilitator of collective creation processes." [5]

I invited the artists to express their thoughts and feelings in two Web-based surveys, one before and the other after the artistic intervention.[6] This process implies that my research instrument adds moments of valuation to the artistic intervention by posing questions that stimulate reflection about the kinds of value the artists expected to find and the value they actually experienced. The logic underpinning the two-step approach is that how people evaluate a situation is shaped in part by how they envisage it beforehand. Therefore, to understand expectations, the pre-experience survey included questions about what they hoped might happen during their intervention, and conversely, what they feared might occur. In order to delve deeper into their implicit success criteria, I also asked what would disappoint them during or after their work with the organization, because expressions of potential sources of disappointment offer insights into tacit expectations of finding or achieving something desirable. The post-experience survey, administered several weeks after the completion of the artistic intervention, elicited what the artists found valuable, surprising, or problematic about the process of intervening in an organizational context. Thus, in this chapter I draw on two moments of reflection: when the artists are on the threshold of starting their intervention, and then again after they have had a few weeks to reflect back on the experience.

[4] The data for this contribution are based on 62 artistic interventions produced by *Conexiones improbables* in 5 editions of short "Creative Pills": EkintzaLab 2011, 2012 and 2013; BidasoaLab 2012; IrunLab 2013; and 2 editions of their longer format: *Conexiones improbables* 2011–12; *Conexiones improbables* 2012–13.

[5] More information is available on the *Conexiones improbables* website, which presents summaries of the projects and images illustrating the process and outcomes in English and Spanish: www.conexionesimprobables.com.

[6] My research takes a multi-stakeholder perspective, so I also send questions to the employees and the managers responsible for the artistic intervention projects, but they are not the focus of this chapter.

Imagining the Organization: Artists' Hopes and Fears Before Embarkation

Aspiring to Processes of Co-Learning

Given that artists often embark on a new venture because "it is precisely the unknown that pushed me to do this project" (CI 2012–13),[7] it is not surprising that some of them chose not to put into words what they hoped would happen during the collaboration. They prefer to "live what happens" (CI 2012–13) rather than to give their hopes any form in advance. Fortunately for my research, however, most of the artists did use the space afforded by the question to formulate what they hoped would happen in the artistic intervention in the organization. The red thread that runs through their responses is the hope for collaborative relationships with people. They write about empathy and personal connection, and about differences in perspectives and language from which they envisage a learning process will grow. They recognize that "there can be opposite ideas but the 'spirit' has to be the same" (BD 2012). They speak of mutual learning, or as some call it, "co-learning," a process to which "all parties bring their needs as well as their competencies" (EK 2013). They formulate high hopes about the people they will encounter in the organizations: "I hope to meet people who are open to listening, reflecting, and taking risks. People who are ready to collaborate with me to seek solutions to their problems. People ready to change in order to improve" (IL 2013).

Working with Emotions

The artists explicitly welcome emotions in the co-learning process, expressing the hope "that we will be immersed in it and co-create something that will be both emotional and interesting" (IL 2013). The kinds of emotions the process should ideally entail include a mix of "surprise, incomprehension, and some resistance" (CI 2011–12). The physicality of the images relating to emotions in co-learning processes is striking. The artist cited above wrote about being immersed with the participants, while another expressed the hope that the participants would be willing "to throw themselves into the abyss of the unknown" (IL 2013). They envisage developing relationships with participants to work with these emotions, and they believe that this co-learning relationship can open people to new ways of seeing things in their setting.

[7] The codes refer to the name and year of the "edition" of artistic interventions produced by *Conexiones improbables*. EK = EkintzaLab; BD = BidasoaLab; IL = IrunLab; CI = Conexiones improbables. Almost all the artists responded in Spanish, a few wrote in Basque or English; I translated the quotations used in this text into English.

The desired interaction is metaphorically captured in the image "That both parties engage in working together like a lock, so that if the key enters, it can open something." The opening can be a releasing of what is already there, but invisible to or undervalued in the organization. Artists refer to revealing and working with the knowledge or competences they believe the employees have. "They certainly have the solutions to their issues and I hope to help them find them, valorize them or rediscover them" (BD 2012). These artists hope to unleash what the employees are capable of by helping them to deal with the fears that are holding them back: "To awaken in them their entrepreneurial and creative capacities, so that they become protagonists of their own process, that they dispel their fears to express and materialize the ideas that occur to them" (EK 2013).

From Seeing Differently to Doing Differently

The artists would like their interaction to engender changes that people consider useful. The interaction should leave traces that extend beyond the actual intervention. Some refer to tangible outcomes, like hoping "that a concrete and replicable model" (CI 2012–13) would result from the collaboration, whereas others use relational terms: "to empower all the employees and the users of the [organization] and improve the sense of belonging to a community that is solid, intergenerational and connected both physically and virtually" (EK 2013). Yet another physical image used for instilling a long-term change is that the experience could "infect them with the artistic attitude of constantly rethinking situations, processes, results, forms" (EK 2011).

Eliciting Implicit Valuation Criteria Ex Negativo

If these are the hopes that artists have before engaging with an organization, what are their concerns? The survey contained two questions: What do you fear might happen? What would disappoint you? With this combination of questions I wanted to stimulate the artists to formulate *ex negativo* the implicit criteria they might use to assess the value of the experience later, beyond those expressed in the form of hopes. There is a certain amount of overlap between these questions, and therefore in answers they generated, but also some interesting differences in emphasis. Although quite a few artists wrote that they had no fears because "both parties have hopes and curiosity; that is the best point of departure" (EK 2012), and because "disagreements can be opportunities to see things from other points of view and opportunities for things to happen that could never have happened without controversies" (CI 2012–13), all the artists identified aspects that would disappoint them. The main fear the artists express is that the people in the organization might not engage with them and the process. The primary source of disappointment would be if the outcome were not interesting or useful.[8]

Fearing Lack of Engagement

The concern about lack of engagement is attached to several elements. The artists fear that people might not be interested or motivated enough, and that they will encounter "indifference, lack of acceptance, delegitimization of what I can bring them" (BD 2012). They realize that stimulating interest is part of their role but they "fear not succeeding in connecting, in getting a common space for experimentation and learning" (EK 2012). Another practical concern they have is time, recognizing that people are under daily pressures, "day-to-day urgencies" (IL 2013), that will compete with the time needed for working together on the project. The difference between the world of the arts and the world of organizations, which is the essential ingredient for the projects, is mentioned as a potential barrier in communication that can undermine engagement. The artists fear that they might "not achieve a minimum of shared language and the ability to play together" (EK 2011) or that "the instruments and language that I will offer them will appear to them to be too strange" (BD 2012). They emphasize the importance of trust, so they fear a situation in which "dynamics of lack of trust are generated" so that "the work gets polarized in two distant poles rather than a single team" (CI 2012–13).

Concern about Lack of Usefulness

The artists addressed their concern about the interaction not being useful from several angles. One is practical impact. They would be disappointed "if nothing we did were put into practice" (BD 2012), if their contribution were to remain "theoretical" (EK 2011). The importance of seeing some kind of impact is evidenced by the potential disappointment "if it does not activate any positive change in the organization and the people" (EK 2011). Another angle is the artists' concern for the sustainability of their contribution, noting that "It would hurt me a lot if there were no continuity after the end of the project" (CI 2012–13).

The Positioning of Art

Several additional elements emerge from the responses to the questions about concerns, revealing qualities that the artists consider to be important in their

[8] I cannot be sure to what extent differences in emphasis in the responses to these two questions relating to potential concerns is attributable to the difference between fears and disappointments. It is possible that the fact the artists were confronted by a second prompt question pushed them to reflect further, and to come up with something more than if they had only had one such question. In future questionnaires or in interviews, it could be interesting to reverse the order of the questions to see whether there is any change in the pattern.

work with an organization. Although they often write about usefulness, and they do not want the artistic intervention to end up being "merely entertainment" (EK 2011), they do not treat usefulness as the exclusively decisive value. The artists care about the content and quality of their projects and they are concerned that their work should not be "banalized" or "used purely instrumentally" (BD 2012). They would be disappointed if "the output is too rigid or fixed and that the idea of 'play' or 'fun' is not visible in the result" (CI 2012–13). They would not be satisfied with a project in which "there were no intensity, enthusiasm, joy, illusion, hope" and in which the participants did not "surpass some limits that they had never passed before" (IL 2013).

In summary, the hopes and concerns expressed by the artists before embarking on an artistic intervention shows that they consider their entry into the world of organizations primarily in terms of the people they will meet there. They expect that these people will have different kinds of knowledge and approaches from those the artists bring in. For the artists these differences are the primary valuable ingredient for the project. They see their own role as both instigating and participating in a learning process as co-learners. Developing good personal relationships while working with people in the organization is important for them, and they refer often to wanting all involved to enjoy the project. However, they do not hope for a smooth ride, emphasizing instead the value of shock, surprise, and resistance. The artists definitely want the engagement to generate something useful, but not at all costs: the outcome should not be banal and the process should not be joyless. By definition, their presence in the organization lasts only a moment, yet they aspire to setting in motion a process that will continue beyond the lifespan of the project.

Postpartum Reflections: Encountering Surprises, Rewards and Problems

We are now ready to listen to the voices of the artists as they reflect on the experience after completing their interventions in the organizations. I asked them whether anything happened that surprised them in the process, because what surprises people can shed light on implicit assumptions that a confrontation with lived reality challenges or disturbs. In addition, asking about surprises taps into emotions: it opens up both how people think and how they feel about the situation. To complement this exploration of the artists' experience in the world of organizations, I asked what they found particularly rewarding and particularly problematic. As a way of summarizing the conclusions the artists draw about the value of engaging in such ventures, I asked whether they would recommend such interventions to other artists and if so, why. This combination of questions provides insights into what they discovered, appreciated and grappled with while working in and with the foreign world of organizations.

Learning from Surprises

The responses contain expressions of surprise when some of the great hopes were fulfilled, as well as when the fears envisioned before entering into the organization were realized. A few artists found that "practically the whole process was accompanied by small surprises" (CI 2011–12), probably because the process was new for all the participants.

Even though they had entered with high expectations, some artists were still positively surprised by the way people responded to them. For example, there was "the huge desire to change the passive attitude of the business" (BD 2012) and "more engagement than we had hoped. None of the suggestions scandalized them" (EK 2013). The ideal situation was achieved when the artist found "the personal treatment, the reception, the motivation and openness to all that I suggested to them" (BD 2012).

Needing to Create Conditions

However, the survey results show that the conditions of high engagement in co-learning that the artists had wished for were not immediately evident when they entered into the organization. Sometimes they needed to discover hidden possibilities, such as unexpected sides of the manager who was responsible for the artistic intervention project: "contrary to the first impression we had, underneath his serious face was a person ready to 'play' with us" (IL 2013). More often it took some time and effort to "create great trust between the company and the artist" (EK 2013) and to "change in attitude from 'wait and see' to 'we have to collaborate to do something different'" (BD 2012). That such a shift was possible sometimes came as a surprise in the last phase of the intervention, when the artists wanted to see their hope of continuity realized: "At the end I was a bit hesitant to propose a game to them that they could continue to use in the process we started—they accepted immediately" (EK 2013).

Dealing with Lack of Organizational Commitment

Although some artists were in the happy position of being surprised by the high level of "motivation and the desire to take the project to a level of reality" (BD 2012), a few had the opposite experience, such as when "the organization dedicated little time and little quality to the collaboration" (EK 2013). A few artists encountered such "fears and resistances to transforming and to opening themselves to participation" that they were puzzled by the organization's decision to work with an artistic intervention. A frustrated artist wondered why the management had agreed to participate in an artistic intervention, asking: "How can you take on a project that you are not willing to do?" (CI 2012–13).

In several cases the engagement was positive at first, but when it came time to shift from developing ideas to actually putting them into practice, barriers became apparent. It came as a surprise, one wrote, because "They felt all our proposals were marvellous—until we suggested implementing them in reality. In that moment, it was all obstacles and difficulties" (IL 2013). In these cases evidently the members of the organization did not succeed in transitioning from talking about new ideas to actually making the necessary changes in the organization. It is unclear from these responses whether the power to block initiatives was located primarily in the hands of the leadership or distributed among employees, but the expression "it was all obstacles and difficulties" suggests that the power of inertia in the organizational culture also played a role in inhibiting the generation of newness from the artistic intervention.

Overall, the artists' responses to the question about what surprised them during their work with the organizations go beyond simply confirming that their hopes and fears about the interaction may come true. Most importantly, the artists often discover that they first have to create the social conditions that they need in order to engage people in co-learning. Comments like "I was surprised by many things, but primarily the degree to which people did not know about the work that others are doing in the organization" (CI 2012–13) indicate how much the artists needed to do in order to build the capacity for learning together. The nature of the problems in the organization can be almost overwhelmingly painful, like the poisoned atmosphere in which "the tone of the meetings was so 'foul'" (EK 2012) and when the impact of the financial crisis made itself felt in day-to-day interactions: "It frightened me to have to deal with a difficult economic situation and great stress it produced" (BD 2012). Of course artists have a great deal of experience in facing the harsh economic climate in their own lives, but dealing with its repercussions on people in an organizational context is a new and different challenge.

Developing the Capacity for Co-learning

Discovering how to create the conditions for co-learning in an organization emerges as the element that the artists value most about the experience. Terms that appear frequently in the responses here are "richness," "inventing," "trying out," "ideas," "perspectives," and "passion." Building the connection requires first "listening and seeing all the characteristics and preoccupations" from which to "detect the point of interest" (IL 2013) in "the encounter between two entities that appear to have nothing in common." (EK 2012)

Several artists refer to learning to send clear signals about how to work. One was particularly adamant about having to set and stick to the rules: "clearly defining the objectives at the beginning of the project and not deviating from them: not taking anything for granted; reflecting on the strategy to follow in order to achieve the objectives" (EK 2013). Another noted that learning "to say the necessary no's" (CI 2011–12) is important to maintain the focus to finish the process successfully. Some artists use strong physical images to get their message about the need for change across: "To insist on changing concrete regulations that asphyxiate the organization and impede them from innovating" (EK 2012). The asphyxiation that artists noticed also isolated the organization from its environment. Some therefore worked on opening the organization to potential stakeholders by "showing the owners that they cannot wait for customers to come, but that they have to go out and find other ways of reaching them" (BD 2012).

Discovering Their Own Value

Working to develop the organizational capacity for learning also helps the artists realize the value of their own abilities, such as "recognizing that I can generate a large volume of ideas in little time. I can keep turning 'the tortilla' around until something sticks in the organization" (IL 2013). The ideas do not have to be used immediately, in fact it is important to have a surplus on which to draw in future, explained the artist who emphasized the value of "generating a bank of ideas as a point of departure for the continuity of the project" (CI 2011–12). The experience also confirms for the artists the value of their process-orientation, which they model in their interaction, namely "being conscious of the importance of 'doing by doing,' to achieve the goal by the process itself" (EK 2013). They discover the practical value of their perspective on innovation, noting, for example, "seeing all the amazing potential that there is that the company is not using" (CI 2012–13). At the same time, they realize that they have to "dose it well in order to have a balance" (EK 2012), and attend to "helping their participants voice take form without imposing my vision" (EK 2013).

Grappling with Uncertainties

Two additional aspects emerge here, the artists' own uncertainties and their roles, both of which may be somewhat related to the double newness at work: artistic interventions are a new means of generating newness in organizations (Berthoin Antal 2009, 2013). As one artist admitted, "Not having done anything like this before we felt a bit insecure at first, but this feeling disappears quickly" (BD 2012).

Some uncertainties revolve around the lack of clarity in the project, a factor that the artists can address with the participants, although, as an artist points out, "absolute clarity on the scope and resources" is difficult to formulate at the outset of an exploratory process (CI 2012–13). They referred to uncertainties about the potential and desired impact of the project as a particular problem. In this regard, some question their own ability, for example, "not knowing whether I could help them in as delicate a situation as they were going through" (EK 2012). In other cases, they question the management commitment, such as "doubting whether the process is really transformational for the WHOLE organization. In other words, whether the engagement of the people who participated is supported by the whole structure, especially by the main decision-makers" (EK 2013).

The artists' willingness to admit to and grapple with uncertainty is a resource. In order to engage people in learning, the artists must "repeatedly address some questions that require deeply rethinking, and calling oneself into question" (EK 2012). As co-learners, the artists cannot remain immune to experiencing uncertainty as well. This dynamic may be one of the most significant features differentiating artistic interventions from the relationships and learning processes in conventional change consultancy projects. "As distinct from an assignment in which you only give answers, here you ask, they ask you, you respond and they respond" (BD 2012).

Since artistic interventions are new situations for all involved, none of the participants know exactly what their roles entail. The artists find themselves engaging in a balancing act with "the difficulty of situating myself between what is expected I should do, what I manage to do, and what we could do" (BD 2012). One aspect of the role definition challenge comes from the fact that people in organizations have a clearer sense of how to work with suppliers and experts than with artists, so there is a risk that they automatically position artists into those kinds of roles, which the artists resent. "At some points I believe they thought I was a supplier for them and they treated me that way" (CI 2012–13). Having been frustrated in an organization that "wanted concrete things from me, like designing a card" an artist advised: "position yourself before the collaboration" (BD 2012). Such advice places the responsibility for role clarification primarily with the artists, which raises the second aspect of the role clarification challenge, namely, how the artists would like to define their role in their own minds. As an artist explained, one of the most problematic aspects of the project was that she found herself "negotiating with myself the ambiguous position between psychologist-person-artist" (BD 2012).

Energizing the Energizers

Do the artists derive enough value from participating in artistic interventions to want to recommend such projects to other artists? If so, what is it that energizes them in the process? When they look back on the journey and respond to the question: "Would you recommend it to other artists?" none of the artists marked "no"; the overwhelming majority (85%) indicated "yes" and the others marked "maybe" because it depends on the interests of the artist and the willingness of the organization to engage. The reasons they give to other artists for embarking on an artistic intervention are somewhat different from the hopes they expressed for themselves at the outset. The difference may be partially attributable to their learning from the process. It is also likely that some of the difference in reasoning is due to the formulation of the question, possibly triggering them to look outwards and think in terms of value in the artworld. The responses provide insights into what energizes the energizers of artistic interventions in organizations.

Engaging with the "Real World"

Whereas at the outset the artists refer often to hoping for "learning" and "mutual/co-learning," the reasons they recommend the experience are that it offers access to "reality/realities," in which to "experiment" and which have proven "enriching." What they find valuable is having had the chance to try out their ideas and methods outside the confines of the art world. Some refer to entering "the real world" as opposed to the "bubble," "white box" or "marble tower" of the art world, while others go beyond dualities and refer to "other realities." The nature of the move between realities is aptly captured by the artist who remembers the initial feelings, "it is very enriching, even if at the outset it all seems very surreal to you" (EK 2013).

Beyond discovering a different context and finding inspiration there, it is also about having an impact, "to transform them" (EK 2013). The series of reasons given in one of the responses is a useful illustration: "Seeing the results of the processes inspires one to create with new possibilities. Visiting different offices from time to time as a source of inspiration. Filling ourselves with new information. Acting directly in society" (EK 2011). There is an excitement in the way some artists formulate their messages to their peers, "Because you have the chance to make some real great work and be useful and make a change for real!" (CI 2011–12). Some express themselves somewhat more poetically, "It is a different experience that allows you to apply your knowledge and practices in a new field and thereby to test them and give them wings" (EK 2013).

Extending and Revalorizing Artists' Methods

The experimentation that the artists value extends beyond trying out art-based methods and ideas in a new setting, to actually doing it in interaction with the people whose world they enter. They recommend artistic interventions because "it allows you to experiment with another type of relations between artist-art-people in the real space" (BD 2012). The reasons the artists offer their peers include elements that might scare off people from other walks of life, such as "It is really interesting to get to know other spheres that conflict with how you think" (EK 2013) and "It is a treatment in humility" (EK 2012).

It is interesting to follow the line of reasoning offered by one of the artists in his/her recommendation to peers. The first sentence explains the essential purpose of many practices from the art world, suggesting their applicability in diverse settings: "Because I believe that many of the tools that we use more or less methodically as artists are work tools that help open questions, review perspectives, ways of seeing, expand or distort the field of the 'symbolic,' establish 'other' relations with what is around you, gain distance." The artist goes on to highlight the additional value they can gain from using their methods outside the art world. It gives new value to the artists' tools: "Furthermore, it helps to revalorize these tools as 'tools' of relational value, beyond artistic production." And engaging the experience reveals to the artists how they create additional value: "It requires work to formulate and share one's ideas which is in turn interesting because it makes us realize that it is part of one's professional skill, not 'divine inspiration' and/or 'genius' " (EK 2011). Thus, working with employees in the world of organizations holds three kinds of attraction for artists: the opportunity to try out new approaches outside the art world, to extend the value of their artistic tools into new domains, and to reposition what it is about being an artist that makes them valuable.

Valuing Practicalities

Two practical reasons for recommending artistic interventions to their peers surface in the responses. A few mention the attractive fact that the artists are paid for the work. And several refer to support of the intermediary *Conexiones improbables*. They value the format of the program and the facilitation that the members of *Conexiones improbables* provide because it gives the artists "a freedom for work and it is possible to achieve better results" (EK 2011). As one artist summarized the recommendation: "It is a valuable space to experiment. If someone wants to apply their art practice to contexts outside of studio or gallery—this is a perfect starting point" (EK 2012).

What lies ahead for artists who are interested in engaging in artistic interventions? Will it become an established professional option? The multiplicity and ambiguity of roles that the artists describe are partially attributable to the newness of artistic interventions as an instrument for stimulating innovation and learning in organizations. The diversity of arts involved and the fact that most artists are entering the field for the first time are additional reasons. Unless or until artistic interventions become an established product/service with a clearly defined professional profile for artist-consultants (Heinsius and Lehikoinen 2013), inventing, experimenting with and negotiating around a combination of roles is likely to remain an integral part of the learning process in every new artistic intervention. But it is also conceivable that the very nature of these interventions, their art-driven uniqueness and unpredictability, will require such fluid role-making even after years of experience.

Returning to the Academy

Stepping back out of the artists' discourse to the world of research, what can organizational scholars learn from artists who have engaged temporarily in being the sand, the oil, and the energy for generating new ways of seeing and doing things in organizations? The responses from the artists about the way they approach artistic interventions in organizations and what they value about the experience suggest that they offer a different model of relations between internal and external sources of knowledge, ideas, and skills for innovation. Whereas the literature portrays artists as culturally foreign beings who embody the scarce resource of creativity and import new ideas and practices into the organization, the artists themselves expect to engage in co-learning. Their model entails revealing, valorizing, and activating potential that exists in the organization.

Expanding the Model of Knowledge Creation

To conceptualize the artists' co-learning model, it is useful to explore how it relates to the model of knowledge creation that Nonaka and his colleagues have developed (Nonaka and Konno 1998; Nonaka and Takeuchi 1995). The "SECI" model foresees *s*ocialization, *e*xternalization, *c*ombination, and *i*nternalization as four modes of creating knowledge from different bodies of explicit and tacit knowledge. The assumption underlying artistic interventions is that organizations can tap into the foreign knowledge of the world of the arts to generate newness. One expectation could be that artists enter into the organization and (1) externalize their knowledge from the world of the arts (i.e., make their tacit knowledge explicitly shareable). Another expectation would be (2) that artists serve as role models when they enter organizations, bringing their artistic ways of knowing and doing into the setting, and thereby offer the employees the opportunity to be temporarily socialized into the world of the arts, absorbing tacit knowledge by observing and interacting with the artists. The externalization and socialization approaches would be followed by (3) internalization, so that the employees can use the knowledge gained from the artists to create newness when they return to their work.

These three processes may well take place during and after artistic interventions, but the artists speak of a different form of knowledge creation. Their point of departure is that people in the organization have valuable knowledge and skills that are different from their own. The possibility of working with these differences is what attracts them to undertake artistic interventions, which suggests that artists are actually interested in the other mode in Nonaka's SECI model, namely (4) combination. In the SECI model, combination refers to bringing together different kinds of explicit knowledge as a matter of inserting, replacing, and connecting elements from different bodies of knowledge and making the new result available in explicit form such as in data banks or manuals. But artists do not speak of simply combining their arts-based knowledge and skills with those of employees. They want to engage in co-learning with the people in the "real world" of organizations, a process they describe quite differently from the technical-sounding activities in this part of the SECI model. The three distinguishing characteristics the artists mention are: the releasing of emotions, the revealing of unvalued knowledge, and the valorization of not-knowing.

Three Distinguishing Characteristics of Co-learning with Artists

Firstly, artists highlight the role of emotions, such as joy and resistance, which they see as resources in the process. Here is where the sand and the oil, the friction and the harmony, and the energy come into play, but differently than expected. The images of oil and sand suggest that artists embody these elements so that their foreign presence should disrupt the situation and facilitate change. Instead the artists seek to awaken and release emotional responses among members of the organization, because therein lies the energy for action. Certainly emotions are powerful motivating forces, but they are much more. They are the invisible side of most organizations. Organizations tend not to provide any legitimate space for emotions to be expressed, seen, experienced openly, and understood. The arts can provide that space—not just to release energy (motivation) but to make it possible for people to choose to bring their full selves, or at least more of their selves into the organization.[9]

Secondly, rather than limiting their sights to the obvious knowledge that is already recognized in its explicitness, the artists sense that there are hidden sources to tap. They note that employees have knowledge, skills, and ideas that are invisible in the organization. The artists do not apply the standard dualism of explicit and tacit knowledge but rather recognize that knowledge, skills and ideas may be hidden because they are not valued in the existing culture or because the employees fear expressing them. Therefore, the artists

[9] I am indebted to Victor Friedman for developing this idea with me.

show how uncovering and valorizing them are crucial to knowledge creation from the different resources of internal and external agents in organizations. Valorizing existing hidden knowledge and enabling it to grow wings and fly, as an artist wrote, is missing from the existing models of knowledge creation in organizations. It is one of the processes that the artists discover they need to undertake in order to build the capacity for learning together.

Thirdly, the artists' reflections on their momentary experience in organizations highlight the value of not-knowing. Whereas the organizational literature focuses on knowledge and knowing as key assets and treats uncertainty and gaps in knowledge as problems to be eradicated, the artists highlight that the capacity to admit uncertainty and to engage with not-knowing is a potential resource for generating newness (Berthoin Antal 2013). In artistic interventions the organization's search for answers is disrupted by the artists undertaking a search for questions in their process of exploring the foreign physical and social context. Their approach can help to make doubt generative by motivating "a search for understanding" because "living doubt is necessary to energize inquiry" (Locke et al. 2008: 908).

Creating "Interspaces" of Possibility

The companion to doubt and not-knowing that the artists stimulate in the organization is imagination, "the power of constructing possible models of human experience" (Frye 1964: 22). Before entering into the organization, artists imagine engaging in co-learning, and when they discover that the conditions for such relationships are not immediately present, they work to build the necessary conditions with people. They challenge the existing mindsets and norms to carve out spaces in which to experiment and play. As organizational scholars are beginning to point out, play is valuable in organizations because it enables humans to "imagine new possibilities for society and become more able to bring those possibilities into lived reality" (Statler et al. 2009: 91). The artists thereby open a new kind of space, or "*ba*" in Nonaka's terminology for knowledge creation (Nonaka and Konno 1998).

I refer to such space as "interspaces" in which members temporarily suspend the cultural norms of the organization and experience the possibility of thinking and working together in new ways (Berthoin Antal and Strauß 2013: 34). Interspaces are by definition ephemeral but they have the potential for changing norms and practices in an organization by extending what is considered desirable and possible in the cultural "toolkit" (Swidler 1986). The artists aspire to having a sustainable impact on the organization, lasting after they have returned to their art world. But the responsibility for following up on the experiential learning by making space for inquiry, experimentation, and play that participants have discovered they are capable of doing together remains with the local population, that is, the employees and the management.

The fact that interspaces for co-learning were not achieved in all the artistic intervention cases in this sample shows that a new approach to organizational learning and innovation cannot always overcome the barriers encountered by other approaches. A few artists commented on what they perceived as lip service: although the management had decided to embark on the project, either they were not willing to get on board when the boat set out on the journey or they essentially took down the sails when it came time to acting upon ideas. The artists attribute the problem to interrelated factors: inappropriate role expectations (especially treating the artist as a supplier), insufficient time invested into the collaboration by the organization, and lack of support from the power structure of the organization for transforming ideas into practice.

Challenges to the World of (Organizational) Research

To seasoned observers of organizational behavior, there is nothing surprising in this gap between espoused theory and action (Argyris and Schön 1978). In order to clarify whether artistic interventions succeed in addressing this gap more or less effectively than other kinds of interventions, and if so, when, comparative research needs to be undertaken. It would be interesting to compare them not only with traditional consulting approaches to organizational development and innovation, and experiential approaches like outward bound activities that have entered the mainstream in the past decades, but also other aesthetically oriented approaches like design-based interventions (Barry and Hansen 2008). Such research could help specify when physical and emotional engagement matter, and when purely cognitive approaches to learning suffice.

The task poses several challenges to the world of organizational research. The most obvious is access to comparable cases: consulting companies are generally reluctant to allow external researchers to accompany their projects and assess the outcomes. Furthermore, cases should have certain similarities (e.g., pursue the same objectives with different means in the same or similar organizations), which does not seem to happen often.

Methods are another challenge: studies should include arts-based ways of collecting and analyzing data, and representing the findings so as to enable the valorization of aesthetic dimensions of organizational processes and outcomes that have remained largely invisible in organizational studies. Some members of the world of organizational research are beginning to embark on experiments of co-learning with members of the world of the arts. And some are expanding their professional identities to scholar-artists (e.g., Nancy Adler and Steve Taylor).

A third challenge to the world of research is structural. Most structures and practices in academia remain solid barriers to newness: peer review processes for journals and research grants, as well as recruitment and promotion criteria in universities, business schools and research centers, make it very difficult to undertake work that does not fit existing categories. Furthermore academic publishers rarely enable the dissemination of findings outside the traditional format of black words on white paper, so material that requires color, motion, or sound is rarely made available to readers.

Researchers and our Stakeholders

In closing, I invite readers to join me in (re-)considering the role of researchers who are undertaking studies of processes of generating newness, and the role of our stakeholders. My studies of artistic interventions in organizations are themselves a form of intervention: by asking the participants to respond to my questions, I insert a moment of reflection and valuation into their process. The questions elicit what they value and give them an opportunity to valorize their work and their perspectives. To what end? The intention is threefold. Firstly, to offer the stakeholders who are directly involved (i.e., the artists, employees, managers, intermediary organizations) material with which they can work to take their learning forward. Secondly, to share insights from these experiments with other stakeholders who are interested in trying new ways of working with people in organizations in order to generate new practices or improve on existing ones (e.g., policy makers and funding bodies). Thirdly, to contribute to conversations among academic colleagues who want to understand the interaction between organizations and society, and the factors that enable or impede learning and social innovation.

These goals are shared by numerous scholars, particularly in the field of action research and participatory action evaluation (Friedman 2001a, 2001b; Friedman and Rogers 2008). But studying and writing about processes that involve the arts in organizations confront researchers with an additional challenge: the pressure to satisfy the demand for certainty in a context where uncertainty is an essential feature. In commenting on an earlier version of this paper, Michael Eriksson pointed out that creating the "illusion of certainty" may be the main differentiator between conventional consulting projects and artistic interventions. In the world of organizations certainty is considered a precondition for planning, therefore consultants often tend to generate a sense of certainty, even if it is just an illusion, whereas uncertainty is considered a precondition for creativity. Research results that create the illusion of certainty around artistic interventions in organizations might encourage some decision-makers who are reluctant to follow McNiff's advice to "trust the process" (1998) when working with the arts, to dare tread into this unknown territory in search of newness. However, such research output is a double-edged sword, because publications claiming certainty in this area risk raising promises of values-added to unrealistic levels, thereby pre-programming failure to meet expectations. The potential cost to the organizational research community of colluding with the demand for certainty in conducting and reporting on studies about artistic interventions is even higher: the interesting insights into processes of generating newness are likely to remain invisible.

Reporting on research without colluding with the demand to provide (the illusion of) certainty entails expecting more from the stakeholders of our research output than they may be accustomed to doing when reading articles that reduce complex realities to "best-practices." In keeping with the theme of artistic interventions, it means challenging readers to engage with uncertainty and develop their "educated imagination" (Frye 1964). After reading an early version of this chapter, Victor Friedman suggested offering readers my findings in the form of pieces of a mosaic to work with. This contribution is therefore an experiment in collaborative sensemaking with my future readers.

References

Argyris, C. and Schön, D. A. (1978). *Organizational Learning: A Theory of Action Perspective*. Reading, MA: Addison-Wesley.

Barry, D. and Hansen, H. (eds.) (2008). *The Sage Handbook of New Approaches in Management and Organization*. Los Angeles, CA: Sage.

Barry, D. and Meisiek, S. (2004). *NyX Innovation Alliances Evaluation Report*. Learning Lab Denmark, Aarhus University http://www.dpb.dpu.dk/dokumentarkiv/Publications/20051215151928/CurrentVersion/rapport.pdf (accessed September 27, 2009).

Barry, D. and Meisiek, S. (2010). Seeing More and Seeing Differently: Sensemaking, Mindfulness and the Workarts. *Organization Studies*, 31(11): 1505–1530.

Berthoin Antal, A. (2009). *A Research Framework for Evaluating the Effects of Artistic Interventions in Organizations*. Gothenburg: TILLT Europe.

Berthoin Antal, A. (2012). Artistic Intervention Residencies and Their Intermediaries: A Comparative Analysis. *Organizational Aesthetics*, 1(1): 44–67.

Berthoin Antal, A. (2013). Art-Based Research for Engaging Not-Knowing in Organizations. In S. McNiff (ed.), *Art as Research*. Chicago, IL: Intellect/University of Chicago Press, 171–180.

Berthoin Antal, A. and Strauß, A. (2013). *Evaluating Artistic Interventions in Organizations: Finding Evidence of Values-Added*. Creative Clash Report. Berlin: Wissenschaftszentrum Berlin für Sozialforschung http://www.wzb.eu/sites/default/files/u30/effects_of_artistic_interventions_final_report.pdf (accessed December 10, 2013).

Biehl-Missal, B. (2011). *Wirtschaftsästhetik—Wie Unternehmen die Kunst als Inspiration und Werkzeug benutzen*. [Business Aesthetics: How Business Ventures Use Arts as Inspiration and Tool]. Wiesbaden: Gabler.

Boyle, M.-E. and Ottensmeyer, E. (2005). Solving Business Problems through the Creative Power of the Arts. *Journal of Business Strategy* 26(5):14–21.

Darsø, L. (2004). *Artful Creation. Learning-Tales of Arts-in-Business*. Frederiksberg, Denmark: Samfundslitteratur.

Dewey, J. (1934/2005). *Art as Experience*. New York, NY: Penguin Group.

Economist (2011). Schumpeter. The Art of Management. Business has much to Learn from the Arts. *Economist*. February 17, 2011 http://www.economist.com/node/18175675 (accessed February 20, 2011).

Friedman, V. J. (2001a). Action Science: Creating Communities of Inquiry in Communities of Practice. In P. B. Reason and H. Bradbury (eds.), *Handbook of Action Research: Participative Inquiry and Practice*. Thousand Oaks, CA: Sage, 159–170.

Friedman, V. J. (2001b). Designed Blindness: An Action Science Perspective on Program Theory Evaluation. *American Journal of Evaluation* 22(2): 161–181.

Friedman, V. J. and Rogers, T. (2008). Action Science: Linking Causal Theory and Meaning Making in Action Research. In P. Reason and H. Bradbury (eds.), *The Sage Handbook of Action Research*. 2nd ed. Thousand Oaks, CA: Sage, 252–265.

Frye, N. (1964). *The Educated Imagination*. Bloomington, IN: Indiana University Press.

Heinsius, J. and Lehikoinen, K. (2013). *Training Artists for Innovation. Competencies for New Contexts*. Helsinki: Theatre Academy of the University of the Arts Helsinki. http://www.cultuur-ondernemen.nl/documents/17339/0/TAFI+book+2013.pdf (accessed December 10, 2013).

Hutter, M. (2013). La Coévolution du Monde de l'Art et du Monde Commercial. Deux Histoires d'Irritation. [The Co-Evolution of Art and Economy]. In P.-J. Benghozi and T. Paris (eds.), *Howard Becker et les Mondes de l'Art. Acte du Colloque de Cerisy*. Palaiseau: Éditions de l'École Polytechnique, 241–254.

Locke, K., Golden-Biddle, K., and Feldman, M. S. (2008). Making Doubt Generative: Rethinking the Role of Doubt in the Research Process. *Organization Science*, 19(6): 907–918.

McNiff, S. (1998). *Trust the Process: An Artist's Guide to Letting Go*. Boston, MA: Shambhala.

Morgan, G. (1986). *Images of Organization*. Newbury Park, CA: Sage.

Nonaka, I. and Konno, N. (1998). The Concept of *Ba*: Building a Foundation for Knowledge Creation. *California Management Review* 40(3): 40–54.

Nonaka, I. and Takeuchi, H. (1995). *The Knowledge-Creating Company: How Japanese Companies Create the Dynamics of Innovation*. New York, NY: Oxford University Press.

Schiuma, G. (2011). *The Value of Arts for Business*. Cambridge: Cambridge University Press.

Seifter, H. and Buswick, T. (eds.) (2005). Arts-Based Learning for Business. Special Issue of *Journal of Business Strategy* 26(5).

Statler, M., Roos, J., and Victor, B. (2009). Ain't Misbehavin': Taking Play Seriously in Organizations. *Journal of Change Management* 9(1): 87–107.

Styhre, A. and Eriksson, M. (2008). Bring in the Arts and Get the Creativity for Free. A Study of the Artists in Residence Project. *Creativity and Innovation Management* 17(1): 47–57.

Swidler, A. (1986). Culture in Action: Symbols and Strategies. *American Sociological Review* 51(2): 273–286.

16

Performing Worth: Shareholder Value and Management Consultancy in Post-Mao China

Kimberly Chong

A partner of Systeo,[1] a global management consultancy, asks a group of junior consultants a simple question:

> What do you tell your parents you do?
>
> A doctor, we are like a doctor. If a client has a problem we go in and help them, volunteers a young female consultant.
>
> The partner wrinkles his face.
>
> Wouldn't it be better to tell people 'I reduce 10 million in inventory? Or I raise revenue by 25%?' Doesn't that sound much better than saying 'well...I make clients these, err, deliverables, these PowerPoints...'

Why the partner should wish to emphasize the reducing of inventory or increasing of revenues, over the actions engendered to achieve these effects, is connected to the central claim of management consultants—that they "create shareholder value" for their clients (Froud et al. 2000). Consultants are paid to generate the right representations—those which reference a strong balance sheet—with the view to increasing a company's share price (Chong 2012). They do this under the guise of "improving efficiency." But this overarching objective and its underlying imperative is not self-evident to junior consultants. Rather, it has to be taught.

As others have noted, since the 1980s the pursuit of shareholder value has become the raison d'etre of the modern corporation, leading to drastic

[1] "Systeo" is a pseudonym. As part of my terms of access, I agreed to anonymize the management consultancy and all of its employees.

changes to work organization and labor practices, which have been implemented, in the main, by management consultants.[2] However, as the above vignette suggests, the logic that connects their everyday work to shareholder value creation is often unclear to management consultants, let alone their clientele. Insofar that consultants are mostly engaged in cost-stripping exercises which, more often than not, fail to improve performance (Fligstein and Shin 2007), we can say that the claim of efficiency is dissonant with the claim of creating shareholder value. This raises the question of: how do experts who propagate shareholder value come to understand the value of their work? Given the considerable attention devoted to exposing the ideological character of shareholder value (e.g. Lazonick and O'Sullivan 2000), it is surprising to find a paucity of answers amongst the existing literature.

Drawing on 16 months of fieldwork inside the China arm of Systeo, this chapter presents ethnographic material of internal training sessions, specifically those which concentrate on client acquisition i.e. "sales." Rather than stressing the efficacy of their products, management consultants instead approach selling by focusing on value. They aim to demonstrate the value of maximizing shareholder value and then use this moral legitimacy as a springboard for their own expertise. This is no straightforward task. I will demonstrate that consultants are trained in various logics of worth each of which attempt to "smoothen" the connections between alternate scales of value. One logic subsumes financial value in an overarching corporate ethics of shareholder value, and another connects valuation to locally relevant and meaningful cultural tropes of ethical behavior which are not subordinated to the creation of economic value but rather are used as models of explication. These multiple logics of worth are not reducible to each other. In agreement with David Stark, I will argue that the rendering of value under a financial logic rests on harnessing the productive friction of multiple evaluative criteria (Stark 2009). In this sense, dissonance can be a productive heuristic in training a value trope characterized by conceptual liability (Guyer 2012) whereby shareholder value "can be variably invoked as cause, consequence or justification" (Froud et al. 2000: 81) of consulting.

Furthermore, I will show that, as an encompassing logic of worth, shareholder value within the consultancy is not always narrated as an ethic. At times it is construed as a product of science, acquiring an objectivity which lends an ideological force to the term, and its ontology is obscured by the explicit ethicizing of shareholder value which implies that shareholder value itself is not an ethic. However, at other times the creation of shareholder value is explicitly tied to business ethics, that creating value for the company's

[2] For discussion on the relationship between the rise of shareholder value and corporate restructuring, see Milberg 2008, Williams 2000, Davis 2009.

shareholders is the right thing to do.[3] I will propose that the success that management consultants have had in popularizing shareholder value as a form of "purposive management" (Froud et al. 2000) rests on the dissonance between multiple modes of valuation (pertaining to companies and labor) and conceptualizations of what value is.

Restructuring and Financial Representations

In China management consulting is still a relatively new industry. It is dominated by Western firms which, prior to market reforms, were prohibited from entering its then closed economy. In the past ten years, however, these Western consultancies have rapidly expanded their operations in China where a host of new clients can be found. The second phase of Deng Xiao Ping's market reforms marked a commitment to the market model and concomitantly a significant acceleration of economic restructuring. Under the policy of "grasping the large and releasing the small" (*zhua da fang xiao*), the state privatized small state-owned enterprises (SOEs) and reorganized large SOEs into corporations, the largest of which were listed on domestic or overseas stock exchanges but with the state retaining control as the majority shareholder. Once socialist work units Chinese SOEs now include some of the largest companies in the world by market capitalization. Implicated in this transformation are the likes of Systeo. Although the state continues to play a strong hand in steering the economy, China's startling growth can also be attributed to the outsourcing of institutional restructuring to Western intermediaries (Steinfeld 2010). Systeo replicates in these organizations the forms of management—the devices, practices and processes—that it first devised in Western settings. Although a market-based system of corporate governance has yet to be fully realized i.e. the stock market has only a weak disciplinary effect on firms (Steinfeld 2004), the hiring of management consultancies and the changes to work organization which they have implemented can be seen as precursors to more aggressive market reforms.

Systeo's consulting work in China mainly consists of the implementation of accounting technologies. That is, installing computer systems in their clients which are designed to improve corporate governance by making visible the individual contributions of workers to shareholder value. These range from small-scale customizations of a client's existing IT system to the installation of comprehensive systems of accountability—enterprise resource planning systems, commonly known as "ERP systems." To underline the global

[3] This is also a key objective for investment bankers, as Karen Ho's ethnography of Wall Street demonstrates so clearly (Ho 2009).

dominance of these technologies, it should be noted that almost every large organization (in the private and public sector) in North America and Europe runs a type of ERP system. In the white papers published by the software company and leading vendor of ERP systems, SAP, it states explicitly that ERP systems are a means of "reengineering" a firm's "culture" under a rationale of shareholder value. For example, in the paper entitled *Enabling Value Based Management,* it states: "To deliver superior value based returns, companies must embed the value approach deep in the organizations' culture—and this requires information systems that support the decision-making process (SAP White Paper 1999: 3)"

In the next paragraph, it poses its own product as the solution: "SAP's next generation solution will facilitate comprehensive value based decision making" (SAP White Paper 1999: 3). ERP systems serve to institutionalize and automate forms of surveillance and control, to produce what has often been termed audit culture (Strathern 2000). They can be used to "prepare" a firm for initial public offering (IPO) on the stock exchange (Reimer 2003), or to bolster the share performance of an existing corporation. In short, what ERP systems do is convey to potential investors that the entity in question has *good management.*[4] By producing "transparency" and "accountability"—corporate virtues which can also be deconstructed (Garsten 2008, Cross 2011)—accounting technologies such as ERP systems aim to convince investors of management quality. We find that management consultants claim to create value by devising means of measuring it (see also Moor and Lury 2011), in doing so they also produce modes of valorizing corporate behavior which can be seen to accord with certain models of the economy.[5] In brief, consulting isn't just the installation of technology, it is also the establishment of value logics that makes this technology desirable, necessary even.

Consultants learn about shareholder value—how it can be represented, created, increased and destroyed—in part through their own experience of being managed. They are evaluated through a performance management system which is operationalized through the same ERP systems which they install in their clients' organizations. Although a number of measures are enmeshed in their targets, consultants quickly learn that the most important metric of all is their billability—the ratio of hours worked that can be billed to clients. For billability is taken as a proxy of their revenue-generating

[4] Good management is one of a number of conviction narratives that can provide the psychological impetus for money managers to make a decision under radical uncertainty (Chong and Tuckett, 2014).

[5] The focus on transparency can be traced to conventional approaches in economics which consider decision-making to be optimized under conditions of perfect information and concomitantly consider "irrational" or sub-optimal behavior to derive from a lack of information. In this model the solution is almost always more information hence the valorizing of transparency.

capacity, and thus a measure of their contribution to Systeo's profitability. And yet this connection with Systeo's financial results often escapes consultants, to the extent that it had to be explicitly elucidated by Liu Xing—the Chairman of, and most senior director in, Systeo's China arm. A common complaint he hears from consultants is "why do you press me so hard on billability?" In order to answer this question "you need to understand Systeo's financial results," he claimed, going on to say that consultants needed to have a "shareholder mindset." This entails three imperatives:

- Solve for Systeo: our clients, our people and our value.
- Little things matter and have a material impact on Systeo's valuation.
- Generate cash flow.

Each bullet point on the PowerPoint slide prescribes a course of action with the goal of elevating Systeo's share price. In making apparent how a "shareholder mindset" can make sense of an evaluation system which reduces all forms of productivity to the measurement of financial value, Liu Xing also, at another level, connects Systeo's work of installing accounting technologies to shareholder value creation. What is unclear, however, are the logics of worth which embellish and justify these connections. These logics are integral to propagating and legitimizing a mode of market valuation whereby a company's worth references culturally constructed evaluations of labor productivity. In the next section I will examine the training of these logics, attending specifically to the top-down narration and presentation of value. The voices of the trainees, whilst undoubtedly an important aspect of acculturation, have not been included. In doing so, I foreground the pedagogical technique of performing worth, looking at how performance is deployed as a method of inculcation and, in the performativity vein, as a mechanism of intervening in social reality.

Creating Value Propositions

In the upper reaches of the organizational hierarchy consultants' responsibilities are less concerned with the day-to-day tasks of installing computer systems, and more about how to sell consulting, internally to—"our people"—and externally to—"our clients." Internally, this entails becoming a "steward" of the company in the form of a career counsellor to their subordinates. In this way they are co-opted into the project of disseminating logics of valuation into the lower reaches of the firm. Externally, they must learn to sell Systeo's system of value-based management, that is, management defined by the ultimate goal of creating shareholder management, as a "win-win" strategy—a win for the client and a win for Systeo.[6] These responsibilities are officially communicated through a set of trainings for each level of management: managers, senior managers and partners. The first I observed was the annual "Manager's Milestone" training, mandatory for all consultants who have been recently promoted to manager level. The title, "Manager's Milestone" is both indicative and misleading. In making it this far, consultants have reached a turning point in their careers. But most of them are not managers. Rather, they have been promoted to the next level up which happens to be called manager, and to mark the occasion they are the recipients of intense corporate training at a five star hotel in Beijing, to teach them *how to create value not how to manage.*

Of fundamental importance is that consultants learn how to construct what is known in Systeo as a "value proposition." In its most basic form, a value proposition is simply a description of an action which is justified by recourse to the production of value. This could be shareholder value, but not necessarily. Indeed, the word "shareholder" is conspicuously absent, lending a conceptual ambiguity to the term. "Value" in this setting cannot be understood solely as social values or economic value (Boltanski and Thévenot 2006). Rather consultants are trained to promulgate a logic of worth within which financial value is subsumed, so collapsing these distinctions of "values" and "value." This is taught progressively with employees being given more and more detailed and technical knowledge about how to construct value propositions as they are promoted up the management hierarchy.

In the Manager's Milestone training consultants are told that in order to sell consulting: "You have to convince yourself first, you have to believe [the value proposition]... You're selling business and personal values."

We see that value is narrated as something that cannot be determined by an apparatus of technical manoeuvres alone, but rather value is something that must resonate with the ethical predilections of Systeo's clients. As will become apparent, this resonance must be created. Chinese partner Edward Zhang lectures the 160 newly promoted "managers" on the importance of relationships, he says that: "Relationships win deals. We carried out a study in FY08 Q1 (the first quarter of the 2008 fiscal year), we found out that 65% of the time 'relationship' was the key factor contributing to the win for Systeo or our competitor."

He tells them that one way of extracting valuable information out of their relationships is by making a "Deal Power Map" when formulating a sales pitch. He tells them that crucial to securing a deal is understanding the organization of their client. But a power map is not an organizational chart [Figure 16.1].

[6] Value-based management is not a term particular to Systeo. Rather it is commonly used in the consulting industry and is also referenced extensively in publications produced by software companies which produce ERP systems, such as the SAP White Paper which I have cited.

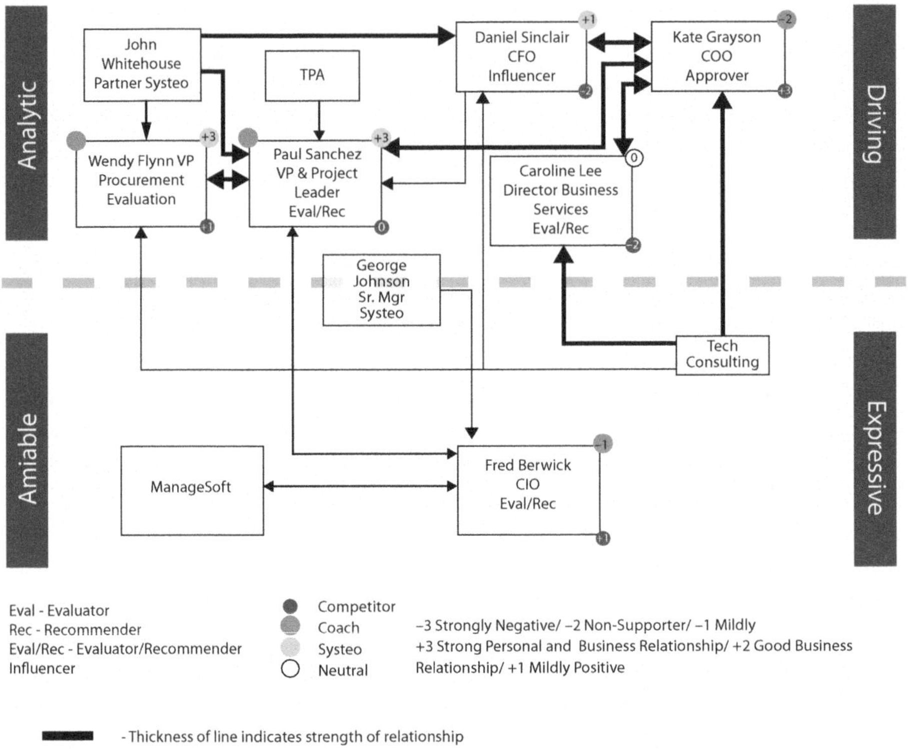

Figure 16.1. Deal Power Map overlaid with SOCIAL STYLES™ model (adapted from fieldwork drawing)

It is a tool which maps out the subjective dimensions of how decisions are made in their clientele. In order to construct the map, you need to find out "who are the key players and what are their roles? Are they evaluators, influencers, recommenders or approvers?" to use Edward's words. This translates into who are the people involved in the decision to buy Systeo's consulting services and how is their relationship to the final outcome structured. For example, do they evaluate the sales proposal, make recommendations or give the final approval. Once their roles are determined, "you gotta find out whether they are analyticals, amiables, drivers or expressives," he continues. This is the terminology of the SOCIAL STYLES™ model which Systeo licenses from the training company Tracom Group. According to this model of human behavior, everyone falls into one of these four categories of social style on the basis of two characteristics—their assertiveness and their responsiveness.[7]

[7] For a more detailed description of this model see the website of Tracom Group: http://www.tracomcorp.com/training-products/model/social-style-model.html (accessed February 10, 2014).

$$\text{Trust} = \frac{\text{Credibility} + \text{Reliability} + \text{Intimacy}}{\text{Self-Orientation}}$$

Figure 16.2. The Trust Equation (Maister et al. 2000)

Edward doesn't bother to describe all four styles, concentrating his efforts on "drivers." He describes this disposition as "I want this tomorrow!" and like many partners I met he says that most consultants are drivers—they produce actions and deliver results. Finding out the social styles of the people who make decisions on the client side is critical, because, according to the SOCIAL STYLES™ model, each "style" trusts differently as modelled by The Trust Equation [Figure 16.2]. Given the costly restructurings they prescribe what consultants certainly do need is to have their clients' trust.

The Trust Equation is the work of management consultants and theorists, David Maister, Charles Green and Robert Galford (2000), and appears in their book "The Trusted Advisor." It is a manual specifically designed for the consulting and advisory industry about how to make the transition "from outside technician to habitué of the inner sanctum." The work of Systeo is to synthesize these two models, SOCIAL STYLES™ and The Trust Equation, which are not their intellectual property, to produce a template for social relationships which will guide their consultants in the sales process. Consultants are told that drivers emphasize credibility to produce trust, amiables reliability, expressives intimacy and analytics self-orientation. If consultants can decipher the social style of their client, then, according to these models, they can work out how to get their client to trust them using the pseudo-scientific means of The Trust Equation.

If we look at the Powermap example which is presented to consultants in their training handout [Figure 16.1], we can see that the approver, Kate Grayson—the person on the client side who makes the final decision as to whether they procure Systeo's services—is a driver. In training this is presented as the typical scenario. Systeo employees are repeatedly told that the decision-makers on both sides of the deal are predominantly drivers—i.e., both consultants and the key client personnel. In fact, the models of SOCIAL STYLES™ and The Trust Equation are utilized, not to emphasize the diversity of subjectivity but rather to emphasize that there is a singular subjectivity associated with consulting. On a slide that's rather tellingly entitled, "How you Want your Client to Feel," we find out that consulting, if directed to drivers, is about delivering, about having "the A team," about saving the client time and money and giving them predictability. In the last slide of Edward's presentation a chart delineates the buzzwords which resonate most strongly

Table 16.1. Matching "SOCIAL STYLES™ and buyer values" (adapted from fieldnotes)

Analytical	Amiable	Driver	Expressive
• Data and Detail • Transparency • Accuracy • Process/Methods	• Teamwork • Consensus • Proven	• Bottom Line • Action Plan • Assets • Experience • Credibility	• Innovation • Vision

with each "style" and consultants are told to use the chart as guidance for tailoring official business communication with their client. In the handout that accompanies the training it states explicitly that these buzzwords summarize and highlight the *values* that each "style" prefers [Table 16.1]. Drivers value and are most responsive to: "(the) bottom line," "action plan," "assets," "experience" and "credibility." Notably, these "values" reference a model of financial valuation in which profitability, return on assets, and credibility of information, are the focus of investor scrutiny. That is to say, if *both* consultants and their clients are drivers then they coalesce around a set of values which are associated with a shareholder value logic of worth. Far from depicting the range of actual behaviors and characteristics of their clients, these tools are designed to teach consultants how clients *should* behave—they *should* be drivers. And thus they *should* be driven by an overarching aim to maximize shareholder value.

The idea that relationships lie at the heart of business is nothing new in China. A whole body of literature talks about the importance of *guanxi* (social relationships) in cultivating economic ties. Trust is built through a system of reciprocal gift giving which binds together Chinese in the mainland and in the diaspora (Yang 1994). But the tools which are presented to the managers suggest a very different model of relationships. Contrary to their narration, these tools are not designed to help consultants manage their relationships between clients; rather they relate a mode of valorizing behavior. We can see that on a broad scale, these tools emphasize the plurality of value registers and contend that, through the calculation of affect, resonance between the consultant and the client can be created. But this promotion of wide-ranging social proficiency is dissonant with the implicit suggestion that shareholder value is the *only point* where clients' needs are commensurate with what consultants can offer. Intuitively this seems implausible. It is obvious that sales also depend on things like intimacy and reliability which are emphasized in the other social styles. However this dissonance need not threaten the legitimacy of these tools nor their capacity to improve sales. Dissonance "allows" these tools to confer a sense of power and expertise (by claiming to cover all manner of subjectivities), and communicate a mode of valorization. Both are required to construct a convincing value proposition.

The Socio-technical Arrangements of Value Creation

"Deal Power Maps," SOCIAL STYLES™ and The Trust Equation are part of a scientific-sounding treatment of "relationships" which governs selling, pandering to a crude alignment of objectivity with truth. This quasi-scientific method is made material by assembling these tools into one unified, IT-enabled, architecture—the QSA tool—where QSA stands for "quality," "strategy" and "assurance." Explicating how to use the QSA tool and why consultants should use it was the main objective of the Sales Effectiveness Workshop. Open only to senior managers and partners, the workshop was the brainchild of the Human Capital Strategy Program—the internal program designed to ensure the effective replication of Systeo's global corporate culture. In practice, this meant the export of value based management models from the West to China, together with the tools that support their implementation.

The QSA tool is a piece of software which has been tailored to fit with Systeo's in-house ERP system. It is designed to standardize working practices and act as a system of surveillance. Nick Carson, an American partner based in the Shanghai office is the "faculty" for the workshop. He tells the participants that using the QSA tool can be the difference between winning and losing a deal. It does this by:

> structuring thinking through a common, single set of Critical Success Factors, 'CSFs,' it tells you the likelihood of closing and winning a deal as you go through the sales cycle, it has a reporting engine that generates consistent output reports for review meetings. It's basically a 'one-stop shop' for all the best practice sales tools.

He runs through the operation of the tool with the comportment of an American game-show host. He darts around the training room rallying the participants to draw on the flipcharts, "be active" and generally "have fun." They struggle to match his meteoric levels of energy and respond hesitantly to his onslaught of questions. Here are is selection of quotes by Nick:

> DING—out of time! The right answer is there are 12 CSFs to complete...
>
> So who do you need for a new business meeting? You got the service center guys in Dalian, and who else? (pause) COME ON GUYS! Legal, HR—you need to involve them otherwise they're gonna kill yer.
>
> You gotta get the IWIKs, you *need* the IWIKs, you're not gonna get to the next level until you find out the IWIKs.

Regarding this last example, I spent over an hour observing the session trying to work out what on earth an IWIK was. There were tabs and buttons in the QSA software labelled with the letters I.W.I.K. There was a whole process of structured thinking dedicated to finding out the IWIKs. Only at the end, Nick flicks the slide and all is revealed—"I. wish. I. knew."

The use of acronyms is just one way that the language of management devices manage to convey a sense of science. It is also by standardizing and dictating practices and then by connecting this set of "correct answers" with the depiction of truth and reality. These two words are used interchangeably by Nick, who tells participants to "Review the QSA tool and find the truth." Groups of participants are given case studies of sales pitches and are asked what they would do in the situation. First they must decipher "the facts" of the case using the QSA tool, and then make decisions such as whether to pursue a particular client, or what coaching they need to give to their team, based on "the truth." In associating truth with the knowledge generated from their sales tools, Systeo training tries to acculturate consultants into selling and defining consulting as the effect of creating value. Deal Power Maps, SOCIAL STYLES™ and The Trust Equation are methods of traversing the conceptual space between ethical values and financial value. These techniques of managing "relationships" construct consulting as the progenitor of multiple registers of value—it creates shareholder value and the ethics which stipulates its domination as *the* value of corporate practice. That is to say, the QSA tool is designed to automate the production of shareholder value. It is the vital socio-technical machinery required to enact shareholder value, the overarching duty to shareholders, as the defining ethical compass of good management.

In this production of worth science is apprehended to depict the "truth" of shareholder value which must, in fact, be *fostered* through the medium of social relationships. But in drawing on social relationships as a functional tool of knowledge, these techniques attempt to empty sociality, between consultants and clients, of its content: the reciprocal obligations and trust which are so constitutive of Chinese economic life. This point was highlighted in Nick's own advice for the senior managers and partners. He makes explicit that these consultants are not being trained in the art of relationships:

> The key secret in China or Asia I've learned [is] if it's gonna be a painful negotiation I don't want Anna [*pointing to partner Anna Wang*] to be involved in negotiation if she's involved in delivery. I want to separate negotiation from the person who is responsible for delivery. You may have different experiences from me, but I have so many experiences in China. Fly someone in from the U.S., get the Global Negotiating Team for ten thousand or twenty thousand US dollars.

By separating the consultants who install the computer systems, i.e. actually do the consulting, from those who sell the value proposition to clients, there is an attempt to expunge social relations from deal making. This construal of value creation, as something that can be aided by the removal of sociality, is at odds with—is dissonant with—the earlier emphasis on "relationships" as the key to winning deals, and which is formalized into the technology of the "Deal Power Map." Here consultants are told that, whilst sociality with the client is an unavoidable aspect of consulting, it can be removed from the selling process. In negotiations where Systeo is struggling to make the profit margin they desire, the removal of consultants who will have prolonged contact with the client, from the task of selling, is an act that is designed to ensure that the creation of shareholder value appears a technical process and a scientific endeavor.

Ethicizing Corporate Praxis

Establishing shareholder value as *the* ethic of business is the beginning of a project of ethical restructuring that new managers are charged with. The next step is for consultants to learn that they have the ability to create value for their clients and for Systeo. In the Manager's Milestone and the senior manager level equivalent, Senior Manager's Milestone, consultants receive extensive training on how they "impact the bottom line" through their sales deals. There are two principle mechanisms: through price and through cost. Senior managers must attend a session where they learn "pricing excellence," narrated as a lesson in "Basic Contract Financials" by the three Chinese partners who conduct the training. Their speeches are peppered with acronyms such as ODE, HSP and CCI, which stand for Original Deal Economics, Hot Skill Premium and Contract Controllable Income, respectively. These are used in their careful explication of the economics of consulting. First of all they tackle price. A slide appears on the projector screen. It has two columns: how pricing our work impacts Systeo's business and how it impacts your rewards. Under the first column is shareholder value, investor perception and client perception; under the second is annual pay increase and bonus program.

Consultants are told that the ability to price well will positively affect Systeo's share prices which will trickle down to individual remuneration. The partners stress that the senior managers must learn to price Systeo's work correctly and learn the time to close, meaning the time to submit the price to the client. Before they explain how to do this they address the second mechanism—costs. Consultants are advised to reduce the costs required to deliver on their promises—the client contract—by lowering the costs of labor. This is achieved using offshore labor found in Systeo's outsourcing division, and by conducting a "pyramid refresh"—technical jargon for changing the structure of their consulting teams to incorporate more lower level staff such as junior consultants in order to reduce costs. This act of costing has to be balanced with the act of pricing to ensure that the profit margin of each contract, technically termed the contract controllable income or "CCI," is at a minimum of 40%.

Senior managers must learn to carefully calibrate the levers of price and cost in order to maintain this golden figure of 40% without compromising Systeo's reputation. Terence Huang, one of the Chinese partners, tells them that, "rule number 1 is that pricing is not cost-plus, instead it is *what you are willing to pay*. So I suggest moving off price and talking about *value*." (emphasis added).

He says that most of the time Systeo's price is higher than their competitors, which he attributes indirectly to the fact that "Systeo does very honest work for clients, our competitors maybe not." What he stresses is that consultants should avoid *only* moving on price since this affects the CCI. But he also incorporates an ethical justification—that it will appear that they are consultants without integrity. Senior Managers are told to avoid simply lowering prices on demand. Instead any change in price must be accompanied by a corresponding reduction in scope—they must reduce what they are willing to provide in the way of consulting services in order to meet the demands of their client. If they lower their price without changing the scope they may become embroiled in a price war with their competitors and it will also appear that the initial price was artificially high. In short, consultants are tasked with continually reinforcing the message that they are giving a fair price. In the Sales workshop, which took place in the depths of the 2008/9 financial crisis, Nick Carson reiterates this point: "you cannot use the economic situation to lower prices, if you do shame on you, it affects our brand. The second you do, you undervalue our firm," he says. His comments draw attention to two ways in which ethics are critical to market valuation. One, he highlights the importance of fostering the right affective ties—the creation of a "brand"—underpinned by a strong ethics. Two, he articulates implicitly who consultants should be creating value for and thus displaying loyalty to. His claim that the lowering of prices, because it culminates in the *undervaluing* of Systeo, is shameful behavior, even in a financial crisis, rests on the assertion that consultants are in the service of Systeo's shareholders and not other constituents such as their clients.

In short, pricing impacts shareholder value for Systeo. By learning how to price, and learning how to cost by managing the composition of their teams, these consultants are invested with the productive power to create value. This differentiates them from their counterparts in Systeo's outsourcing division. Knowledge workers in the company's offshore service center in Dalian are constantly reminded of the costs which they impart and which make their jobs vulnerable to global competition. In contrast, senior managers in Beijing's consulting practice are told that each Systeo Strategy consultant brings in an average of 239,000 yuan (£23,900) of revenue each year. In Systeo consultants are deemed to create shareholder value by generating revenues, whilst outsourcing workers is deemed to destroy shareholder value by generating cost. This split in the firm is a crucial part of a particular logic of worth. The separation of cost from revenue in pursuit of profit is necessary to stratify the economy into those who can restructure it and those who must bear the consequences of the restructuring i.e. it legitimizes outsourcing (Chong 2012). By imbuing themselves with the power to create shareholder value, consultants construct themselves as experts of the knowledge economy, an ethical responsibility which finds official expression in the discourse of value-based leadership. As the similarity in terminology suggests, it goes hand-in-hand with the value-based management that consultants preach. If value-based management is the creation of shareholder value for Systeo and its clients, then value-based leadership is the explicit ethicizing of that very process. In the Senior Manager's Milestone, the Chairman of Systeo China, Liu Xing, draws on China's extensive tradition of moral philosophy to explain the concept of value-based leadership. Quoting Confucius and Mencius, he describes his own leadership as being guided by the principles of: *Hefa, heli, heqing* (Be lawful, reasonable and fair), *buxiao wu liangjiang* (To be unfilial is to be unscrupulous). He then recounts a conversation he once had with a professor at Beijing University, who conducts research on the "best practices" of Chinese employees. The professor told him of an entrepreneur who had a particularly interesting way of evaluating and rewarding his employees. He would match the remittances that his employees would send back to their family at Chinese New Year. The entrepreneur justified his actions, saying that if he doesn't see his employees taking care of their parents, he won't respect them. "That's his principle," says Liu Xing, "I'm not saying you have to have the same principle but you have to have *a principle* to guide your decisions."

This training of the explicit ethicization of corporate praxis is particular to the cultivation of managers. Lower down the organizational hierarchy training focuses mainly on the development of soft-skills: of presenting, organizing meetings, and professional appearance. It is only when consultants are promoted to manager level and above, that the development of an ethical consciousness of value creation is considered paramount. It is couched as "leadership." Furthermore, it is a morality which draws on the resources of local culture. The training of shareholder value in the scientific treatment of "relationships" was deemed to conflict with Chinese modes of economic operation, by seeking to expunge the social and the cultural so fundamental to orthodox Chinese business practice. In contrast, value-based leadership is conceptualized as the expression of cultural modes of being. The meeting of value-based leadership and value-based management is narrated as the point where Chinese culture is commensurable with shareholder value. By extension I would argue that Systeo is engaged with the discursive and material restructuring that seeks to make China's principles, more generally, commensurable with shareholder value creation. When this does not materialize, expatriate consultants generally do not, however, question the coherence of shareholder value logic.

Rather, the failure of shareholder value creation to become hegemonic in China is portrayed by expatriate partners as a critical shortcoming of post-Mao Chinese culture, which is seen as still bearing the hallmarks of the old socialist command economy. They see their role as ensuring the effective transportation and replication of Systeo's business practices in China. In an interview with one American partner I was told

> China is missing the leaders. If analysts and consultants are the foot soldiers, the managers and senior managers are front-line infantry, then partners are air cover—ready to shoot down any problems if not covered by the people on the ground. Currently China needs more air cover, it needs more local leaders with skills.
>
> I spend a lot of time coaching, explaining things that I wouldn't have to do elsewhere. I enjoy helping local talent which is good cos there's a lot of people here who don't "get" business operations. Or at least it feels like there's more of a lack here.

Like many expatriate staff based in China, this partner saw Systeo's methods of business as "best practice" and in this way normalized the systems of value-based management which they propagate. He justifies the need for foreign expertise, in the way of expatriate partners, in terms of their "Systeoness"—their acculturation into the logics of worth that underpin their central claim to create shareholder value. Chinese consultants are exposed to these archetypes of Systeoness through formal management training, as I have described here, but also through their daily work which often involves contact with expatriate managers and partners. Whilst Chinese culture must be made commensurable with shareholder value, Systeo's corporate culture is already isomorphic with value creation, being constructed from the very logics that inform the trope of "maximizing shareholder value."

Conclusion

Corporate training seeks to acculturate consultants into a logic of worth which both collapses and references financial value to ethical values. This chapter has shown how establishing a mode of valorization which encourages consultants to imbibe the notion that producing financial value is, in and of itself, ethical behavior is central to this process. Using the tools of SOCIAL STYLES™, The Trust Equation, and the Deal Power Map, consultants learn that the scientific management of relationships is a requisite for the creation of value. These tools enable consultants to dig beneath the surface of social relationships, to uncover "the facts." In other words it is only by getting past the hazy fuzz of sociality that consultants can find out that what their client really wants is shareholder value, and that this demand is what drives and defines consulting. Armed with "the facts" consultants can then construct value propositions. They are engaged in the tautological process of uncovering shareholder value as the real need of clients, and simultaneously constructing their sales pitches to convince clients of the primacy of shareholder value as the cornerstone of corporate governance.

Partners often confessed that they had difficulty achieving this end-goal. One American partner told me that: "Chinese clients are often principle based rather than objective based. So rather than aiming to reduce costs, improve capital charge, they have principles like become an innovator or maintain social harmony." To use the language of SOCIAL STYLES™, Chinese clients are not easily constructed as drivers and as such "are not on the same page" as consultants, to use consultancy-speak. It is only through the alignment of values and value, where shareholder value is ingrained as the hegemonic ethic, that consulting can be seen to "add value." If Chinese clients do not place shareholder value as the ultimate objective of restructuring, the value of consultants' claim to "add value" is somewhat diminished. Thus, one could imagine that the task of selling value would include attempts to make Chinese businesses commensurable with shareholder value by, for example, expressly teaching consultants how to frame cost-stripping exercises and the implementation of accounting technologies in terms of localized "Chinese" business ethics. However, this does not take place. (In the last section we saw how consultants were encouraged to draw on localized ethics such as Confucianism to inform their own management praxis, but they were not encouraged to draw on Confucianism as a resource for sales.) As I have repeatedly demonstrated, the performance and narration of value in consulting—whilst attempting to peg different value registers together—does not obviate or preclude the existence of dissonance. Let us take the just-mentioned statement of the American partner. In contrast to the narratives and practices of internal training he implies that shareholder value is an "objective" (seemingly uninformed by an ethics) rather than an ethical imperative, an objective that Chinese clients do not pursue, their corporate strategy is instead guided by "principles" (which by his phrasing we can take to not include the pursuit of shareholder value). We see here an obvious dissonance of representations, which, however, is not necessarily problematic. In fact it is by problematizing dissonance that this partner goes on to formulate an imperative to modernize Chinese SOEs (i.e. the fact that they are not oriented to shareholder value is all the more reason to chase their business). That is to say, dissonance in this instance legitimizes Systeo's project of ethicizing Chinese companies through a large dose of consultancy. Far from posing a threat to performed discourses, dissonance can be productive for a profession which is all about selling a logic of worth.

References

Boltanski, L. and Thévenot, L. (2006). *On Justification: Economies of Worth*. Princeton, NJ: Princeton University Press.

Chong, K. (2012). *The Work of Financialization: An Ethnography of a Global Management Consultancy in Post-Mao China*. Unpublished PhD Dissertation, London School of Economics and Political Science.

Chong, K. and Tuckett, D. (2014). Constructing Conviction through Action and Narrative: How Money Managers Manage Uncertainty and the Consequences for Financial Market Functioning. *Socio-Economic Review*. Http://ser.oxfordjournals.org/content/early/2014/05/01/ser.mwu020.short?rss=1 (accessed, July 17, 2014)

Cross, J. (2011). Detachment as a Corporate Ethic: Materializing CSR in the Diamond Supply Chain. *Focaal: Journal of Global and Historical Anthropology* 60: 34–46.

Davis, G. F. (2009). *Managed by the Markets: How Finance Re-Shaped America*. Oxford and New York: Oxford University Press.

Fligstein, N. and Shin, T. (2007). Shareholder Value and the Transformation of the US Economy, 1984–2001. *Sociological Forum* 22(4): 399–424.

Froud, J., Haslam, C., Johal, S., and Williams, K. (2000). Shareholder Value and Financialization: Consultancy Promises, Management Moves. *Economy and Society* 29(1): 80–110.

Garsten, C. (2008). The Corporate Gaze: Transparency and Other Corporate Visions. *Ethnographic Praxis in Industry Conference Proceedings*. 2008 (1): 1–7.

Guyer, J. (2012). Terms of Debate versus Words in Circulation: Some Rhetorics of the Crisis. In J. Carrier (ed.), *A Handbook of Economic Anthropology* (2nd ed.). Cheltenham, UK: Edward Elgar Publishing.

Ho, K. (2009). *Liquidated: An Ethnography of Wall Street*. Durham and London: Duke University Press.

Lazonick, W. and O'Sullivan, M. (2000). Maximizing Shareholder Value: A New Ideology for Corporate Governance. *Economy and Society* 29(1): 13–35.

Maister, D. H., Green, C., and Galford, R. (2000). *The Trusted Advisor*. London: Free Press.

Milberg, W. (2008). Shifting Sources and Uses of Profits: Sustaining US Financialization with Global Value Chains. *Economy and Society* 37(3): 420–451.

Moor, L. and Lury, C. (2011). Making and Measuring Value: Comparison, Singularity and Agency in Brand Valuation Practice. *Journal of Cultural Economy* 4(4): 439–454.
Reimer, K. (2003). International Examples of Large-Scale Systems—Theory and Practice II: Implementing ERP Systems in China. *Communications of the Association for Information Systems* 11(20): 1–34.
SAP White Paper (1999). *Enabling Value Based Management.* http://www.juergendaum.com/finance/sap_vbm_wp.pdf (accessed August 30, 2013).
Stark, D. (2009). *The Sense of Dissonance: Accounts of Worth in Economic Life.* Princeton, NJ: Princeton University Press.
Steinfeld, E. (2004). Market Visions: The Interplay of Ideas and Institutions in Chinese Financial Restructuring. *Political Studies* 52(4): 643–663.
Steinfeld, E. (2010). *Playing our Game: Why China's Economic Rise Doesn't Threaten the West.* Oxford and New York: Oxford University Press.
Strathern, M. (2000). Introduction. In M. Strathern (ed.), *Audit Cultures: Anthropological Studies in Accountability, Ethics and the Academy.* London: Routledge, 1-18.
Williams, K. (2000). From Shareholder Value to Present-Day Capitalism. *Economy and Society* 29(1): 1–12.
Yang, M. (1994). *Gifts, Favors, and Banquets: The Art of Social Relationships in China.* Ithaca, NY: Cornell University Press.

Index of Names

Index of Subjects

Numbers in **bold** refer to figures and tables

The manufacturer's authorised representative in the EU for product safety is Oxford University Press España S.A. of El Parque Empresarial San Fernando de Henares, Avenida de Castilla, 2 - 28830 Madrid (www.oup.es/en or product.safety@oup.com). OUP España S.A. also acts as importer into Spain of products made by the manufacturer.
Printed and bound by CPI Group (UK) Ltd, Croydon, CR0 4YY
14/09/2026
02216307-0015